Oral Medicine and Immunity

Oral Medicine and Immunity

Editor

Kenichi Kumagai

MDPI

Basel • Beijing • Wuhan • Barcelona • Belgrade • Novi Sad • Cluj • Manchester

Editor
Kenichi Kumagai
Oral and Maxillofacial
Surgery
Tokyo University
Tokyo
Japan

Editorial Office
MDPI
St. Alban-Anlage 66
4052 Basel, Switzerland

This is a reprint of articles from the Special Issue published online in the open access journal *International Journal of Molecular Sciences* (ISSN 1422-0067) (available at: www.mdpi.com/journal/ijms/special_issues/oral_immunity).

For citation purposes, cite each article independently as indicated on the article page online and as indicated below:

Lastname, A.A.; Lastname, B.B. Article Title. *Journal Name* **Year**, *Volume Number*, Page Range.

ISBN 978-3-7258-1066-6 (Hbk)
ISBN 978-3-7258-1065-9 (PDF)
doi.org/10.3390/books978-3-7258-1065-9

Contents

Preface

The oral cavity is the port of entry to the gastrointestinal and respiratory tracts, characterized by the juxtaposition of soft and hard tissues. It is continuously subject to challenges by the external environment, such as foreign antigens or material.

Oral microbial communities are now seen as the fundamental etiological agent in oral diseases through their interface with host inflammatory responses. Thus, it is essential to understand the molecular mechanisms and pathogenicity of oral diseases from immunological and microbiological perspectives. Oral medicine concerns with the diagnosis and management of oral diseases, including oral mucosal disease, oral cancer, salivary gland disorders, temporomandibular disorders and oral manifestations of systemic and infectious diseases. This Special Issue aims to outline all areas of oral medicine in order to depict various oral diseases from immunological and microbiological perspectives.

Kenichi Kumagai
Editor

International Journal of *Molecular Sciences*

MDPI

Review

Function of Innate Lymphoid Cells in Periodontal Tissue Homeostasis: A Narrative Review

Zhiyu Ma [1], Jinsong Wang [1,2], Lei Hu [1,3,4,*] and Songlin Wang [1,2,4,5,6,*]

1 Beijing Laboratory of Oral Health, School of Basic Medicine, School of Stomatology, Capital Medical University, Beijing 100050, China
2 Department of Biochemistry and Molecular Biology, School of Basic Medicine, Capital Medical University Beijing 100070, China
3 Department of Prosthodontics, School of Stomatology, Capital Medical University, Beijing 100050, China
4 Immunology Research Center for Oral and Systemic Health, Beijing Friendship Hospital, Capital Medical University, Beijing 100070, China
5 Laboratory for Oral and General Health Integration and Translation, Beijing Tiantan Hospital, Capital Medical University, Beijing 100070, China
6 Research Unit of Tooth Development and Regeneration, Chinese Academy of Medical Sciences, Beijing 100700, China
* Correspondence: hulei@ccmu.edu.cn (L.H.); slwang@ccmu.edu.cn (S.W.)

Abstract: Periodontitis is an irreversible inflammatory response that occurs in periodontal tissues. Given the size and diversity of natural flora in the oral mucosa, host immunity must strike a balance between pathogen identification and a complicated system of tolerance. The innate immune system, which includes innate lymphoid cells (ILCs), certainly plays a crucial role in regulating this homeostasis because pathogens are quickly recognized and responded to. ILCs are a recently discovered category of tissue-resident lymphocytes that lack adaptive antigen receptors. ILCs are found in both lymphoid and non-lymphoid organs and are particularly prevalent at mucosal barrier surfaces, where they control inflammatory response and homeostasis. Recent studies have shown that ILCs are important players in periodontitis; however, the mechanisms that govern the innate immune response in periodontitis still require further investigation. This review focuses on the intricate crosstalk between ILCs and the microenvironment in periodontal tissue homeostasis, with the purpose of regulating or improving immune responses in periodontitis prevention and therapy.

Keywords: innate immunity; periodontitis (PD); innate lymphoid cells (ILC)

Citation: Ma, Z.; Wang, J.; Hu, L.; Wang, S. Function of Innate Lymphoid Cells in Periodontal Tissue Homeostasis: A Narrative Review. *Int. J. Mol. Sci.* **2023**, *24*, 6099. https://doi.org/10.3390/ijms24076099

Academic Editor: Christopher W Cutler

Received: 15 February 2023
Revised: 16 March 2023
Accepted: 18 March 2023
Published: 23 March 2023

1. Introduction

Periodontitis, a multifactorial disease, is associated with dysbiosis of the oral microbiota, aberrant inflammatory immune responses, and genetic predisposition. Microbial dysbiosis, initial oral-soft-tissue inflammation, and innate and adaptive immunity work together to protect periodontal tissue throughout the immune response, involving inflammation, regression, and healing. Although the adaptive immune system has been widely researched in the pathogenesis of periodontitis, the innate immune system in general and innate lymphoid cells, in particular have, received less attention.

Innate lymphoid cells (ILCs) are a heterogeneous population of newly discovered lymphoid cells that have recently been identified. They play central roles within mucosal barrier function, microbiome homeostasis, and control of inflammation in the early defense stage. ILCs are divided into five distinct ILC subsets, namely helper ILCs (i.e., ILC1s, ILC2s, ILC3s), natural killer (NK) cells, and lymphoid tissue inducer (LTi) cells, which are characterized by the production of different cytokines and functions through signature cytokines and transcription factors involved in their development [1–3]. ILC1s contributes to protective immune response directed against intracellular bacteria or viruses, as well as tumors through interferon-γ (IFN) [4]; ILC2s contribute to response to infections with extracellular parasites or microbes after being stimulated by allergens through interleukin (IL)-5, IL-9, and IL-13 [5]; ILC3s act against extracellular microorganisms (i.e., bacteria and fungi) through IL-22 and IL-17 [3,6]; and LTi cells play a crucial role in lymphoid tissue development during fetal life, and they may also help repair tissue in secondary lymphoid organs in adults.

ILCs in different organs and tissues exhibit different immunophenotypes and functions, adapting to particular tissue microenvironments [7]. Research in the past decade has increased our comprehension of how ILCs affects the tumor and inflammation microenvironment. NK-cell-based immunotherapy has produced immune checkpoint inhibitors (ICIs) for the production of ILCs, and these molecules may regulate ILCs to control their activity in tumor microenvironments [8]. Although ILCs in health and chronic inflammatory diseases have been explored and their role in some chronic diseases, such as allergy and inflammatory bowel disease (IBD), have been proposed in recent years, the evidence from animal studies is far from conclusive. This study provides an overview of the phenotypic and developmental origin of ILC subsets and discusses what has come to be recognized as the functions of ILCs in periodontal tissue.

2. Innate Lymphoid Cells

Innate lymphoid cells (ILCs) can be divided into helper ILCs and cytotoxic ILCs (natural killer [NK] cells) based on cytokines and transcription factors [9] (Table 1).There are three distinct subpopulations of helper ILCs: ILC1s are T-bet-dependent and secrete interferon-γ (IFN-γ); ILC2s are dependent on GATA3 and secrete type 2 cytokines, such as interleukin (IL)-5 and IL-13; and ILC3s are dependent on retinoic acid receptor-related orphan receptor (ROR)-γt and are further classified as lymphoid tissue-induced (LTi) cells and natural cytotoxicity receptor (NCR)-negative cells or NCR-positive cells [10]. The diversity of ILC subgroups within groups is extensive, such as ILC1a-d, ILC2a-d, and ILC3a-e subgroups [11]. Regulatory ILCs (ILCregs), similar to regulatory T-cells (Tregs), were discovered in the intestine in, addition to these three conventional subgroups, secreting TGF-β and inhibiting ILC1s and ILC3s in an IL-10-dependent manner [10]. Circulating ILCs do not appear to fit into this scheme despite their phenotypic similarity to their tissue-resident counterparts; hence, they are not discussed further in this article [12].

Various organ-specific subpopulations of ILCs are found in multiple non-lymphoid tissues, including lung, skin, intestine, liver, and adipose tissue [13]. Intraepithelial ILC1s were enriched in the oral mucosa, and ILC2s were predominantly enriched in oral draining lymph nodes (dLNs). Compared to the draining mesenteric lymph nodes, ILC populations are more abundant in the lower GI tract's mucosal surface [14]. The data reported here suggest that a similar phenomenon happens in the oral mucosa because the gingivae have an enrichment of ILCs and a more diverse ILC population than oral dLNs [15]. It is hypothesized that the increase in the variety and amount of ILCs in gingiva should react to ongoing microbial disturbance in the mouth to preserve oral homeostasis. In periodontal inflammatory tissues, the number and frequency of different ILC subpopulations increase, thereby promoting inflammation [15,16]. However, this phenomenon further validates the above speculation. The ratio of ILC2s and ILC3s in the gingiva of mice and humans differ significantly. Human gingiva only has a modest percentage of IL-17^{+}ILC3, and no ILC2 population has been found [16]. The gingiva of mice has roughly similar amounts of ILC1 and ILC2 [15]. This variation may be due to environmental effects brought on by different microbes in different racial groups or between some non-pathogenic mice and humans.

Table 1. Characteristics of innate lymphoid cells in mouse and human.

ILC Group	Subsets	Stimuli	Mediators	Phenotype: Mouse	Phenotype: Human	Transcription Factors	References
Group 1	ILC1s	tumors, intracellular microbe (bacteria, virus, parasites)	IFN-r, granzymes, perforin	CD49a, NK1.1, NKp46, CD122, CD127, CD200R, TRAIL	NKp46, CD122, CD127, CD200R, TRAIL	T-bet EOMES^{hig}	[17–21]
	NK cells			CD49b, NK1.1, NKp46, CD122, KLRG1	NKp46, NKp30, NKp80, CD56, CD16, CD122, CD127, KLRG1	T-bet EOMES	[20,22–24]
Group 2	ILC2s	extracellular parasites and allergens	IL-4, IL-5, IL-9, IL-13, AREG	CD127, CD25, ST2, KLRG1	NKp46, CD127	GATA3 RORα	[25–27]
Group 3	NCR$^+$ILC3s	extracellular microbes (bacteria, fungi)	IL-22, IL-17, GM-CSF, Lympho-toxin	NKp46, CD127	NKp46, NKp44, CD127, CCR6, CD56	RORγt AhR	[9,28,29]
	NCR$^-$ILC3s			NKp46, NKp44, CD127, CD56, CCR6	CD127, CCR6		[30,31]
	Lti	mesenchymal organizer cells (CXCL13, RANKL)	RANK, Lym-photoxin, TNF, IL-17, IL-22	CD127, CCR6	CD127, CCR6, CD7		[31,32]

3. The Development of ILCs

Mouse and human ILCs differ significantly in several key aspects, such as surface markers, phenotype, and several genes, including Id2, NFIL3, Zbtb16, GATA3, and Tox [10,15,33]. The developmental lineage of mouse ILCs is being explored using genetically modified animal models, but knowledge of human ILC development is very limited. The first reason for this is the lack of comparable genetic and tracing techniques in animal models of lymphoid development. The second is the variation in how progenitor cells are defined, lineage identification that is not reliant solely on cytokine production, and other indicators.

Similar to other lymphocytes, mouse ILCs originate from a common lymphoid progenitor (CLP), with ILCs predominantly present in the $\alpha4\beta7$-positive portion of the CLP population ($\alpha4\beta7^+$ CLP). Downstream of the murine CLPs, two different progenitors have been discovered, each with limited ILC potential. These include common ILC precursors (CILPs) and early innate lymphoid progenitor cells (EILPs) [34–36]. It has also been suggested that CILP may serve as an intermediary step of maturation between CLP and EILCP given the significant decrease in CILPs, along with EILCPs in Tcf7-deficient mice [37]. EILPs have lower levels of Zbtb16, IL-7Rα, and Id2 than CILPs. Integrins $\alpha4\beta7$ and CXCR6 are markers for CILPs. CILPs can differentiate into at least two different precursors: NK progenitors (NKPs), which can give rise to cNK cells; and common helper ILC progenitors (CHILPs), which can differentiate into all ILC subpopulations [38]. CHILPs express Id2 and CD127 and contain two groups: the group of cells that express PLZF (promyelocytic leukemia zinc finger), called ILC precursor cells (ILCps), and a population of RORγt+ PLZF− cells, which highly express TOX and not GATA3, called lymphoid tissue inducer precursors (LTips) [39]. LTips differentiate into LTi cells, and ILCps can further differentiate into ILC1s, ILC2s, and ILC3s, depending on the transcription factors T-bet, GATA-3, and RORγt [40]. Unfortunately, while it is understood that ILCregs do not originate from ILCps in the intestine tissue, the specific origin of ILCregs from CHILPs remains unknown [10] (Figure 1).

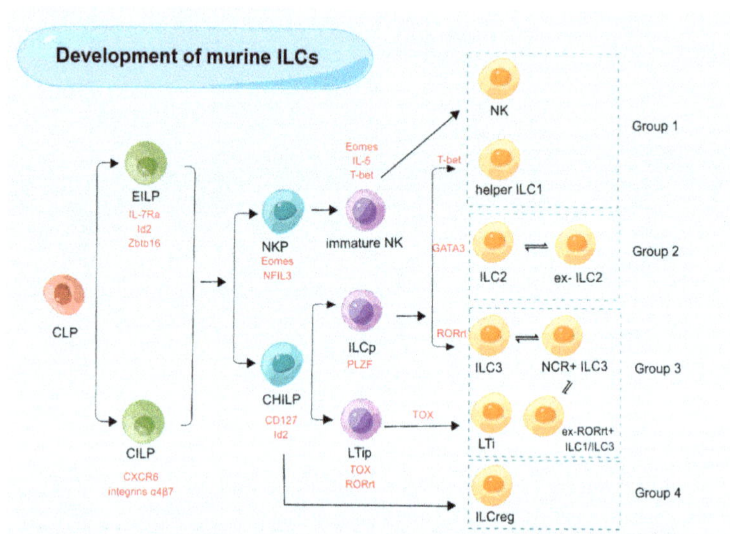

Figure 1. The development of murine innate lymphoid cells (ILCs). CLP: common lymphoid progenitor; EILP: early innate lymphoid progenitor; CILP: common ILC precursor; CHILP: common helper ILC progenitor; NKP: natural killer progenitor; ILCp: ILC precursor; LTip: lymphoid tissue inducer precursor; LT: lymphoid tissue inducer; ILCreg: regulatory ILC.

In human, early tonsillar progenitors (ETPs), which were once referred to as stage 1 and stage 2 NK cell developmental intermediates (NKDIs) and are currently detected in secondary lymphoid tissues (SLTs), are the earliest ILC progenitors that have been identified. Specific classifications of ETPs include Lin$^-$CD34$^+$CD10$^+$CD117$^-$ ETP1 and Lin-CD34$^+$CD10$^-$CD117$^+$ ETP2 [33]. The expression of IL-1R1 is variable in ETP2s. IL-1R1+ ETP2s are ILC-restricted, characterized as ID2$^+$RORγt$^+$RAG1, while IL-1R1$^-$ETP2s show some T-cell and DC potential [41]. IL-1R1$^+$ ETP2s are the earliest committed human CILCP that has been found so far. Stage 3 NKID has restricted potential and ILC generation, which has the phenotypic characteristics of a Lin$^-$CD34$^-$CD7$^+$CD127$^+$CD117$^+$CRTH2$^-$ ILC progenitor (ILCp) [42]. The expression of NKp46, CD56, and killer cell lectin-like receptor subfamily G member 1 (KLRG1) can be used to distinguish between several ILCp populations with restricted differentiation potential. The ability of CD56-positive ILCps to differentiate into NK cells, ILC1s, and ILC3s is limited, whereas NKp46-positive ILCps mostly differentiate into ILC3s, and KLRG1-positive ILC precursors primarily develop into ILC2s [43,44] (Figure 2).

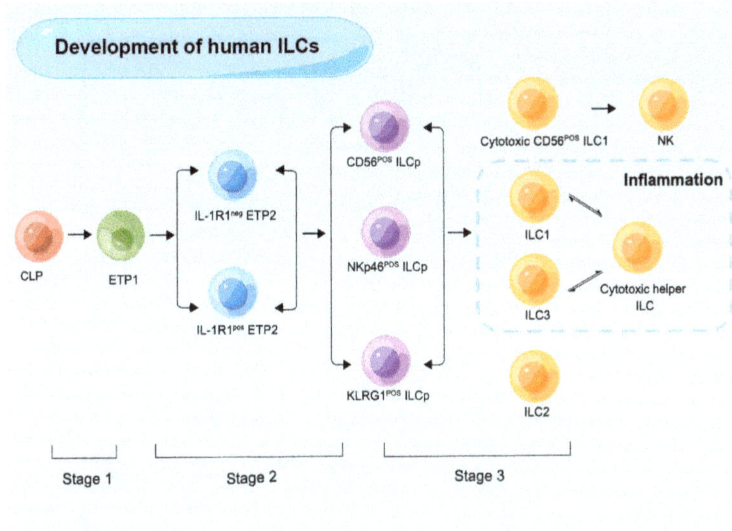

Figure 2. The development of human innate lymphoid cells (ILCs). CLP: common lymphoid progenitor; ETP: early tonsillar progenitors; ILCp: ILC precursor; KLRG: killer cell lectin-like receptor subfamily G member 1.

4. Plasticity of ILC Subsets

Effective innate immune responses depend heavily on the heterogeneity and plasticity of ILCs. ILCs interact with epithelial cells, stromal cells, myeloid cells, and adaptive immune cells in the surrounding environment to modify their activities and phenotypes in response to environmental cues, known as plasticity. The phenotype of ILCs is relatively flexible because the local tissue microenvironment in inflamed tissue can differ from that of healthy tissue. ILC subsets are selectively activated and accumulate, and the equilibrium between cytokines and chemokines is frequently disturbed. Finally, a subset of ILCs with a new specialized function forms through a process termed as transdifferentiation. Transdifferentiation increases the immune system's effectiveness because it avoids the need for progenitor cell differentiation.

In 2010, it was discovered, for the first time, that ILCs have plasticity comparable to T-cells in mice and humans [45–47]. Environmental cues can trigger ILCs to become increasingly plastic and interconvert their phenotypes, even if the phenotype of ILC subsets is well defined. ILCs precursors develop from $CD34^+$ myeloid precursors to circulating naïve ILCs that exist in peripheral blood and undergo further maturation in tissue, known as tissue-resident ILCs. Before the terminal differentiation of ILC subsets, circulating naïve ILCs undergo a plasticity phase [48], and both extrinsic and intrinsic divers of ILCs plasticity have been documented [49].

In inflammation-related diseases, ILCs plasticity is a ubiquitous phenomenon involved in the regulation of immune response and inflammation. Based on the transcription factor RORγt, ILC1 differentiated to ILC3 in the presence of IL-2, IL-23, and IL-1β, and this process was accelerated in the presence of retinoic acid [50]. Furthermore, transdifferentiation is a reversible process. A middle ILC3-ILC1 cluster with a high directional preference for ILC1s in human tonsils and intestine was discovered by RNA velocity analysis [51]. Changes in the epigenetic landscape result in the transition of ILC2 into ILC1 in an IL-1β-receptor and IL-12-IL-12R pathway-dependent manner. Although their transdifferentiation in oral mucosal still needs to be explored, this occurrence is generally associated with Crohn's disease and chronic obstructive pulmonary disease (COPD) [52–54]. Inflammatory ILC2 cells (iILC2 cells) can transform into ILC3-like cells and develop the capacity to generate

IL-17, contributing to immunity to both fungi and helminths, during Th17 polarization circumstances in vitro or Candida albicans infection in vivo [55]. ILC2s exhibit phenotypic plasticity during the psoriasis inflammatory process. ScRNA-seq validated the dynamic changes in the status of the original or static ILC2, as well as the metastasis of the effector ILC2 [56]. In response to TNF-α, IL-1β, and IL-23, ILC2s transform into ILC3-like cells that produce IL-17. Meanwhile, IL-4 has the power to reverse this event [57]. This process has also been noted in nasal mucosal, and this transdifferentiation is abrogated by vitamin D3 and IL-4 [58]. However, further research is required to confirm this process in the oral mucosa. Notably, there is currently no published research on human ILC2-ILC3 plasticity. A subset of NCR$^+$ ILC3s suppressed the conversion of RORγt into ILC1-like cells during oral microbial infection. ILC1-like cells are phenotypically and functionally similar to ILC1s, although they are not actual ILC1 subpopulations, and are hence referred to as "ex-RORγt ILCs". ILC3s convert from RORγt$^+$ ILC3 to ex-RORγt$^+$ ILC3 in chronic inflammatory situations to adjust to environmental cytokines and preserve tissue homeostasis.

ILCs infiltrate the tumor microenvironment and respond rapidly to inhibit tumor growth. Transforming growth factor-β (TGF-β) and IFN-γ produced by ILC1s has an important role in the immune response to tumors. TGF-β signaling triggers the conversion of ILC3 to ILCreg by scRNA-seq to promote tumor growth [59]. TGF-β signaling also promotes the conversion of NK cells to intermediate ILC1s (intILC1s) to promote immune escape and tumorigenesis. NK cells transform into ILC1-like cells in TGF-β-rich tumor environment [60]. Oral squamous cell carcinoma (SqCC) is a major type of oral cancer. ILC1 to ILC3 conversion by SqCC tumor cells results in the production of IL-23, which encourages IL-17-mediated tumor cell proliferation [61].

The plasticity of ILCs is intimately correlated with the development of illness, and the development of disease can be somewhat managed by regulating cell plasticity. The immunological microenvironment is altered due to the illness process, although it is yet known whether the plasticity of ILCs is a cause or an effect of disease onset.

5. ILCs in Periodontal Homeostasis

The modulation of the activation state and quantity of local ILCs, which involves three main aspects, is what determines the quality and strength of the local immune response in the oral mucosa, as follows: (1) activation or inhibition of tissue-resident ILCs; (2) plasticity and differentiation of specific ILC subpopulations; and (3) tissue-specific migration and regional accumulation of peripheral ILCs [27]. It is significant to highlight that ILCs play a crucial function in the oral mucosa given that all three aspects of ILCs are pertinent to immune response. They are a potential therapeutic target for maintaining periodontal homeostasis because, while they play a protective role in periodontal tissues, they may also initiate and amplify locally detrimental immune projections (Figure 3).

Under the condition of infection or chronic inflammation, the number of ILCs on the local oral mucosal surface [62] and draining lymph nodes is significantly increased to effectively regulate the innate and adaptive immune response of oral microorganisms [63]. This population is partly supplemented by a small amount of circulating precursor cells and circulating mature ILCs [64], as well as the migration of specific subpopulations of ILCs from the mucosal surface in the presence of chemokines and homing receptors [14]. However, the mode and mechanism of ILC migration from the gingiva to the oral drainage lymph nodes remains to be investigated. Although the majority of oral draining lymph node and gingival populations are CD117$^-$ NKp46$^-$ ILC subpopulations, it has been shown that the gingival epithelium and oral draining lymph nodes produce INF-γ and IL-5 ILC subpopulations in close proportions [15]. This may be related to the potential plasticity and heterogeneity of ILCs.

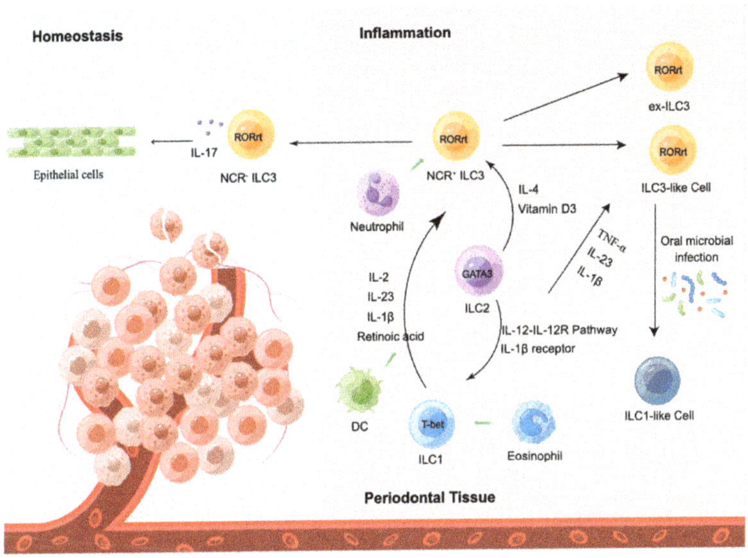

Figure 3. ILC subsets maintain periodontal homeostasis. At steady state, ILC subsets reside within the basal layers of the oral mucosal epithelium, contacting the basement membrane and slowly moving across the mucosal epithelium via contacting and squeezing between mucosal epithelial cells. At inflamed state, they generate cytokines and interact with other immune cells, such as neutrophils and eosinophils, to control ILCs' plasticity during homeostasis via a complex network of transcription factors, resulting in the establishment of an antiviral/inflamed state of the mucosal epithelium.

5.1. ILC1

ILC1s are a heterogeneous mixed-cell population, a class of helper ILC1 (also known as tissue-resident ILC1), characterized as $lin^{-}CD127^{+}CD117^{-}CRTH2^{-}NKp44^{-}$, with the ability to express T-bet and produce TNF-α and IFN-γ. It has been determined that human gingival tissue is dominated by the ILC1 subgroup [65], with the largest number of ILC1s in the periodontium and gingiva (approximately 50% of total ILC1s) [66]. In addition, a small subset of RANKL-expressing ILC1s was also found, which may play an immune regulatory role in gingivitis and periodontitis lesions [65].

Another classical member of this population is the NK cell, which expresses CD62L, CCR7, and S1PR, and highly expresses granzyme and perforin. NK cells have a distinct origin in comparison to ILC1. More in-depth studies utilizing polychromic receptor mice have revealed that ILC precursors, which were previously believed to be unable to develop into NK cells, exhibit significant NK cell precursor activity. As a result, the difference between ILC1s and NK cells is considerably less apparent because certain subsets have been described that combine ILC1 and NK cell characteristics. NK cells represent innate $CD8^{+}$ CTL, which interacts directly with periodontopathogenic microorganisms [67], such as *Aggregatibacter actinomycetemcomitans*, *Porphyromonas gingivalis*, and *Fusobacterium nucleatum*. This interaction induces NK cells to form immune synapses, followed by polarization of their microtubule organizing centers (MTOCs) and secretory lysosomes towards the lysis synapses [68]. Prior to fusing with the plasma membrane, secretory lysosomes settle and increase IFN-c and TNF-α levels [69], ultimately leading to periodontal tissue destruction and alveolar bone resorption. In addition, NK cells have long been demonstrated to be crucial in the prevention of human and monkey immunodeficiency virus (HIV/SIV) by preventing viral multiplication and transmission in the oral cavity [70].

Helper ILC1 subpopulations and NK cells contribute to the IFN-γ-mediated proinflammatory release of IL-12, IL-15, and IL-18. NK cells share a large degree of similarity in their properties and functions with helper ILC1s, which also produce IFN-γ. IFN-γ is expressed at higher levels in periodontitis than in gingivitis, indicating that it may be implicated in the pathogenesis of oral mucosal inflammation through the aberrant secretion of IFN-γ, which induces chemokine production by epithelial cells. In the uninfected oral epithelium, ILC1s cross the basal epithelium to maintain barrier integrity and homeostasis. Numerous studies have shown that ILC1s can secrete IFN-γ to induce a local antiviral state during early phases of viral infection [69,71–73]. First, the source of IFN-γ is determined to clarify how helper ILC1s, rather than NK cells, play a role in restricting viral propagation using Rag2$^{-/-}$ IL-2rg$^{-/-}$ animals (lacking all ILC subgroups) or mice treated with NK1.1 (lack of NK cells and ILC1) [72]. Depletion of ILC1 during VACV infection and IFN-γ-mediated neutralization of viral load in uninfected tissues provide additional evidence that ILC1-mediated antiviral effects depend on steady state or early secreted IFN-γ during VACV infection. IFN-γ-regulating genes (IRGs), including IRF7, were significantly upregulated in the oral mucosa following viral infection [69,72,73]. Prior to viral infection and near to the ILC1 area, IRF7 was expressed concentrically in the uninfected mucosa [71]. It has certainly been demonstrated that ILC1s secrete INF-γ to create a local antiviral state in the oral mucosa before viral infection in order to stop viral reproduction and transmission after infection has occurred. In addition, IFN-γ is associated with bone through a complex and context-dependent mechanism and is thought to regulate osteoclast differentiation and bone resorption in periodontitis.

5.2. ILC2

The ILC2 subpopulation highly expresses transcription factor GATA3 and is characterized as lin$^-$KLRG1$^+$IL-7R$^+$CD117$^-$IL33R$^+$ IL1R2$^+$. Two functionally and phenotypically unique subpopulations of ILC2 in mice have been identified. Inflammatory ILC2s (iILC2s), which are exclusively produced by IL-25 activation in an inflammatory state, and natural ILC2s (nILC2s) are defined as a homeostatic and IL-33-responsive cell [55]. There is still no published evidence that ILC2 subpopulations exist in humans, and single-cell RNA sequencing was unable to identify transcriptionally different subpopulations of human tonsillar ILC2 [74].

ILC2s are distinguished by the expression of transcription factor Bcl11b, which regulates mature ILC2s' identity and function [75], the IL-33 receptor, prostaglandin D2 receptor 2 (CRTH2), and by varied levels of c-Kit, all of which are important for ILC2s' location and function. ILC2s can also be identified by the expression of the killer cell lectin-like receptor subfamily G member 1 (KLRG1), a co-inhibitory receptor that has previously been discovered on T and NK cells [76]. KLRG1 is increased in response to IL-25 during infection and binds to members of the cadherin family [55].

ILC2s participate in several functions, such as lipid metabolism, parasite defense, and accumulation during type 2 inflammation. A previous study showed that ILC2s had a significant effect on periodontitis. In a ligation-induced periodontitis mouse model, AMPK knockdown promoted the release of IL-33 from ILC2s, a novel compensatory mechanism for suppressing inflammation [62]. ILC2s have been attributed to the regulation of inflammatory and immune responses by IL-33 in asthma, allergic dermatitis, and allergic lung inflammation [77,78]. The synthesis of ILC2s, which secrete substantial levels of IL-5 and IL-13, can be activated and encouraged by IL-33 derived from epithelial cells. IL-33 promotes STAT3 translocation into mitochondria via MAPK-induced STAT3 phosphorylation at the S727 site. This, in turn, controls electron transport chain activity and cellular respiration and lessens ILC2-mediated inflammation in acute asthmatic lungs [79]. In contrast, IL-33 promotes ILC2 growth and activation, and activated ILC2s recruit adaptive immune cells through high expression of CCL5 to exert anti-tumor immune effects [80].

5.3. ILC3

ILC3s are the most diverse mimics of RORγt-dependent Th17 cells available, including LTi cells. They are characterized as lin$^-$CD127$^+$CD117$^+$ and express transcription factors RORγt and AHR [9]. NKp46, NKG20, and NK1.1 are three RORγt NCRs found in the ILC3 population. ILC3s are finely separated into NCR$^+$ ILC3 (4%) and NCR$^-$ ILC3 (18%) in periodontal inflammatory tissues [65]. While NCR$^-$ ILC3s primarily produce IL-17A, NCR$^+$ ILC3s can produce IL-17A, IL-22, and IFN-γ [81]. NCR$^+$ ILC3s are the exclusive source of IL-22 in adult tonsils. ILC3s are the primary source of IL-17A in periodontal inflammation and fungal infection, despite the fact that Th17 cells are the primary source of IL-17A in other tissues [17]. In addition to these two subpopulations, two other subpopulations of ILC3s have been identified in human tonsils: HLA-DR$^+$ILC3, which account for approximately 10% of all ILC3s in tonsils; and CD62L+ILC3, which express CD45RA but do not secrete IL-22 and IL-17, and are therefore presumed to be naïve ILC3s [74,82]. ILC3s are underrepresented in the pool of circulating lymphocytes, as opposed to tissues. In contrast to mature ILC3s seen in secondary lymphoid organs, the majority of circulating ILC3s exhibit low amounts of RORγt. Circulating ILC3s are multipotent ILC precursors (ILCPs), which nonetheless have the capacity to give birth to functionally mature helper ILC subsets and NK cells.

Lymphoid tissue inducer (LTi) cells are regarded as belonging to a distinct ILC lineage. When the ILC3 group was first described in 1997 [83], LTi cells were its first members. In the developing fetus, LTi cells contribute to the organogenesis of secondary lymphoid tissues. In fact, LTi cells produce IL-22, IL-17A, and IL-17F as a result of RORγt expression and are involved in the creation of secondary lymphoid organs during embryogenesis [84] and maturity, as well as in their restoration after infection [85]. LTi cells in the mouse develop in a manner distinct from helper ILC3 cells. However, it has been challenging to identify a specific LTi population in humans. In humans, NRP1+ ILC3 generates substantial amounts of IL-22 and IL-17A [86]. These cells could therefore be a representation of LTi cells in humans. It is yet uncertain, though, whether these cells develop differently from helper ILC3s than they do in mice.

The dual role of ILC3s depends on cytokine levels. Intestinal epithelial cells are induced by IL-22 to produce antimicrobial peptides quickly, such as α- and β- defensins, and IL-22 also enhances epithelial cell renewal to maintain epithelial barrier integrity. IL-17A and IFN-γ trigger the pathogenesis of periodontitis by inducing TNF-α and RANKL production to induce periodontal tissue destruction [17]. However, ILC3s reduce the production of IFN-γ and IL-17 by decreasing CCR6 expression in response to IL-23 stimulation while increasing IL-22 production [87], thereby suppressing inflammatory response. In addition, ILC3s may limit the multiplication of periodontal microorganisms and their toxic products to avoid entering the systemic immune system and causing systemic diseases, such as Alzheimer's disease. It is possible that ILCs encourage endostasis by promoting the development of isolated lymphoid follicles and epithelial cell foci with the aid of dendritic cells and Tregs, or by triggering the production of GM-CSF through the flora.

6. Conclusions

The developmental lineage and characterization of ILCs has provided new insights into immune response in periodontal tissues. This new group of innate immune cells in periodontal tissues contributes to resistance to pathogen invasion and viral replication, promotes local oral tissue repair and inflammation, regulates periodontal tissue homeostasis, and maintains the integrity of the oral mucosal barrier. However, further investigation of the molecular mechanisms that make these processes possible is required.

ILCs can adjust to shifting periodontal tissue circumstances owing to their plasticity, which may be crucial for fine-tuning responses to various pathologic stimuli. Given that the majority of canonical ILC subsets are tissue resident under homeostatic conditions, their plasticity enables an immediate response to alterations in the microenvironment brought on by pathogens without the requirement for de novo recruitment of ILC subsets. Restoration of the original ILC subset composition may occur along with the resolution of inflammation. Future research should pay close attention to the unraveling of metabolic programs and signaling cascades that control ILC differentiation and plasticity. It may be possible to better understand the role that ILCs play in maintaining periodontal homeostasis by pursuing further research on ILC heterogeneity and plasticity, which may also lead to the discovery of new biomarkers and therapeutic targets for cutting-edge, individualized treatments.

Author Contributions: Conceptualization: Z.M., J.W. and L.H.; writing—original draft preparation: Z.M. and L.H.; writing—review and editing; supervision: Z.M., J.W., S.W. and L.H. All authors have read and agreed to the published version of the manuscript.

Funding: This study was supported by the National Natural Science Foundation of China (82030031, 82001067); grants from the Chinese Research Unit of Tooth Development and Regeneration, Academy of Medical Sciences (2019-12M-5-031); the Beijing Municipal Science and Technology Commission (Z181100001718208); the Beijing Municipal Education Commission (119207020201); Beijing Advanced Innovation Center for Big Data-based Precision Medicine (PXM2021_014226_000026) and grants from the Beijing Municipal Government (Beijing Scholar program—PXM2020_014226_000005; PXM2021_014226_000020); Innovation Research Team Project of Beijing Stomatological Hospital, Capital Medical University (CXTD202201). Beijing Municipal Administration of Hospitals' Youth Program (QML20191504), Scientific Research Common Program of Beijing Municipal Commission of Education (KM202110025009) and Beijing Talents Fund (2018000021469G285).

Institutional Review Board Statement: Not applicable.

Informed Consent Statement: Not applicable.

Data Availability Statement: Not applicable.

Conflicts of Interest: The authors declare that they have no conflict of interest.

References

1. Artis, D.; Spits, H. The biology of innate lymphoid cells. *Nature* **2015**, *517*, 293–301. [CrossRef] [PubMed]
2. Colonna, M. Innate Lymphoid Cells: Diversity, Plasticity, and Unique Functions in Immunity. *Immunity* **2018**, *48*, 1104–1117. [CrossRef] [PubMed]
3. Goc, J.; Lv, M.; Bessman, N.J.; Flamar, A.L.; Sahota, S.; Suzuki, H.; Teng, F.; Putzel, G.G.; Eberl, G.; Withers, D.R.; et al. Dysregulation of ILC3s unleashes progression and immunotherapy resistance in colon cancer. *Cell* **2021**, *184*, 5015–5030.e5016. [CrossRef] [PubMed]
4. Bernink, J.H.; Peters, C.P.; Munneke, M.; te Velde, A.A.; Meijer, S.L.; Weijer, K.; Hreggvidsdottir, H.S.; Heinsbroek, S.E.; Legrand, N.; Buskens, C.J.; et al. Human type 1 innate lymphoid cells accumulate in inflamed mucosal tissues. *Nat. Immunol.* **2013**, *14*, 221–229. [CrossRef]
5. Sonnenberg, G.F.; Artis, D. Innate lymphoid cells in the initiation, regulation and resolution of inflammation. *Nat. Med.* **2015**, *21*, 698–708. [CrossRef]
6. Martin, J.C.; Chang, C.; Boschetti, G.; Ungaro, R.; Giri, M.; Grout, J.A.; Gettler, K.; Chuang, L.S.; Nayar, S.; Greenstein, A.J.; et al. Single-Cell Analysis of Crohn's Disease Lesions Identifies a Pathogenic Cellular Module Associated with Resistance to Anti-TNF Therapy. *Cell* **2019**, *178*, 1493–1508.e1420. [CrossRef]
7. Kansler, E.R.; Li, M.O. Innate lymphocytes-lineage, localization and timing of differentiation. *Cell. Mol. Immunol.* **2019**, *16*, 627–633. [CrossRef]
8. Pesce, S.; Trabanelli, S.; Di Vito, C.; Greppi, M.; Obino, V.; Guolo, F.; Minetto, P.; Bozzo, M.; Calvi, M.; Zaghi, E.; et al. Cancer Immunotherapy by Blocking Immune Checkpoints on Innate Lymphocytes. *Cancers* **2020**, *12*, 3504. [CrossRef]
9. Spits, H.; Artis, D.; Colonna, M.; Diefenbach, A.; Di Santo, J.; Eberl, G.; Koyasu, S.; Locksley, R.; McKenzie, A.; Mebius, R.; et al. Innate lymphoid cells—A proposal for uniform nomenclature. *Nat. Rev. Immunol.* **2013**, *13*, 145–149. [CrossRef]
10. Wang, S.; Xia, P.; Chen, Y.; Qu, Y.; Xiong, Z.; Ye, B.; Du, Y.; Tian, Y.; Yin, Z.; Xu, Z.; et al. Regulatory Innate Lymphoid Cells Control Innate Intestinal Inflammation. *Cell* **2017**, *171*, 201–216.e218. [CrossRef]
11. Guia, S.; Fenis, A.; Vivier, E.; Narni-Mancinelli, E. Activating and inhibitory receptors expressed on innate lymphoid cells. *Semin. Immunopathol.* **2018**, *40*, 331–341. [CrossRef] [PubMed]
12. Bennstein, S.B.; Scherenschlich, N.; Weinhold, S.; Manser, A.R.; Noll, A.; Raba, K.; Kögler, G.; Walter, L.; Uhrberg, M. Transcriptional and functional characterization of neonatal circulating Innate Lymphoid Cells. *Stem. Cells Transl. Med.* **2021**, *10*, 867–882. [CrossRef] [PubMed]
13. Gasteiger, G.; Fan, X.; Dikiy, S.; Lee, S.; Rudensky, A. Tissue residency of innate lymphoid cells in lymphoid and nonlymphoid organs. *Science* **2015**, *350*, 981–985. [CrossRef] [PubMed]
14. Kim, C.H.; Hashimoto-Hill, S.; Kim, M. Migration and Tissue Tropism of Innate Lymphoid Cells. *Trends Immunol.* **2016**, *37*, 68–79. [CrossRef]
15. Brown, J.; Campbell, L.; Malcolm, J.; Adrados Planell, A.; Butcher, J.; Culshaw, S. Enrichment of Innate Lymphoid Cell Populations in Gingival Tissue. *J. Dent. Res.* **2018**, *97*, 1399–1405. [CrossRef]

16. Dutzan, N.; Konkel, J.E.; Greenwell-Wild, T.; Moutsopoulos, N.M. Characterization of the human immune cell network at the gingival barrier. *Mucosal Immunol.* **2016**, *9*, 1163–1172. [CrossRef]
17. Cortez, V.S.; Fuchs, A.; Cella, M.; Gilfillan, S.; Colonna, M. Cutting edge: Salivary gland NK cells develop independently of Nfil3 in steady-state. *J. Immunol.* **2014**, *192*, 4487–4491. [CrossRef]
18. Schuster, I.S.; Wikstrom, M.E.; Brizard, G.; Coudert, J.D.; Estcourt, M.J.; Manzur, M.; O'Reilly, L.A.; Smyth, M.J.; Trapani, J.A.; Hill, G.R.; et al. TRAIL+ NK cells control CD4+ T cell responses during chronic viral infection to limit autoimmunity. *Immunity* **2014**, *41*, 646–656. [CrossRef]
19. Tessmer, M.S.; Reilly, E.C.; Brossay, L. Salivary gland NK cells are phenotypically and functionally unique. *PLoS Pathog.* **2011**, *7*, e1001254. [CrossRef]
20. Gordon, S.M.; Chaix, J.; Rupp, L.J.; Wu, J.; Madera, S.; Sun, J.C.; Lindsten, T.; Reiner, S.L. The transcription factors T-bet and Eomes control key checkpoints of natural killer cell maturation. *Immunity* **2012**, *36*, 55–67. [CrossRef]
21. Mackay, L.K.; Minnich, M.; Kragten, N.A.; Liao, Y.; Nota, B.; Seillet, C.; Zaid, A.; Man, K.; Preston, S.; Freestone, D.; et al. Hobit and Blimp1 instruct a universal transcriptional program of tissue residency in lymphocytes. *Science* **2016**, *352*, 459–463. [CrossRef] [PubMed]
22. Caligiuri, M.A. Human natural killer cells. *Blood* **2008**, *112*, 461–469. [CrossRef] [PubMed]
23. Hudspeth, K.; Donadon, M.; Cimino, M.; Pontarini, E.; Tentorio, P.; Preti, M.; Hong, M.; Bertoletti, A.; Bicciato, S.; Invernizzi, P.; et al. Human liver-resident CD56(bright)/CD16(neg) NK cells are retained within hepatic sinusoids via the engagement of CCR5 and CXCR6 pathways. *J. Autoimmun.* **2016**, *66*, 40–50. [CrossRef]
24. Diefenbach, A.; Colonna, M.; Koyasu, S. Development, differentiation, and diversity of innate lymphoid cells. *Immunity* **2014**, *41*, 354–365. [CrossRef] [PubMed]
25. Salimi, M.; Barlow, J.L.; Saunders, S.P.; Xue, L.; Gutowska-Owsiak, D.; Wang, X.; Huang, L.C.; Johnson, D.; Scanlon, S.T.; McKenzie, A.N.; et al. A role for IL-25 and IL-33-driven type-2 innate lymphoid cells in atopic dermatitis. *J. Exp. Med.* **2013**, *210*, 2939–2950. [CrossRef]
26. Bal, S.M.; Bernink, J.H.; Nagasawa, M.; Groot, J.; Shikhagaie, M.M.; Golebski, K.; van Drunen, C.M.; Lutter, R.; Jonkers, R.E.; Hombrink, P.; et al. IL-1β, IL-4 and IL-12 control the fate of group 2 innate lymphoid cells in human airway inflammation in the lungs. *Nat. Immunol.* **2016**, *17*, 636–645. [CrossRef]
27. Schulz-Kuhnt, A.; Wirtz, S.; Neurath, M.F.; Atreya, I. Regulation of Human Innate Lymphoid Cells in the Context of Mucosal Inflammation. *Front. Immunol.* **2020**, *11*, 1062. [CrossRef]
28. Ishizuka, I.E.; Chea, S.; Gudjonson, H.; Constantinides, M.G.; Dinner, A.R.; Bendelac, A.; Golub, R. Single-cell analysis defines the divergence between the innate lymphoid cell lineage and lymphoid tissue-inducer cell lineage. *Nat. Immunol.* **2016**, *17*, 269–276. [CrossRef]
29. An, Z.; Flores-Borja, F.; Irshad, S.; Deng, J.; Ng, T. Pleiotropic Role and Bidirectional Immunomodulation of Innate Lymphoid Cells in Cancer. *Front. Immunol.* **2019**, *10*, 3111. [CrossRef]
30. Zhong, C.; Zheng, M.; Cui, K.; Martins, A.J.; Hu, G.; Li, D.; Tessarollo, L.; Kozlov, S.; Keller, J.R.; Tsang, J.S.; et al. Differential Expression of the Transcription Factor GATA3 Specifies Lineage and Functions of Innate Lymphoid Cells. *Immunity* **2020**, *52*, 83–95.e84. [CrossRef]
31. Vivier, E.; Artis, D.; Colonna, M.; Diefenbach, A.; Di Santo, J.P.; Eberl, G.; Koyasu, S.; Locksley, R.M.; McKenzie, A.N.J.; Mebius, R.E.; et al. Innate Lymphoid Cells: 10 Years On. *Cell* **2018**, *174*, 1054–1066. [CrossRef] [PubMed]
32. Montaldo, E.; Juelke, K.; Romagnani, C. Group 3 innate lymphoid cells (ILC3s): Origin, differentiation, and plasticity in humans and mice. *Eur. J. Immunol.* **2015**, *45*, 2171–2182. [CrossRef] [PubMed]
33. Scoville, S.D.; Freud, A.G.; Caligiuri, M.A. Cellular pathways in the development of human and murine innate lymphoid cells. *Curr. Opin. Immunol.* **2019**, *56*, 100–106. [CrossRef] [PubMed]
34. O'Sullivan, T.E.; Geary, C.D.; Weizman, O.E.; Geiger, T.L.; Rapp, M.; Dorn, G.W., 2nd; Overholtzer, M.; Sun, J.C. Atg5 Is Essential for the Development and Survival of Innate Lymphocytes. *Cell Rep.* **2016**, *15*, 1910–1919. [CrossRef]
35. Yu, X.; Wang, Y.; Deng, M.; Li, Y.; Ruhn, K.; Zhang, C.; Hooper, L. The basic leucine zipper transcription factor NFIL3 directs the development of a common innate lymphoid cell precursor. *eLife* **2014**, *3*, e04406. [CrossRef]
36. Yu, Y.; Tsang, J.C.; Wang, C.; Clare, S.; Wang, J.; Chen, X.; Brandt, C.; Kane, L.; Campos, L.S.; Lu, L.; et al. Single-cell RNA-seq identifies a PD-1(hi) ILC progenitor and defines its development pathway. *Nature* **2016**, *539*, 102–106. [CrossRef]
37. Seillet, C.; Mielke, L.A.; Amann-Zalcenstein, D.B.; Su, S.; Gao, J.; Almeida, F.F.; Shi, W.; Ritchie, M.E.; Naik, S.H.; Huntington, N.D.; et al. Deciphering the Innate Lymphoid Cell Transcriptional Program. *Cell Rep.* **2016**, *17*, 436–447. [CrossRef]
38. Lim, A.I.; Verrier, T.; Vosshenrich, C.A.; Di Santo, J.P. Developmental options and functional plasticity of innate lymphoid cells. *Curr. Opin. Immunol.* **2017**, *44*, 61–68. [CrossRef]
39. Curio, S.; Belz, G.T. The unique role of innate lymphoid cells in cancer and the hepatic microenvironment. *Cell. Mol. Immunol.* **2022**, *19*, 1012–1029. [CrossRef] [PubMed]
40. Constantinides, M.G.; McDonald, B.D.; Verhoef, P.A.; Bendelac, A. A committed precursor to innate lymphoid cells. *Nature* **2014**, *508*, 397–401. [CrossRef]
41. Scoville, S.D.; Mundy-Bosse, B.L.; Zhang, M.H.; Chen, L.; Zhang, X.; Keller, K.A.; Hughes, T.; Chen, L.; Cheng, S.; Bergin, S.M.; et al. A Progenitor Cell Expressing Transcription Factor RORγt Generates All Human Innate Lymphoid Cell Subsets. *Immunity* **2016**, *44*, 1140–1150. [CrossRef] [PubMed]

42. Bar-Ephraim, Y.E.; Koning, J.J.; Burniol Ruiz, E.; Konijn, T.; Mourits, V.P.; Lakeman, K.A.; Boon, L.; Bögels, M.; van Maanen, J.P.; Den Haan, J.M.M.; et al. CD62L Is a Functional and Phenotypic Marker for Circulating Innate Lymphoid Cell Precursors. *J. Immunol.* **2019**, *202*, 171–182. [CrossRef] [PubMed]
43. Chen, L.; Youssef, Y.; Robinson, C.; Ernst, G.F.; Carson, M.Y.; Young, K.A.; Scoville, S.D.; Zhang, X.; Harris, R.; Sekhri, P.; et al. CD56 Expression Marks Human Group 2 Innate Lymphoid Cell Divergence from a Shared NK Cell and Group 3 Innate Lymphoid Cell Developmental Pathway. *Immunity* **2018**, *49*, 464–476.e464. [CrossRef] [PubMed]
44. Mazzurana, L.; Czarnewski, P.; Jonsson, V.; Wigge, L.; Ringnér, M.; Williams, T.C.; Ravindran, A.; Björklund, Å.K.; Säfholm, J.; Nilsson, G.; et al. Tissue-specific transcriptional imprinting and heterogeneity in human innate lymphoid cells revealed by full-length single-cell RNA-sequencing. *Cell Res.* **2021**, *31*, 554–568. [CrossRef] [PubMed]
45. Cella, M.; Otero, K.; Colonna, M. Expansion of human NK-22 cells with IL-7, IL-2, and IL-1beta reveals intrinsic functional plasticity. *Proc. Natl. Acad. Sci. USA* **2010**, *107*, 10961–10966. [CrossRef]
46. Crellin, N.K.; Trifari, S.; Kaplan, C.D.; Satoh-Takayama, N.; Di Santo, J.P.; Spits, H. Regulation of cytokine secretion in human CD127(+) LTi-like innate lymphoid cells by Toll-like receptor 2. *Immunity* **2010**, *33*, 752–764. [CrossRef]
47. Vonarbourg, C.; Mortha, A.; Bui, V.L.; Hernandez, P.P.; Kiss, E.A.; Hoyler, T.; Flach, M.; Bengsch, B.; Thimme, R.; Hölscher, C.; et al. Regulated expression of nuclear receptor RORγt confers distinct functional fates to NK cell receptor-expressing RORγt(+) innate lymphocytes. *Immunity* **2010**, *33*, 736–751. [CrossRef] [PubMed]
48. Shih, H.Y.; Sciumè, G.; Mikami, Y.; Guo, L.; Sun, H.W.; Brooks, S.R.; Urban, J.F., Jr.; Davis, F.P.; Kanno, Y.; O'Shea, J.J. Developmental Acquisition of Regulomes Underlies Innate Lymphoid Cell Functionality. *Cell* **2016**, *165*, 1120–1133. [CrossRef]
49. Pelletier, A.; Stockmann, C. The Metabolic Basis of ILC Plasticity. *Front. Immunol.* **2022**, *13*, 858051. [CrossRef]
50. Bernink, J.H.; Krabbendam, L.; Germar, K.; de Jong, E.; Gronke, K.; Kofoed-Nielsen, M.; Munneke, J.M.; Hazenberg, M.D.; Villaudy, J.; Buskens, C.J.; et al. Interleukin-12 and -23 Control Plasticity of CD127(+) Group 1 and Group 3 Innate Lymphoid Cells in the Intestinal Lamina Propria. *Immunity* **2015**, *43*, 146–160. [CrossRef]
51. Cella, M.; Gamini, R.; Sécca, C.; Collins, P.L.; Zhao, S.; Peng, V.; Robinette, M.L.; Schettini, J.; Zaitsev, K.; Gordon, W.; et al. Subsets of ILC3-ILC1-like cells generate a diversity spectrum of innate lymphoid cells in human mucosal tissues. *Nat. Immunol.* **2019**, *20*, 980–991. [CrossRef] [PubMed]
52. Lim, A.I.; Menegatti, S.; Bustamante, J.; Le Bourhis, L.; Allez, M.; Rogge, L.; Casanova, J.L.; Yssel, H.; Di Santo, J.P. IL-12 drives functional plasticity of human group 2 innate lymphoid cells. *J. Exp. Med.* **2016**, *213*, 569–583. [CrossRef] [PubMed]
53. Ohne, Y.; Silver, J.S.; Thompson-Snipes, L.; Collet, M.A.; Blanck, J.P.; Cantarel, B.L.; Copenhaver, A.M.; Humbles, A.A.; Liu, Y.J. IL-1 is a critical regulator of group 2 innate lymphoid cell function and plasticity. *Nat. Immunol.* **2016**, *17*, 646–655. [CrossRef] [PubMed]
54. Silver, J.S.; Kearley, J.; Copenhaver, A.M.; Sanden, C.; Mori, M.; Yu, L.; Pritchard, G.H.; Berlin, A.A.; Hunter, C.A.; Bowler, R.; et al. Inflammatory triggers associated with exacerbations of COPD orchestrate plasticity of group 2 innate lymphoid cells in the lungs. *Nat. Immunol.* **2016**, *17*, 626–635. [CrossRef]
55. Huang, Y.; Guo, L.; Qiu, J.; Chen, X.; Hu-Li, J.; Siebenlist, U.; Williamson, P.R.; Urban, J.F., Jr.; Paul, W.E. IL-25-responsive, lineage-negative KLRG1(hi) cells are multipotential 'inflammatory' type 2 innate lymphoid cells. *Nat. Immunol.* **2015**, *16*, 161–169. [CrossRef]
56. Wallrapp, A.; Burkett, P.R.; Riesenfeld, S.J.; Kim, S.J.; Christian, E.; Abdulnour, R.E.; Thakore, P.I.; Schnell, A.; Lambden, C.; Herbst, R.H.; et al. Calcitonin Gene-Related Peptide Negatively Regulates Alarmin-Driven Type 2 Innate Lymphoid Cell Responses. *Immunity* **2019**, *51*, 709–723.e706. [CrossRef]
57. Bernink, J.H.; Ohne, Y.; Teunissen, M.B.M.; Wang, J.; Wu, J.; Krabbendam, L.; Guntermann, C.; Volckmann, R.; Koster, J.; van Tol, S.; et al. c-Kit-positive ILC2s exhibit an ILC3-like signature that may contribute to IL-17-mediated pathologies. *Nat. Immunol.* **2019**, *20*, 992–1003. [CrossRef]
58. Golebski, K.; Ros, X.R.; Nagasawa, M.; van Tol, S.; Heesters, B.A.; Aglmous, H.; Kradolfer, C.M.A.; Shikhagaie, M.M.; Seys, S.; Hellings, P.W.; et al. IL-1β, IL-23, and TGF-β drive plasticity of human ILC2s towards IL-17-producing ILCs in nasal inflammation. *Nat. Commun.* **2019**, *10*, 2162. [CrossRef]
59. Wang, S.; Qu, Y.; Xia, P.; Chen, Y.; Zhu, X.; Zhang, J.; Wang, G.; Tian, Y.; Ying, J.; Fan, Z. Transdifferentiation of tumor infiltrating innate lymphoid cells during progression of colorectal cancer. *Cell Res.* **2020**, *30*, 610–622. [CrossRef]
60. Cortez, V.S.; Ulland, T.K.; Cervantes-Barragan, L.; Bando, J.K.; Robinette, M.L.; Wang, Q.; White, A.J.; Gilfillan, S.; Cella, M.; Colonna, M. SMAD4 impedes the conversion of NK cells into ILC1-like cells by curtailing non-canonical TGF-β signaling. *Nat. Immunol.* **2017**, *18*, 995–1003. [CrossRef]
61. Koh, J.; Kim, H.Y.; Lee, Y.; Park, I.K.; Kang, C.H.; Kim, Y.T.; Kim, J.E.; Choi, M.; Lee, W.W.; Jeon, Y.K.; et al. IL23-Producing Human Lung Cancer Cells Promote Tumor Growth via Conversion of Innate Lymphoid Cell 1 (ILC1) into ILC3. *Clin. Cancer Res.* **2019**, *25*, 4026–4037. [CrossRef] [PubMed]
62. Qin, X.; Hoda, M.N.; Susin, C.; Wheeler, J.N.; Marshall, B.; Perry, L.; Saad, N.; Yin, L.; Elsayed, R.; Elsalanty, M.; et al. Increased Innate Lymphoid Cells in Periodontal Tissue of the Murine Model of Periodontitis: The Role of AMP-Activated Protein Kinase and Relevance for the Human Condition. *Front. Immunol.* **2017**, *8*, 922. [CrossRef] [PubMed]
63. Klose, C.S.; Artis, D. Innate lymphoid cells as regulators of immunity, inflammation and tissue homeostasis. *Nat. Immunol.* **2016**, *17*, 765–774. [CrossRef] [PubMed]

64. Ercolano, G.; Wyss, T.; Salomé, B.; Romero, P.; Trabanelli, S.; Jandus, C. Distinct and shared gene expression for human innate versus adaptive helper lymphoid cells. *J. Leukoc. Biol.* **2020**, *108*, 723–737. [CrossRef]
65. Kindstedt, E.; Koskinen Holm, C.; Palmqvist, P.; Sjostrom, M.; Lejon, K.; Lundberg, P. Innate lymphoid cells are present in gingivitis and periodontitis. *J. Periodontol.* **2019**, *90*, 200–207. [CrossRef]
66. Li, C.; Liu, J.; Pan, J.; Wang, Y.; Shen, L.; Xu, Y. ILC1s and ILC3s Exhibit Inflammatory Phenotype in Periodontal Ligament of Periodontitis Patients. *Front. Immunol.* **2021**, *12*, 708678. [CrossRef] [PubMed]
67. Seillet, C.; Brossay, L.; Vivier, E. Natural killers or ILC1s? That is the question. *Curr. Opin. Immunol.* **2021**, *68*, 48–53. [CrossRef]
68. Wilensky, A.; Chaushu, S.; Shapira, L. The role of natural killer cells in periodontitis. *Periodontol 2000* **2015**, *69*, 128–141. [CrossRef]
69. Lujan, R.A.; Vrba, S.M.; Hickman, H.D. Antiviral Activities of Group I Innate Lymphoid Cells. *J. Mol. Biol.* **2022**, *434*, 167266. [CrossRef]
70. Li, H.; Reeves, R.K. Functional perturbation of classical natural killer and innate lymphoid cells in the oral mucosa during SIV infection. *Front. Immunol.* **2012**, *3*, 417. [CrossRef]
71. Nabekura, T.; Shibuya, A. ILC1: Guardians of the oral mucosa against enemy viruses. *Immunity* **2021**, *54*, 196–198. [CrossRef] [PubMed]
72. Shannon, J.P.; Vrba, S.M.; Reynoso, G.V.; Wynne-Jones, E.; Kamenyeva, O.; Malo, C.S.; Cherry, C.R.; McManus, D.T.; Hickman, H.D. Group 1 innate lymphoid-cell-derived interferon-gamma maintains anti-viral vigilance in the mucosal epithelium. *Immunity* **2021**, *54*, 276–290.e275. [CrossRef] [PubMed]
73. Weizman, O.E.; Adams, N.M.; Schuster, I.S.; Krishna, C.; Pritykin, Y.; Lau, C.; Degli-Esposti, M.A.; Leslie, C.S.; Sun, J.C.; O'Sullivan, T.E. ILC1 Confer Early Host Protection at Initial Sites of Viral Infection. *Cell* **2017**, *171*, 795–808 e712. [CrossRef] [PubMed]
74. Björklund, Å.K.; Forkel, M.; Picelli, S.; Konya, V.; Theorell, J.; Friberg, D.; Sandberg, R.; Mjösberg, J. The heterogeneity of human CD127(+) innate lymphoid cells revealed by single-cell RNA sequencing. *Nat. Immunol.* **2016**, *17*, 451–460. [CrossRef] [PubMed]
75. Califano, D.; Cho, J.J.; Uddin, M.N.; Lorentsen, K.J.; Yang, Q.; Bhandoola, A.; Li, H.; Avram, D. Transcription Factor Bcl11b Controls Identity and Function of Mature Type 2 Innate Lymphoid Cells. *Immunity* **2015**, *43*, 354–368. [CrossRef] [PubMed]
76. Nagasawa, M.; Heesters, B.A.; Kradolfer, C.M.A.; Krabbendam, L.; Martinez-Gonzalez, I.; de Bruijn, M.J.W.; Golebski, K.; Hendriks, R.W.; Stadhouders, R.; Spits, H.; et al. Correction: KLRG1 and NKp46 discriminate subpopulations of human CD117(+)CRTH2(-) ILCs biased toward ILC2 or ILC3. *J. Exp. Med.* **2019**, *216*, 2221–2222. [CrossRef] [PubMed]
77. Moro, K.; Kabata, H.; Tanabe, M.; Koga, S.; Takeno, N.; Mochizuki, M.; Fukunaga, K.; Asano, K.; Betsuyaku, T.; Koyasu, S. Interferon and IL-27 antagonize the function of group 2 innate lymphoid cells and type 2 innate immune responses. *Nat. Immunol.* **2016**, *17*, 76–86. [CrossRef]
78. Karagiannis, F.; Masouleh, S.K.; Wunderling, K.; Surendar, J.; Schmitt, V.; Kazakov, A.; Michla, M.; Holzel, M.; Thiele, C.; Wilhelm, C. Lipid-Droplet Formation Drives Pathogenic Group 2 Innate Lymphoid Cells in Airway Inflammation. *Immunity* **2020**, *52*, 620–634 e626. [CrossRef]
79. Fu, L.; Zhao, J.; Huang, J.; Li, N.; Dong, X.; He, Y.; Wang, W.; Wang, Y.; Qiu, J.; Guo, X. A mitochondrial STAT3-methionine metabolism axis promotes ILC2-driven allergic lung inflammation. *J. Allergy Clin. Immunol.* **2022**, *149*, 2091–2104. [CrossRef]
80. Moral, J.A.; Leung, J.; Rojas, L.A.; Ruan, J.; Zhao, J.; Sethna, Z.; Ramnarain, A.; Gasmi, B.; Gururajan, M.; Redmond, D.; et al. ILC2s amplify PD-1 blockade by activating tissue-specific cancer immunity. *Nature* **2020**, *579*, 130–135. [CrossRef]
81. Konya, V.; Czarnewski, P.; Forkel, M.; Rao, A.; Kokkinou, E.; Villablanca, E.J.; Almer, S.; Lindforss, U.; Friberg, D.; Hoog, C.; et al. Vitamin D downregulates the IL 23 receptor pathway in human mucosal group 3 innate lymphoid cells. *J. Allergy Clin. Immunol.* **2018**, *141*, 279–292. [CrossRef] [PubMed]
82. Lim, A.I.; Li, Y.; Lopez-Lastra, S.; Stadhouders, R.; Paul, F.; Casrouge, A.; Serafini, N.; Puel, A.; Bustamante, J.; Surace, L.; et al. Systemic Human ILC Precursors Provide a Substrate for Tissue ILC Differentiation. *Cell* **2017**, *168*, 1086–1100.e1010. [CrossRef]
83. Mebius, R.E.; Rennert, P.; Weissman, I.L. Developing lymph nodes collect CD4+CD3- LTbeta+ cells that can differentiate to APC, NK cells, and follicular cells but not T or B cells. *Immunity* **1997**, *7*, 493–504. [CrossRef] [PubMed]
84. Sonnenberg, G.F.; Monticelli, L.A.; Elloso, M.M.; Fouser, L.A.; Artis, D. CD4(+) lymphoid tissue-inducer cells promote innate immunity in the gut. *Immunity* **2011**, *34*, 122–134. [CrossRef] [PubMed]
85. van de Pavert, S.A. Lymphoid Tissue inducer (LTi) cell ontogeny and functioning in embryo and adult. *Biomed. J.* **2021**, *44*, 123–132. [CrossRef]
86. Shikhagaie, M.M.; Björklund, Å.K.; Mjösberg, J.; Erjefält, J.S.; Cornelissen, A.S.; Ros, X.R.; Bal, S.M.; Koning, J.J.; Mebius, R.E.; Mori, M.; et al. Neuropilin-1 Is Expressed on Lymphoid Tissue Residing LTi-like Group 3 Innate Lymphoid Cells and Associated with Ectopic Lymphoid Aggregates. *Cell Rep.* **2017**, *18*, 1761–1773. [CrossRef]
87. Buonocore, S.; Ahern, P.P.; Uhlig, H.H.; Ivanov, I.I.; Littman, D.R.; Maloy, K.J.; Powrie, F. Innate lymphoid cells drive interleukin-23-dependent innate intestinal pathology. *Nature* **2010**, *464*, 1371–1375. [CrossRef]

International Journal of
Molecular Sciences

Review

Genetic/Protein Association of Atopic Dermatitis and Tooth Agenesis

Wanlu Ouyang [1], Charlene E. Goh [2], Wei Bo Ng [3], Fook Tim Chew [4], Eric Peng Huat Yap [5] and Chin-ying Stephen Hsu [2,*]

[1] Department of Orthodontics, Shanghai Xuhui District Dental Disease Prevention and Control Institute, No. 500, Fenglin Road, Shanghai 200032, China
[2] Faculty of Dentistry, National University of Singapore, Singapore 119085, Singapore
[3] Faculty of Science, National University of Singapore, Singapore 117543, Singapore
[4] Allergy and Molecular Immunology Laboratory, Lee Hiok Kwee Functional Genomics Laboratories, Department of Biological Sciences, Faculty of Science, National University of Singapore, Block S2, Level 5, 14 Science Drive 4, Lower Kent Ridge Road, Singapore 117543, Singapore
[5] Lee Kong Chian School of Medicine, Nanyang Technological University, Singapore 636921, Singapore
* Correspondence: denhsus@nus.edu.sg

Abstract: Atopic dermatitis and abnormalities in tooth development (including hypomineralization, hypodontia and microdontia) have been observed to co-occur in some patients. A common pathogenesis pathway that involves genes and protein interactions has been hypothesized. This review aims to first provide a description of the key gene mutations and signaling pathways associated with atopic dermatitis and tooth agenesis (i.e., the absence of teeth due to developmental failure) and identify the possible association between the two diseases. Second, utilizing a list of genes most commonly associated with the two diseases, we conducted a protein–protein network interaction analysis using the STRING database and identified a novel association between the Wnt/β-catenin signaling pathway (major pathway responsible for TA) and desmosomal proteins (component of skin barrier that affect the pathogenesis of AD). Further investigation into the mechanisms that may drive their co-occurrence and underlie the development of the two diseases is warranted.

Keywords: atopic dermatitis; tooth agenesis; skin barrier; gene–protein interaction

Citation: Ouyang, W.; Goh, C.E.; Ng, W.B.; Chew, F.T.; Yap, E.P.H.; Hsu, C.-y.S. Genetic/Protein Association of Atopic Dermatitis and Tooth Agenesis. *Int. J. Mol. Sci.* **2023**, *24*, 5754. https://doi.org/10.3390/ijms24065754

Academic Editor: Kenichi Kumagai

Received: 31 December 2022
Revised: 7 March 2023
Accepted: 15 March 2023
Published: 17 March 2023

1. Introduction

Atopic dermatitis (AD), also known as eczema and atopic eczema, is the most common chronic inflammatory skin disease [1] and is estimated to have the highest disease burden among all skin diseases [2].

Interestingly, some studies have shown epidemiological commonalities between AD and dental caries, and dental structural abnormalities such as hypomineralization and hypodontia (developmentally missing teeth). For example, in the GUSTO birth cohort study in Singapore, children diagnosed with AD in the first 18 months of life had a 3-times higher risk of developing tooth decay by age 3, despite controlling for several potential confounders [3]. A similar association was observed in adults in a nationwide cross-sectional study of 21,606 Korean adults, finding higher odds for having experienced caries in those with AD compared with those with no AD [4]. Another large population-based survey of Korean adolescents also showed significantly higher odds of AD among participants with oral symptoms, including sensitive teeth, toothache, etc. [5].

Due to the shared ectodermal tissue origin of the teeth and skin, an "ectodermal subclinical development defect" has been suggested, whereby genetic mutations associated with AD share a common pathogenic pathway with abnormalities in tooth development and can cause structural defects in the tooth, such as hypo-mineralization of the enamel. This structural defect in turn increases the tooth's susceptibility to dental caries and may have resulted in the AD-caries associations observed. A longitudinal cohort of 6-year-old twins, which demonstrated moderate to strong associations between hypo-mineralization of the second molars (HSPM) and infantile eczema [6], provides further support for the link between AD and abnormalities in tooth development.

Tooth agenesis (TA) is an extreme case of abnormality in tooth development, where there is an absence of teeth due to developmental failure, and is one of the most prevalent dental and craniofacial malformations in humans [7]. TA can be categorized into the following three groups: hypodontia (less than 6 missing teeth), oligodontia (6 or more missing teeth) and anodontia (complete absence of dentition).

Non-syndromic TA is TA that is not associated with any other systemic abnormalities or genetic syndromes. However, as the main cause of tooth agenesis is genetic, TA may involve other organs or tissues, as the involved networks of signaling molecules and transcription factors in TA and epithelial–mesenchymal interactions play essential and extensive roles during embryogenesis [8,9].

A few studies offer direct evidence for the association between AD and hypodontia. While only allergy (allergic rhinitis and pollinosis) was significantly associated, atopy (which includes atopic dermatitis) and asthma were also among the top conditions experienced by patients with hypodontia [10]. A 2017 Italian study found that 13/90 (14.4%) of children with atopic dermatitis had anatomical dental abnormalities, including agenesis and hypoplasias [11]. Our team also recently reported a significant association between severe–moderate AD and hypodontia and microdontia [12].

While the published review articles have focused on either AD or TA [13–17], there is an emerging need for a review that focuses on the overlapping areas and decodes the potential common signaling pathways and/or genes involved in the development of both AD and TA.

Therefore, the aim of this review is to provide a description of the key gene mutations and signaling pathways associated with atopic dermatitis and tooth agenesis, respectively, identify the possible associations between the two diseases, and to propose exploratory hypotheses and mechanisms to narrow down the possible shared pathogenic pathways for future research to interrogate.

2. Genes Associated with Atopic Dermatitis (AD)

The two major pathophysiological pathways in AD are abnormalities of epidermal structure and function, and cutaneous inflammation due to inappropriate immune responses to antigens encountered in the skin [18]. Both pathways may influence each other and cause a systemic T helper type 2 (Th2) inflammatory pathway and a Th17/Th22 cell response, which may in turn affect epidermal structure and function. From the genetic point of view, the disease is inherited and multifactorial [19]. In this review, we will focus on the genes involved in epidermal barrier dysfunction, as both AD and TA are potentially caused by structural defects during ectodermal tissue development.

2.1. Mutations in Genes Related to Epidermal Barrier

The epidermal barrier is the first line of defense between the host organism and the environment. As illustrated in Figure 1, the skin barrier resides primarily in the *stratum corneum (SC)*, which consists of corneocytes surrounded by intercellular lipid lamellae and attached by *corneodesmosomes* [20]. The *tight junctions* attached to lateral walls of keratinocytes in the upper stratum granulosum (SG) have also been included in the basic skin barrier structure. Keratin filaments form macrofibrils by cross-linking with the *cornified envelope (CE)* of corneocytes. The SC lipid layer is covalently attached to the external surface of CE proteins, forming the cornified lipid envelope (CLE) [21].

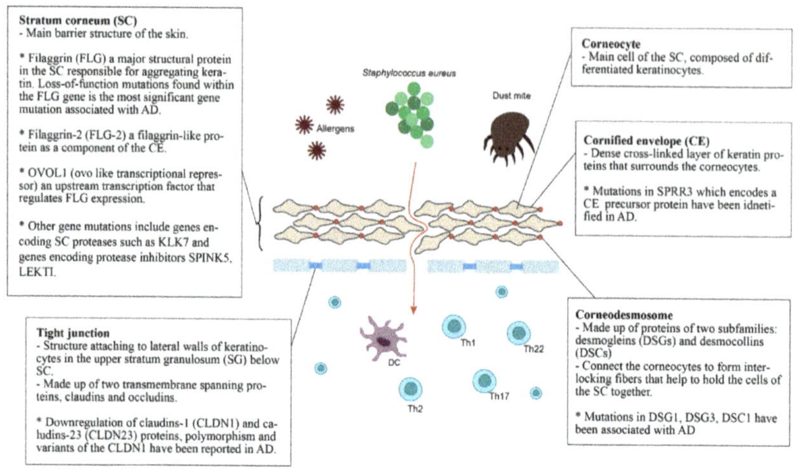

Figure 1. Defective epidermal barrier in atopic dermatitis (AD). The genes that affect SC may have an effect on both corneocytes and CE. For example, the filaggrin protein can be found in corneocytes but it aggregates keratin, affecting the whole SC. Therefore, genes with specific localization of action are placed in the small boxes, while the others are placed in the larger box of SC.

Stratum corneum (SC): The most significant gene variants associated with AD are the loss-of-function mutations found in the filaggrin (FLG) gene. An estimated 27.5% of Caucasian Americans, 48% of Europeans, 31.4% of Chinese and 20% of Japanese populations with AD present mutations in the FLG gene [22]. Filaggrin is a protein found in the corneocytes responsible for aggregating keratin in the formation of the stratum corneum [23]. It is produced from a precursor, pro-filaggrin. Studies have confirmed that FLG null mutations increase the risk of AD, impairing skin barrier function [24,25]. Homozygous mutations in the FLG gene are associated with an increased risk of severe AD, with earlier onset, longer duration, and increased skin infections [26–28]. In addition, filaggrin has a broad range of immunomodulatory effects [28–31].

Filaggrin-2 (FLG-2) is a filaggrin-like protein, and is part of the corneocyte envelope [32]. The expression of the FLG2 protein is reported to be decreased in patients with AD [33]. A link between polymorphisms in the FLG2 gene and more persistent AD in African American populations has been found [34]. However, a recent study in Brazil found no correlation between AD patients and polymorphisms in the FLG2 gene [35].

OVOL1 (ovo-like transcriptional repressor) is an upstream transcription factor that regulates FLG expression. FLG, OVOL1 and IL13 are reported to be the three genes most significantly associated with AD among the 31 susceptible gene loci reported in a meta-analysis of genome-wide association studies [36].

Other gene mutations that result in epidermal barrier dysfunction include genes that encode SC proteases such as KLK7, and genes that encode protease inhibitors SPINK5 and LEKTI. Kallikreins are a family of 15 trypsin- or chymotrypsin-like secreted serine proteases (KLK1-KLK15). The expression of KLK5-8, KLK10, KLK13 and KLK14 is significantly increased in AD patients [37], and the elevation of KLK7 is predominant in the SC [37]. A 4-bp insertion in the 3′-untranslated region of the KLK7 gene was found to have a significant association with AD [38].

The multidomain serine protease inhibitor Kazal-type 5 (SPINK5), otherwise known as the lympho-epithelial Kazal-type-related inhibitor (LEKTI), plays a role in keratinocyte differentiation during skin and hair morphogenesis, and the protective barrier function of skin by inhibiting the activity of KLKs in the epidermis [39,40]. Loss of KLK regulation through SPINK5 mutation could cause excessive KLK activity, resulting in permeable barrier defects [40,41]. AD has been associated with SPINK5 mutations in certain populations, specifically eastern Asians [40,42–46].

Cornified envelope (CE): The corneocyte envelope (CE) serves as a scaffold for lipids to attach and provides a supportive force to the corneocytes. The envelope is formed from structural proteins, including involucrin, loricrin, and the small proline-rich (SPRR) proteins [47]. An extra 24-bp defect in the central domain and additional in-frame deletions and insertions of the SPRR3 gene have been associated with AD [48]. Levels of FLG, FLG2, and SPRR3 mRNAs and proteins were also found to be reduced in AD skin [49].

Corneodesmosomes and tight junctions: Mutations within genes that express corneodesmosomal proteins (desmogleins and desmocollins) and tight junction proteins (claudins and ocludins) also contribute to the progression of AD [43]. Desmoglein-1 (DSG1), Claudin-1 (CLDN1) and Claudin-23 (CLDN23) have been reported to cause downregulation in AD [50]. CLDN1 haplo-type-tagging single nucleotide polymorphisms reveal linkage to AD in two North American populations [51]. The risk of eczema herpeticum in AD subjects is also associated with variants in the CLDN1 gene [51]. DSG3-/- mice appear to have traumatized skin that displays a distinct separation of desmosomes under electron microscopy [52]. Mice lacking desmocollin 1 (DSC1) have a fragile and flaky epidermis with acanthosis in the stratum granulosum [53].

2.2. Gene Polymorphisms in Inflammation and Immunity

Genetic variants associated with these immunological events contribute to the aberrant inflammatory and immune response in AD. Mutations in pattern-recognition receptors (PRRs) have been observed to be related to AD; these include polymorphisms in toll-like receptors (TLRs)—TLR2, TLR4, TLR6, TLR9—and gene polymorphisms in nucleotide-binding oligomerization domain-like receptors (NLRs)—CARD4, CARD12, CARD15, NALP1, NALP12, and NOD1; several SNPs of the human β-defensin 1 (DEFB1) gene have also been found in AD patients [19]. Mutations in IL-1 family cytokines and receptors genes that induce systemic Th2-type inflammatory responses, e.g., the susceptibility loci 2q12, which contain the receptors of the IL-1 family cytokines (IL1RL1, IL18R1, and IL18RAP) and the IL-18 gene play key roles in innate immunity and contribute to the pathogenesis of AD. Mutations in genes implicated in the vitamin D metabolism and synthesis of its receptors (CYP27A1, CYP2R1 and VDR) have been reported to be associated with AD. Mutations in interleukin genes produced by keratinocytes, including IL-25, TSLP, IL-33 and IL-7RA, were found in the epidermis in lesions of AD exposed to stress, e.g., UV or mechanical trauma. The adaptive immune response in AD is associated with the increased expression of the Th2 cytokines (IL-4, IL-5, IL-13, and IL-31) and the Th22 cytokine IL-22 during the acute phase of AD [54,55]. Several distinct polymorphisms of IL-4, IL-5, IL-13, IL-4 receptor alpha (IL-4RA), IL-5 receptor alpha (IL-5RA), and IL-13 receptor alpha (IL-13RA) have been found to influence the susceptibility to AD in different populations. Genetic variants in IL-12 and IL-12R [54,55], IFNG and IFNGR1, as well as interferon regulatory factor (IRF)-2, were significantly associated with AD and eczema herpeticum (EH) [56,57]. Other cytokine and receptor variants were also identified in AD, including IL-2, IL-6, IL-9, and IL-10 [19]. Correlations between AD and genetic polymorphisms of FcεRIα—the alpha-chain of high-affinity IgE receptors—have also been observed [19].

3. Genes Associated with Tooth Agenesis (TA)

Tooth development is a series of genetically regulated processes with successive and reciprocal interactions of the epithelium and mesenchyme (Figure 2). Four major signaling pathways (Fgf, Wnt, Bmp and Shh) and numerous transcription factors are key to tooth development. Disturbances at any stage or alterations in any pathway may lead to tooth agenesis [58].

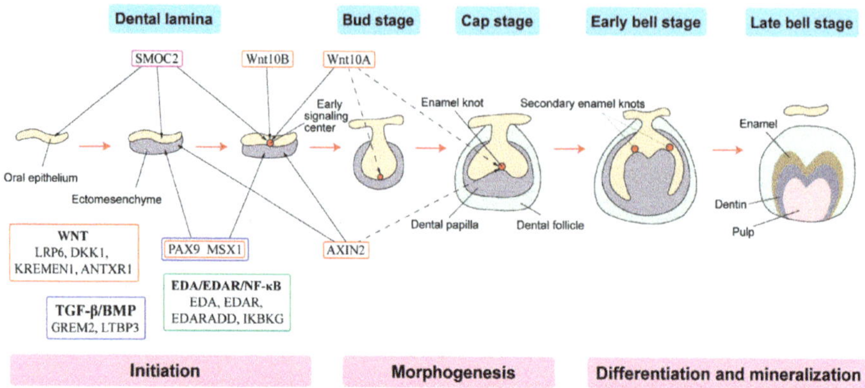

Figure 2. Schematic representation of tooth development stages. (Modified from the work of Nanci A. in 2013. Development of the tooth and its supporting tissues [59]) Main groups of signaling pathways and genes associated with tooth agenesis (TA). Gene mutations in Wnt pathway (orange box), TGF–β/BMP pathway (blue box), EDA/EDAR/κB pathway (green box) and SPARC family (purple box) affecting the initiation stage of tooth development results in tooth agenesis.

3.1. Paired Box Gene 9 (PAX9)

PAX9 is the most prevalent gene for non-syndromic TA [60]. It encodes a member of the paired box family of transcription factors and is expressed in the mesenchyme to induce odontogenic signals and initiate dental development [61]. PAX9 is important for organogenesis, as it induces the activation of Wnt and TGF-β/BMP signaling pathways [62]. Mutations in PAX9 often lead to the absence of second molars [61,63,64], and are associated with a high risk of maxillary lateral incisor agenesis [65]. Second premolars were also sometimes affected [7,66].

3.2. Muscle Segment Homeobox 1 (MSX1)

MSX1, a member of the homeobox genes, encodes for a protein that acts as a transcriptional repressor during embryogenesis and is critical for the development of teeth [67]. The Wnt/β-catenin signaling may increase MSX1 expression which subsequently activates the TGF-β/BMP cascade for odontogenesis [68]. Mutations in MSX1 have been associated with severe forms of hypodontia, oligodontia with cleft lip, and non-syndromic TA, usually missing mandibular central incisors, upper lateral incisors, second premolars, and third molars [61,69–71]. Regulation of BMP4 expression can also be affected through the synergistic interaction between MSX1 and PAX9 [72].

3.3. Axis Inhibitor 2 (AXIN2)

AXIN2 encodes an intracellular inhibitor of Wnt/β-catenin signaling and has been associated with lower incisor agenesis [73,74]. AXIN2 missense mutants were found to enhance β-catenin degradation and reduced Wnt activation, whereas the truncated mutants seemed to heighten the activation of Wnt/β-catenin [75]. A nonsense mutation in the AXIN2 gene was reported as etiologic for familial TA and predisposes patients to colorectal cancer [76]. AXIN2 is highly expressed in the enamel knot and underlying mesenchyme during tooth formation in mice [76].

3.4. Ectodysplasin A (EDA) and Relevant Genes

Ectodysplasin A, a protein of the tumor necrosis factor family, plays an important role during the development of ectodermal organs and teeth by activating the IKBKG-NF-κB signaling pathway. Mutations in EDA and EDA receptor genes have been reported to affect sporadic hypodontia in families [77]. Most of the mutations in EDA are identified to cause

X-linked hypohidrotic ectodermal dysplasia (HED) [78]. In addition, some of these EDA mutations have also been associated with missing maxillary lateral incisor cases [64].

Most mutations in EDAR and EDARADD are associated with ectodermal dysplasia, with a few associated with non-syndromic TA. Several mutations of IKBKG will lead to incontinentia pigmenti and ectodermal dysplasia (HGMD); therefore, NF-κB activity is likely to be affected by these mutations in the aforementioned genes [60].

In addition, single nucleotide polymorphisms in the EDAR gene have also been associated with other dental malformations. For example, the presence/absence of the V370A allele of the EDAR gene has been correlated with modern human shovel-shaped incisors [79]. The 1540C allele of EDAR was also found to be strongly associated with the presence of incisor shoveling and hair thickness [80].

3.5. Other Genes Related to Wnt Signaling Pathway

Recently, the genetic link between the Wnt pathway and TA was highlighted once again through whole exome and Sanger sequencing, with the observation of many mutations in genes that encode for Wnt ligands and its receptors [60]. The reported genetic mutations include genes that encode for Wnt ligands such as WNT10A and WNT10B, and associated receptors such as LRP6. For individuals with non-syndromic TA, WNT10A is one of the most commonly mutated genes [60]. It is expressed in the dental epithelium at the dental lamina and bud stage and in the enamel knot during the cap stage [81]. Mutations in WNT10A account for more than half of the isolated hypodontia and oligodontia cases [81,82] and have been identified with odonto-onycho-dermal dysplasia [83–85].

Wnt10B, a structurally related protein, is also expressed in the dental epithelium during the early bud and cap stages of tooth development. Similar to Wnt10A, genetic mutations in Wnt10B have been found in dental anomalies, such as TA and oligodontia [86,87]. Impaired odontoblastic differentiation and vasculogenesis of dental stem cells can result from these Wnt10B mutants, as they are unable to activate Wnt signaling pathways [87].

LRP6 is important for cell differentiation and proliferation, as it encodes a protein of the Wnt-Fzd-LRP5-LRP6 complex, which triggers the Wnt/β-catenin signaling cascade. LRP6 mutants are unable to activate β-catenin, therefore preventing the activation of Wnt signaling [87]. These LRP6 mutations have been reported in those with non-syndromic TA [88].

Other genes implicated in isolated TA or oligodontia accompanied with ectodermal dysplasia include mutations in DKK1 and associated KREMEN1; while mutations in ANTXR1 were also implicated with syndromic TA [89–91]. DKK1, Dickkopf Wnt signaling pathway inhibitor 1, is involved in the regulation of embryonic and vascular development when it binds to the transmembrane receptor, KREMEN1, and co-receptor, LRP6, to inhibit Wnt/β-catenin signaling. Through its interaction with LRP6, ANTXR1 plays an important role in modulating Wnt signaling and stabilizing β-catenin [60].

3.6. Other Genes Related to TGF-β/BMP Signaling Pathway

The implicated TGF-β/BMP-associated genes include GREM2—for mutations associated with TA, taurodontism, short tooth roots, and microdontia—and LTBP3—for mutations associated with inherited dental anomalies and isolated oligodontia [92–94]. GREM2 is involved in the regulation of embryonic morphogenesis, specifically TGF-β signaling in tooth development, as it encodes for a BMP antagonist protein [91]. LTBP3 encodes for a protein that regulates the assembly, secretion, and targeting activity of the TGF-β molecules, through the formation of a complex [93,94].

3.7. SMOC2 Gene

Secreted protein acidic and rich in cysteine (SPARC)-related modular calcium binding 2 (SMOC2) is an early dental developmental gene in human beings, supported by its high expression in areas such as the oral ectoderm and dental epithelium [95]. This gene encodes a member of the SPARC family protein, which promotes matrix assembly and stimulates endothelial cell proliferation and migration, as well as angiogenic activity [96]. Dental anomalies such as dental dysplasia, severe oligodontia and extreme microdontia have been reported with SMOC2 gene mutations [97].

4. Protein–Protein Interaction Network Functional Enrichment Analysis in AD and TA

From the above review of the gene mutations associated with epidermal barrier defects in AD and TA, no direct overlap of the genes involved was observed. Nevertheless, it is possible that there are shared biological pathways or processes, or indirect interactions through intermediary molecules between the two diseases. Recent research has shown that protein–protein interactions (PPI) are crucial for most biological activities, and examining the protein interaction networks can help to identify key proteins that may be involved in both diseases or that act as "hubs" linking multiple pathways. For example, PPIs between Parkinson's disease and periodontitis have been found to indicate new candidate molecular mechanisms [98]. Therefore, we aim to conduct a protein interaction analysis between AD and TA to explore new potential targets for research purposes.

4.1. Methods

The protein–protein interaction networks for TA and AD were investigated using version 11.5 of the STRING (Search Tool for the Retrieval of Interacting Genes/Proteins) database (https://string-db.org/, accessed on 15 November 2022), together with association and analysis methods. The STRING database contains known and predicted protein–protein interactions, stemming from computational prediction, knowledge transfer between organisms, and interactions aggregated from other (primary) databases. In general, PPIs in STRING are derived from the following five main domains: Genomic Context Predictions, High-throughput Lab Experiments, (Conserved) Co-Expression, Automated Textmining and Previous Knowledge in Databases (https://cn.string-db.org/cgi/about?footer_active_subpage=content, accessed on 15 November 2022). Apart from in-house predictions and homology transfers, STRING also relies on many resources maintained elsewhere (https://cn.string-db.org/cgi/credits?footer_active_subpage=datasources, accessed on 15 November 2022). The methodological details of the STRING database and network analysis have been reported in a recent paper [99].

The lists of proteins associated with AD and TA as summarized above and in Figures 1 and 2 were input and a "high confidence" cutoff of 0.7 was set, as is the case in previously published studies [98,100,101]. A PPI network was then generated [102], as shown in Figure 3.

4.2. Results

In the resulting protein–protein interaction network map, proteins are presented as nodes, which are connected by color-coded lines representing potential protein–protein associations. We identified potential protein–protein interactions (PPIs) between the proteins known to be associated with TA and AD. Catenin beta-1 (CTNNB1) presents interactions with TA-associated genes, including AXIN2, WNT10A, WNT10B and LRP6, as a key downstream component of the canonical Wnt signaling pathway. Importantly, there is evidence that indicates functional links between CTNNB1 and desmosomal proteins (DSC1 and DSG3), which play an important role in the maintenance of skin barrier function, and thus affect the pathogenesis of AD (Figure 3). There are "experimentally determined" interactions (denoted by a pink line) between CTNNB1 and DSC1, and other interactions derived from "text-mining", meaning "co-mentioned in PubMed abstracts" (denoted by a light green line) and co-expression of genes (denoted by a black line). Furthermore, between CTNNB1 and DSG3, there are known "experimentally determined" interactions and "text-mining" interactions.

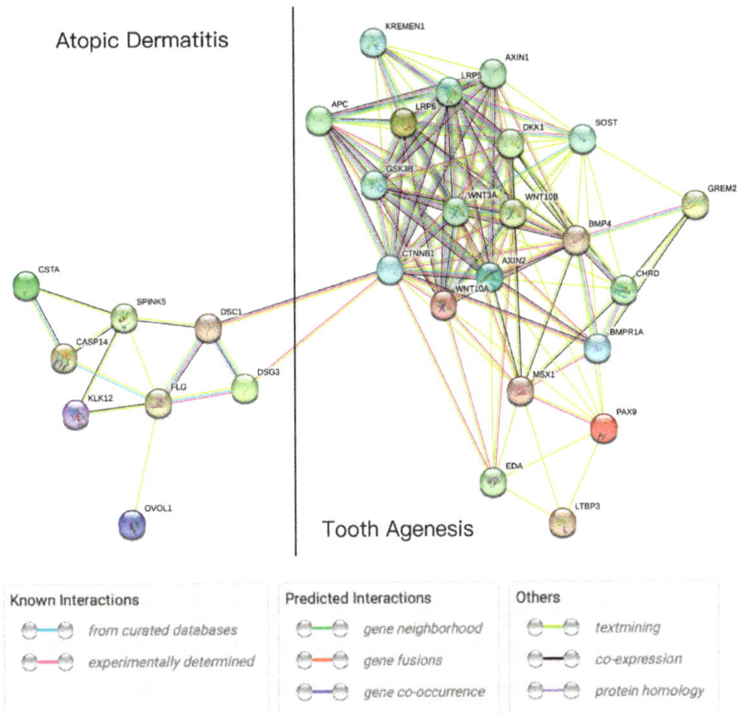

Figure 3. Search Tool for the Retrieval of Interacting Genes/Proteins (STRING) analysis reveals protein interaction networks between tooth agenesis and atopic dermatitis proteins. The color-coded nodes presented proteins, while color-coded lines represent protein–protein associations. Light blue lines represent associations in curated databases; pink lines represent experimentally determined associations derived from experimental/biochemical data; light green lines represent text-mining associations meaning "co-mentioned in PubMed abstracts"; black lines represent co-expression of genes in homo sapiens or other organisms; light purple lines represent putative homologs. There were no predicted interactions (green, red and blue lines) in the results.

5. Discussion and Limitations

While the Wnt/b-catenin pathway to junctional/desmosomal proteins interaction may have been previously determined experimentally (as illustrated by the purple/pink connection line), this interaction has not been proposed to be involved in the mechanistic pathways through which AD and TA co-occur. The identification of the interaction of these proteins that are associated with the known gene mutations for these two diseases is the novel result of this database search.

The β-catenin protein encoded by the CTNNB1 gene is part of a complex of proteins that constitute adherens junctions (AJs). AJs are necessary for the creation and maintenance of epithelial cell layers by regulating cell growth and adhesion between cells. Meanwhile, β-catenin is an integral part of the canonical Wnt signaling pathway.

Tooth development is a dynamic process that goes through the bud, cap and bell stages, root development and tooth eruption [103]. The Wnt/β-catenin signaling pathway is involved in embryonic development in many aspects and is active at all stages in various regions of tooth-forming, playing a key role in odontogenesis [104]. The machinery of the Wnt pathway includes extracellular secreted glycoproteins (19 Wnt ligands at the human level), seven transmembrane-spanning receptors (Frizzled and LRP5/6), cytoplasmic proteins (DVL, APC, AXIN, GSK3β and β-catenin, etc.), nuclear transcription factors (TCF/LEF), and several related molecules (MSX1, DKK1, KREMEN1, and ANTXR1) [60]. The Wnt and Wnt-associated pathways are demonstrated to play a major role in the molecular pathogenesis of the non-syndromic TA [60].

Research on the relationship between desmosomal components and β-catenin signaling has been carried out. It was shown that DSC3 regulated β-catenin in suprabasal keratinocytes by inducing β-catenin stabilization and transgene-mediated DSC3a and DSC3b expression in differentiating keratinocytes enhances β-catenin signaling [105]. The abnormal expression of DSC3, DSG3, and β-catenin was found in oral carcinomas and that the reduced or absent expression of β-catenin had a positive correlation with reduced or absent expression of DSC3 in 24 patients with lymph node metastasis [106]. Silencing DSG3 was reported to inhibit the activation of the Wnt/β-catenin signaling pathway in mice with chronic rhinosinusitis [107]. Additionally, Sawa et al. found that human dental pulp fibroblasts did not express desmoplakin (DPK, cytoplasmic membrane-associated protein in the epithelium) until they were cultured in the differentiation medium, whereas odontoblasts expressed vimentin-binding DPK-1 [108].

The canonical Wnt/β-catenin pathway is a fundamental mechanism that accounts for various biological activities, including cell proliferation, differentiation and development. Distinct from the non-canonical Wnt pathways that are independent of β-catenin, the canonical Wnt/β-catenin pathway has β-catenin as its typical characteristic [109]. In the inactive state, the β-catenin protein is degraded by a destruction complex composed of AXIN, glycogen synthase kinase 3β (GSK3β) and adenomatous polyposis coli (APC) [109]. Upon Wnt ligand receptor/coreceptor (LRP5/6) binding, this complex becomes inactivated, leading to cytoplasmic accumulation and subsequent nuclear translocation of β-catenin [110]. In the nucleus, β-catenin stimulates the transcription of target genes in cooperation with T-cell-factor/lymphoid enhancer-binding factors to regulate the expression of downstream target genes [111]. The effects of the Wnt/β-catenin inhibitor ICG-001 in an AD-like murine model generated by repeated topical application of the hapten oxazolone (Ox) were examined, and ICG-001 was found to attenuate epidermal permeability barrier function in Ox-AD mice [111]. A longitudinal birth cohort study that involved 1699 children in Korea found that children with the AD phenotype in early life were closely related to the development of asthma only in the cases of accompanying food allergy (FA) [112]. Ingenuity pathway analysis (IPA) of the colonocyte transcriptome revealed that the differentiation of FA with AD was best described by the genes in 'Wnt/β-catenin Signaling', specifically AXIN1, CCND1, FZD2, and WNT6, indicating that this mechanism might be regulated by Wnt signaling [112]. This pathway also has a strong link with TA.

In summary, PAX9 and MSX1 are the most prevalent genes for non-syndromic TA. Mutated genes that encode the components in the canonical Wnt/β-catenin pathway and Wnt-associated genes account for the highest genetic risk for isolated TA compared to mutated genes involved in several other pathways. TGF-β/BMP and EDA/EDAR/NF-κB signaling pathways also contribute to TA. Skin epithelial function and immune responses are the two major biologic pathways responsible for AD and they interactionally affect each other. Genetic mutations that cause structural defects in the epidermal barrier include genes that encode epidermal barrier structural proteins, stratum corneum proteases and protease inhibitors.

We identified protein interactions between desmosomal proteins and β-catenin. Desmosomes and tight junctions are essential components of the skin barrier, affecting the pathogenesis of AD. β-catenin is a key component of the Wnt/β-catenin signaling pathway, which is the major pathway responsible for TA. While experimental studies have shown that the DSG and DSC levels have a positive correlation with β-catenin expression and affect the Wnt signaling pathway, this protein interaction has not been previously proposed as one of the mechanistic pathways through which AD and TA co-occur. The specific mechanism by which these two diseases interact through this pathway remains unclear and more evidence is needed, for example, the GWAS risk of disease and EQTL analysis. Nevertheless, our findings help to narrow down the possible shared pathogenic pathways for future research to interrogate in this novel field, and help support the hypothesis that shared genetic mutations in the epidermal structure could increase the risk of tooth agenesis, thus linking structural defects in the skin barrier and tooth formation. Further studies may explore the mechanisms of gene variation in desmosomes and adherens in epithelial membranes, and their interactions with the Wnt signaling pathway linked to TA, in particular in the context of patients with TA and AD.

Author Contributions: Conceptualization, W.O., C.E.G., F.T.C. and C.-y.S.H.; methodology, W.O., C.E.G., F.T.C. and C.-y.S.H.; software, W.O. and C.E.G.; validation, W.O., C.E.G., F.T.C., E.P.H.Y. and C.-y.S.H.; formal analysis, W.O.; investigation, W.O., C.E.G., W.B.N., F.T.C. and C.-y.S.H.; resources, W.O., C.E.G., F.T.C. and C.-y.S.H.; writing—original draft preparation, W.O.; writing—review and editing, W.O., C.E.G., W.B.N., F.T.C., E.P.H.Y. and C.-y.S.H.; visualization, W.O., C.E.G. and C.-y.S.H.; supervision, C.E.G., F.T.C., E.P.H.Y. and C.-y.S.H.; project administration, W.O., C.E.G., F.T.C. and C.-y.S.H. All authors contributed to the paper and approved the submitted draft. All authors have read and agreed to the published version of the manuscript.

Funding: This project was partially supported by the Academic Research Funds of Singapore Ministry of Education via the National University of Singapore (A-0002960-00-00) and the Faculty of Dentistry, National University of Singapore (A-0002960-01 00).

Institutional Review Board Statement: Not applicable.

Informed Consent Statement: Not applicable.

Data Availability Statement: Not applicable.

Conflicts of Interest: The authors declare no conflict of interest.

References

1. Kim, J.; Kim, B.E.; Leung, D.Y.M. Pathophysiology of atopic dermatitis: Clinical implications. *Allergy Asthma Proc.* **2019**, *40*, 84–92. [CrossRef] [PubMed]
2. Laughter, M.R.; Maymone, M.B.C.; Mashayekhi, S.; Arents, B.W.M.; Karimkhani, C.; Langan, S.M.; Dellavalle, R.P.; Flohr, C. The global burden of atopic dermatitis: Lessons from the Global Burden of Disease Study 1990–2017. *Br. J. Dermatol.* **2021**, *184*, 304–309. [CrossRef] [PubMed]
3. Kalhan, T.A., Loo, E.X.L.; Kalhan, A.C.; Kramer, M.S.; Karunakaran, B.; Lam, C.U.; Van Bever, H.; Shek, L.P.-C.; Goh, A.; Chong, Y.S.; et al. Atopic dermatitis and early childhood caries: Results of the GUSTO study. *J. Allergy Clin. Immunol.* **2017**, *139*, 2000–2003. [CrossRef] [PubMed]
4. Park, H.J.; Choi, M.; Park, H.-J.; Haw, S. Dental Caries in Adults with Atopic Dermatitis: A Nationwide Cross-Sectional Study in Korea. *Ann. Dermatol.* **2021**, *33*, 154–162. [CrossRef]
5. Shim, J.-S.; Yang, M.-S. Identification of oral symptoms associated with atopic dermatitis in adolescents: Results from the Korea national representative survey 2009–2017. *Sci. Rep.* **2020**, *10*, 19461. [CrossRef]
6. Silva, M.J.; Kilpatrick, N.M.; Craig, J.M.; Manton, D.J.; Leong, P.; Burgner, D.; Scurrah, K.J. Etiology of Hypomineralized Second Primary Molars: A Prospective Twin Study. *J. Dent. Res.* **2018**, *98*, 77–83. [CrossRef]
7. Nieminen, P. Genetic basis of tooth agenesis. *J. Exp. Zool. Part B Mol. Dev. Evol.* **2009**, *312B*, 320–342. [CrossRef] [PubMed]
8. Lan, Y.; Jia, S.; Jiang, R. Molecular patterning of the mammalian dentition. *Semin. Cell Dev. Biol.* **2014**, *25–26*, 61–70. [CrossRef]
9. Balic, A.; Thesleff, I. Tissue Interactions Regulating Tooth Development and Renewal. *Curr. Top. Dev. Biol.* **2015**, *115*, 157–186. [CrossRef]
10. Yamaguchi, T.; Tomoyasu, Y.; Nakadate, T.; Oguchi, K.; Maki, K. Allergy as a possible predisposing factor for hypodontia. *Eur. J. Orthod.* **2008**, *30*, 641–644. [CrossRef]
11. Perugia, C.; Saraceno, R.; Ventura, A.; Lorè, B.; Chiaramonte, C.; Docimo, R.; Chimenti, S. Atopic dermatitis and dental manifestations. *G. Ital. Dermatol. Venereol.* **2017**, *152*, 122–125. [CrossRef]
12. Tan, S.; Leong, S.M.; Hsu, C.-Y.; Chandran, N. Association of moderate–severe atopic dermatitis with dental anomalies. *Indian J. Dermatol.* **2022**, *67*, 539–542. [CrossRef]
13. Sroka-Tomaszewska, J.; Trzeciak, M. Molecular Mechanisms of Atopic Dermatitis Pathogenesis. *Int. J. Mol. Sci.* **2021**, *22*, 4130. [CrossRef]
14. Maas, R.; Bei, M. The Genetic Control of Early Tooth Development. *Crit. Rev. Oral Biol. Med.* **1997**, *8*, 4–39. [CrossRef]
15. Vastardis, H. The genetics of human tooth agenesis: New discoveries for understanding dental anomalies. *Am. J. Orthod. Dentofac. Orthop.* **2000**, *117*, 0650–0656. [CrossRef]
16. Thesleff, I. Genetic basis of tooth development and dental defects. *Am. J. Med. Genet. Part A* **2000**, *58*, 191–194. [CrossRef]
17. De Coster, P.J.; Marks, L.A.; Martens, L.C.; Huysseune, A. Dental agenesis: Genetic and clinical perspectives. *J. Oral Pathol. Med.* **2009**, *38*, 1–17. [CrossRef]

18. Weidinger, S.; Novak, N. Atopic dermatitis. *Lancet* **2016**, *387*, 1109–1122. [CrossRef]
19. Liang, Y.; Chang, C.; Lu, Q. The Genetics and Epigenetics of Atopic Dermatitis—Filaggrin and Other Polymorphisms. *Clin. Rev. Allergy Immunol.* **2016**, *51*, 315–328. [CrossRef]
20. Choi, E.H. Aging of the skin barrier. *Clin. Dermatol.* **2019**, *37*, 336–345. [CrossRef]
21. Kezic, S.; Novak, N.; Jakasa, I.; Jungersted, J.M.; Simon, M.; Brandner, J.M.; Middelkamp-Hup, M.A.; Weidinger, S. Skin barrier in atopic dermatitis. *Front. Biosci.* **2014**, *19*, 542–556. [CrossRef]
22. Polcari, I.; Becker, L.; Stein, S.L.; Smith, M.S.; Paller, A.S. Filaggrin Gene Mutations in African Americans with Both Ichthyosis Vulgaris and Atopic Dermatitis. *Pediatr. Dermatol.* **2014**, *31*, 489–492. [CrossRef]
23. Sandilands, A.; Sutherland, C.; Irvine, A.D.; McLean, W.H.I. Filaggrin in the frontline: Role in skin barrier function and disease. *J. Cell Sci.* **2009**, *122*, 1285–1294. [CrossRef]
24. Kaufman, B.P.; Guttman-Yassky, E.; Alexis, A.F. Atopic dermatitis in diverse racial and ethnic groups-Variations in epidemiology, genetics, clinical presentation and treatment. *Exp. Dermatol.* **2018**, *27*, 340–357. [CrossRef]
25. Irvine, A.D.; McLean, W.H.I.; Leung, D.Y. Filaggrin Mutations Associated with Skin and Allergic Diseases. *N. Engl. J. Med.* **2011**, *365*, 1315–1327. [CrossRef]
26. Kim, B.E.; Leung, D.Y. Epidermal Barrier in Atopic Dermatitis. *Allergy, Asthma Immunol. Res.* **2012**, *4*, 12–16. [CrossRef]
27. Brown, S.J.; McLean, W.H.I. One Remarkable Molecule: Filaggrin. *J. Investig. Dermatol.* **2012**, *132*, 751–762. [CrossRef]
28. Marwah, I.; Wang, X.; Chan, H.; Ogg, G.S.; Gutowska-Owsiak, D. Filaggrin-insuciency in keratinocytes influences responsiveness of allergen-specific T cells to cognate antigen and compounds barrier function deficiency. *Clin. Immunol.* **2014**, *153*, 153–155. [CrossRef]
29. Jarrett, R.; Salio, M.; Lloyd-Lavery, A.; Subramaniam, S.; Bourgeois, E.; Archer, C.; Cheung, K.L.; Hardman, C.; Chandler, D.; Salimi, M.; et al. Filaggrin inhibits generation of CD1a neolipid antigens by house dust mite–derived phospholipase. *Sci. Transl. Med.* **2016**, *8*, 325ra18. [CrossRef]
30. Lee, K.H.; Cho, K.-A.; Kim, J.-Y.; Kim, J.-Y.; Baek, J.-H.; Woo, S.-Y.; Kim, J.-W. Filaggrin knockdown and Toll-like receptor 3 (TLR3) stimulation enhanced the production of thymic stromal lymphopoietin (TSLP) from epidermal layers. *Exp. Dermatol.* **2011**, *20*, 149–151. [CrossRef]
31. Leitch, C.S.; Natafji, E.; Yu, C.; Abdul-Ghaar, S.; Madarasingha, N.; Venables, Z.C.; Chu, R.; Fitch, P.M.; Muinonen-Martin, A.J.; Campbell, L.E.; et al. Filaggrin-null mutations are associated with increased maturation markers on Langerhans cells. *J. Allergy Clin. Immunol.* **2016**, *138*, 482–490. [CrossRef]
32. Wu, Z.; Hansmann, B.; Meyer-Hoffert, U.; Gläser, R.; Schröder, J.-M. Molecular Identification and Expression Analysis of Filaggrin-2, a Member of the S100 Fused-Type Protein Family. *PLoS ONE* **2009**, *4*, e5227. [CrossRef]
33. Makino, T.; Mizawa, M.; Yamakoshi, T.; Takaishi, M.; Shimizu, T. Expression of filaggrin-2 protein in the epidermis of human skin diseases: A comparative analysis with filaggrin. *Biochem. Biophys. Res. Commun.* **2014**, *449*, 100–106. [CrossRef]
34. Margolis, D.J.; Gupta, J.; Apter, A.J.; Ganguly, T.; Hoffstad, O.; Papadopoulos, M.; Rebbeck, T.R.; Mitra, N. Filaggrin-2 variation is associated with more persistent atopic dermatitis in African American subjects. *J. Allergy Clin. Immunol.* **2014**, *133*, 784–789. [CrossRef]
35. Hertz, A.; Azulay-Abulafia, L.; Nascimento, A.P.D.; Ohara, C.Y.; Kuschnir, F.C.; Porto, L.C. Analysis of filaggrin 2 gene polymorphisms in patients with atopic dermatitis. *An. Bras. Dermatol.* **2020**, *95*, 173–179. [CrossRef]
36. Nedoszytko, B.; Reszka, E.; Gutowska-Owsiak, D.; Trzeciak, M.; Lange, M.; Jarczak, J.; Niedoszytko, M.; Jablonska, E.; Romantowski, J.; Strapagiel, D.; et al. Genetic and Epigenetic Aspects of Atopic Dermatitis. *Int. J. Mol. Sci.* **2020**, *21*, 6484. [CrossRef]
37. Komatsu, N.; Saijoh, K.; Kuk, C.; Liu, A.C.; Khan, S.; Shirasaki, F.; Takehara, K.; Diamandis, E.P. Human tissue kallikrein expression in the stratum corneum and serum of atopic dermatitis patients. *Exp. Dermatol.* **2007**, *16*, 513–519. [CrossRef]
38. Vasilopoulos, Y.; Cork, M.J.; Murphy, R.; Williams, H.C.; Robinson, D.A.; Duff, G.W.; Ward, S.J.; Tazi-Ahnini, R. Genetic association between an AACC insertion in the 3′UTR of the stratum corneum chymotryptic enzyme gene and atopic dermatitis. *J. Investig. Dermatol.* **2004**, *123*, 62–66. [CrossRef]
39. Walley, A.J.; Chavanas, S.; Moffatt, M.F.; Esnouf, R.M.; Ubhi, B.; Lawrence, R.; Wong, K.; Abecasis, G.R.; Jones, E.Y.; Harper, J.I.; et al. Gene polymorphism in Netherton and common atopic disease. *Nat. Genet.* **2001**, *29*, 175–178. [CrossRef]
40. Hachem, J.-P.; Wagberg, F.; Schmuth, M.; Crumrine, D.; Lissens, W.; Jayakumar, A.; Houben, E.; Mauro, T.M.; Leonardsson, R.; Brattsand, M.; et al. Serine Protease Activity and Residual LEKTI Expression Determine Phenotype in Netherton Syndrome. *J. Investig. Dermatol.* **2006**, *126*, 1609–1621. [CrossRef]
41. Fortugno, P.; Furio, L.; Teson, M.; Berretti, M.; El Hachem, M.; Zambruno, G.; Hovnanian, A.; D'Alessio, M. The 420K LEKTI variant alters LEKTI proteolytic activation and results in protease deregulation: Implications for atopic dermatitis. *Hum. Mol. Genet.* **2012**, *21*, 4187–4200. [CrossRef]
42. Nishio, Y.; Noguchi, E.; Shibasaki, M.; Kamioka, M.; Ichikawa, E.; Ichikawa, K.; Umebayashi, Y.; Otsuka, F.; Arinami, T. Association between polymorphisms in the SPINK5 gene and atopic dermatitis in the Japanese. *Genes Immun.* **2003**, *4*, 515–517. [CrossRef]
43. Kusunoki, T.; Okafuji, I.; Yoshioka, T.; Saito, M.; Nishikomori, R.; Heike, T.; Sugai, M.; Shimizu, A.; Nakahata, T. SPINK5 polymorphism is associated with disease severity and food allergy in children with atopic dermatitis. *J. Allergy Clin. Immunol.* **2005**, *115*, 636–638. [CrossRef]

44. Lan, C.-C.E.; Tu, H.-P.; Wu, C.-S.; Ko, Y.-C.; Yu, H.-S.; Lu, Y.-W.; Li, W.-C.; Chen, Y.-C.; Chen, G.-S. Distinct SPINK5 and IL-31 polymorphisms are associated with atopic eczema and non-atopic hand dermatitis in Taiwanese nursing population. *Exp. Dermatol.* **2011**, *20*, 975–979. [CrossRef]
45. Kato, A.; Fukai, K.; Oiso, N.; Hosomi, N.; Murakami, T.; Ishii, M. Association of SPINK5 gene polymorphisms with atopic dermatitis in the Japanese population. *Br. J. Dermatol.* **2003**, *148*, 665–669. [CrossRef]
46. Zhao, L.P.; Di, Z.; Zhang, L.; Wang, L.; Ma, L.; Lv, Y.; Hong, Y.; Wei, H.; Chen, H.D.; Gao, X.H. Association of SPINK5 gene polymorphisms with atopic dermatitis in Northeast China. *J. Eur. Acad. Dermatol. Venereol.* **2012**, *26*, 572–577. [CrossRef]
47. Cabral, A.; Voskamp, P.; Cleton-Jansen, A.-M.; South, A.; Nizetic, D.; Backendorf, C. Structural Organization and Regulation of the Small Proline-rich Family of Cornified Envelope Precursors Suggest a Role in Adaptive Barrier Function. *J. Biol. Chem.* **2001**, *276*, 19231–19237. [CrossRef]
48. Kelsell, D.P.; Byrne, C. SNPing at the Epidermal Barrier. *J. Investig. Dermatol.* **2011**, *131*, 1593–1595. [CrossRef]
49. Trzeciak, M.; Sakowicz-Burkiewicz, M.; Wesserling, M.; Dobaczewska, D.; Gleń, J.; Nowicki, R.; Pawelczyk, T. Expression of Cornified Envelope Proteins in Skin and Its Relationship with Atopic Dermatitis Phenotype. *Acta Derm. Venereol.* **2017**, *97*, 36–41. [CrossRef]
50. De Benedetto, A.; Rafaels, N.M.; McGirt, L.Y.; Ivanov, A.I.; Georas, S.N.; Cheadle, C.; Berger, A.E.; Zhang, K.; Vidyasagar, S.; Yoshida, T.; et al. Tight junction defects in patients with atopic dermatitis. *J. Allergy Clin. Immunol.* **2011**, *127*, 773–786.e7. [CrossRef]
51. Herrmann, N.; Koch, S.; Leib, N.; Bédorf, J.; Wilms, H.; Schnautz, S.; Fimmers, R.; Bieber, T. TLR2 down-regulates FcεRI and its transcription factor PU.1 in human Langerhans cells. *Allergy* **2013**, *68*, 621–628. [CrossRef]
52. Koch, P.J.; Mahoney, M.G.; Ishikawa, H.; Pulkkinen, L.; Uitto, J.; Shultz, L.; Murphy, G.F.; Whitaker-Menezes, D.; Stanley, J.R. Targeted Disruption of the Pemphigus Vulgaris Antigen (Desmoglein 3) Gene in Mice Causes Loss of Keratinocyte Cell Adhesion with a Phenotype Similar to Pemphigus Vulgaris. *J. Cell Biol.* **1997**, *137*, 1091–1102. [CrossRef]
53. Chidgey, M.; Brakebusch, C.; Gustafsson, E.; Cruchley, A.; Hail, C.; Kirk, S.; Merritt, A.; North, A.; Tselepis, C.; Hewitt, J.; et al. Mice lacking desmocollin 1 show epidermal fragility accompanied by barrier defects and abnormal differentiation. *J. Cell Biol.* **2001**, *155*, 821–832. [CrossRef]
54. Tsunemi, Y.; Saeki, H.; Nakamura, K.; Sekiya, T.; Hirai, K.; Fujita, H.; Asano, N.; Kishimoto, M.; Tanida, Y.; Kakinuma, T.; et al. Interleukin-12 p40 gene (IL12B) 3′-untranslated region polymorphism is associated with susceptibility to atopic dermatitis and psoriasis vulgaris. *J. Dermatol. Sci.* **2002**, *30*, 161–166. [CrossRef]
55. Takahashi, N.; Akahoshi, M.; Matsuda, A.; Ebe, K.; Inomata, N.; Obara, K.; Hirota, T.; Nakashima, K.; Shimizu, M.; Tamari, M.; et al. Association of the IL12RB1 promoter polymorphisms with increased risk of atopic dermatitis and other allergic phenotypes. *Hum. Mol. Genet.* **2005**, *14*, 3149–3159. [CrossRef]
56. Leung, D.Y.; Gao, P.-S.; Grigoryev, D.N.; Rafaels, N.M.; Streib, J.E.; Howell, M.D.; Taylor, P.A.; Boguniewicz, M.; Canniff, J.; Armstrong, B.; et al. Human atopic dermatitis complicated by eczema herpeticum is associated with abnormalities in IFN-γ response. *J. Allergy Clin. Immunol.* **2011**, *127*, 965–973.e5. [CrossRef]
57. Gao, P.-S.; Leung, D.Y.; Rafaels, N.M.; Boguniewicz, M.; Hand, T.; Gao, L.; Hata, T.R.; Schneider, L.C.; Hanifin, J.M.; Beaty, T.H.; et al. Genetic Variants in Interferon Regulatory Factor 2 (IRF2) Are Associated with Atopic Dermatitis and Eczema Herpeticum. *J. Investig. Dermatol.* **2012**, *132*, 650–657. [CrossRef]
58. Al-Ani, A.H.; Antoun, J.S.; Thomson, W.M.; Merriman, T.R.; Farella, M. Hypodontia: An Update on Its Etiology, Classification, and Clinical Management. *BioMed Res. Int.* **2017**, *2017*, 9378325. [CrossRef]
59. Nanci, A. *Ten Cate's Oral Histology: Development, Structure, and Function*, 8th ed.; Elsevier Mosby: St. Louis, MO, USA, 2013; pp. 70–94.
60. Yu, M.; Wong, S.W.; Han, D.; Cai, T. Genetic analysis: Wnt and other pathways in nonsyndromic tooth agenesis. *Oral Dis.* **2019**, *25*, 646–651. [CrossRef]
61. Wong, S.-W.; Han, D.; Zhang, H.; Liu, Y.; Zhang, X.; Miao, M.Z.; Wang, Y.; Zhao, N.; Zeng, L.; Bai, B.; et al. Nine Novel *PAX9* Mutations and a Distinct Tooth Agenesis Genotype-Phenotype. *J. Dent. Res.* **2018**, *97*, 155–162. [CrossRef]
62. Jia, S.; Zhou, J.; Fanelli, C.; Wee, Y.; Bonds, J.; Schneider, P.; Mues, G.; D'Souza, R.N. Small-molecule Wnt agonists correct cleft palates in *Pax9* mutant mice *in utero*. *Development* **2017**, *144*, 3819–3828. [CrossRef] [PubMed]
63. Kapadia, H.; Frazier-Bowers, S.; Ogawa, T.; D'Souza, R.N. Molecular characterization of a novel *PAX9* missense mutation causing posterior tooth agenesis. *Eur. J. Hum. Genet.* **2006**, *14*, 403–409. [CrossRef] [PubMed]
64. Suda, N.; Ogawa, T.; Kojima, T.; Saito, C.; Moriyama, K. Non-syndromic Oligodontia with a Novel Mutation of *PAX9*. *J. Dent. Res.* **2011**, *90*, 382–386. [CrossRef]
65. Alves-Ferreira, M.; Pinho, T.; Sousa, A.; Sequeiros, J.; Lemos, C.; Alonso, I. Identification of Genetic Risk Factors for Maxillary Lateral Incisor Agenesis. *J. Dent. Res.* **2014**, *93*, 452–458. [CrossRef]
66. Matalova, E.; Fleischmannova, J.; Sharpe, P.T.; Tucker, A.S. Tooth Agenesis: From Molecular Genetics to Molecular Dentistry. *J. Dent. Res.* **2008**, *87*, 617–623. [CrossRef]
67. MacKenzie, A.; Ferguson, M.W.; Sharpe, P.T. Expression patterns of the homeobox gene, Hox-8, in the mouse embryo suggest a role in specifying tooth initiation and shape. *Development* **1992**, *115*, 403–420. [CrossRef]
68. Yin, W.; Bian, Z. The Gene Network Underlying Hypodontia. *J. Dent. Res.* **2015**, *94*, 878–885. [CrossRef] [PubMed]

69. Vastardis, H.; Karimbux, N.; Guthua, S.W.; Seidman, J.G.; Seidman, C.E. A human MSX1 homeodomain missense mutation causes selective tooth agenesis. *Nat. Genet.* **1996**, *13*, 417–421. [CrossRef]
70. Cobourne, M.T.; Sharpe, P.T. Diseases of the tooth: The genetic and molecular basis of inherited anomalies affecting the dentition. *Wiley Interdiscip. Rev. Dev. Biol.* **2013**, *2*, 183–212. [CrossRef]
71. Satokata, I.; Maas, R.L. Msx1 deficient mice exhibit cleft palate and abnormalities of craniofacial and tooth development. *Nat. Genet.* **1994**, *6*, 348–356. [CrossRef]
72. Ogawa, T.; Kapadia, H.; Feng, J.Q.; Raghow, R.; Peters, H.; D'Souza, R.N. Functional Consequences of Interactions between *Pax9* and Msx1 Genes in Normal and Abnormal Tooth Development. *J. Biol. Chem.* **2006**, *281*, 18363–18369. [CrossRef]
73. Kuchler, C.; Lips, A.; Tannure, P.N.; Ho, B.; Costa, M.C.; Granjeiro, J.M.; Vieira, A.R. Tooth Agenesis Association with Self-reported Family History of Cancer. *J. Dent. Res.* **2013**, *92*, 149–155. [CrossRef]
74. Callahan, N.; Modesto, A.; Meira, R.; Seymen, F.; Patir, A.; Vieira, A.R. Axis inhibition protein 2 (AXIN2) polymorphisms and tooth agenesis. *Arch. Oral Biol.* **2009**, *54*, 45–49. [CrossRef] [PubMed]
75. Yue, H.; Liang, J.; Yang, K.; Hua, B.; Bian, Z. Functional analysis of a novel missense mutation in *AXIN2* associated with non-syndromic tooth agenesis. *Eur. J. Oral Sci.* **2016**, *124*, 228–233. [CrossRef]
76. Lammi, L.; Arte, S.; Somer, M.; Järvinen, H.; Lahermo, P.; Thesleff, I.; Pirinen, S.; Nieminen, P. Mutations in *AXIN2* Cause Familial Tooth Agenesis and Predispose to Colorectal Cancer. *Am. J. Hum. Genet.* **2004**, *74*, 1043–1050. [CrossRef]
77. Bergendal, B.; Klar, J.; Stecksén-Blicks, C.; Norderyd, J.; Dahl, N. Isolated oligodontia associated with mutations in *EDARADD*, *AXIN2*, *MSX1*, and *PAX9* genes. *Am. J. Med. Genet. Part A* **2011**, *155*, 1616–1622. [CrossRef] [PubMed]
78. Galluccio, G.; Castellano, M.; La Monaca, C. Genetic basis of non-syndromic anomalies of human tooth number. *Arch. Oral Biol.* **2012**, *57*, 918–930. [CrossRef]
79. Hlusko, L.J.; Carlson, J.P.; Chaplin, G.; Elias, S.A.; Hoffecker, J.F.; Huffman, M.; Jablonski, N.G.; Monson, T.A.; O'Rourke, D.H.; Pilloud, M.A.; et al. Environmental selection during the last ice age on the mother-to-infant transmission of vitamin D and fatty acids through breast milk. *Proc. Natl. Acad. Sci. USA* **2018**, *115*, E4426–E4432. [CrossRef] [PubMed]
80. Kimura, R.; Yamaguchi, T.; Takeda, M.; Kondo, O.; Toma, T.; Haneji, K.; Hanihara, T.; Matsukusa, H.; Kawamura, S.; Maki, K.; et al. A Common Variation in EDAR Is a Genetic Determinant of Shovel-Shaped Incisors. *Am. J. Hum. Genet.* **2009**, *85*, 528–535. [CrossRef]
81. van den Boogaard, M.-J.; Créton, M.; Bronkhorst, Y.; van der Hout, A.; Hennekam, E.; Lindhout, D.; Cune, M.; Ploos van Amstel, H.K. Mutations in *WNT10A* are present in more than half of isolated hypodontia cases. *J. Med. Genet.* **2012**, *49*, 327–331. [CrossRef]
82. Biedziak, B.; Firlej, E.; Dąbrowska, J.; Bogdanowicz, A.; Zadurska, M.; Mostowska, A. Novel Candidate Genes for Non-Syndromic Tooth Agenesis Identified Using Targeted Next-Generation Sequencing. *J. Clin. Med.* **2022**, *11*, 6089. [CrossRef]
83. Adaimy, L.; Chouery, E.; Mégarbané, H.; Mroueh, S.; Delague, V.; Nicolas, E.; Belguith, H.; de Mazancourt, P.; Mégarbané, A. Mutation in *WNT10A* Is Associated with an Autosomal Recessive Ectodermal Dysplasia: The Odonto-onycho-dermal Dysplasia. *Am. J. Hum. Genet.* **2007**, *81*, 821–828. [CrossRef]
84. Bohring, A.; Stamm, T.; Spaich, C.; Haase, C.; Spree, K.; Hehr, U.; Hoffmann, M.; Ledig, S.; Sel, S.; Wieacker, P.; et al. *WNT10A* Mutations Are a Frequent Cause of a Broad Spectrum of Ectodermal Dysplasias with Sex-Biased Manifestation Pattern in Heterozygotes. *Am. J. Hum. Genet.* **2009**, *85*, 97–105. [CrossRef]
85. Kantaputra, P.; Sripathomsawat, W. *WNT10A* and isolated hypodontia. *Am. J. Med. Genet. Part A* **2011**, *155*, 1119–1122. [CrossRef]
86. Kantaputra, P.N.; Hutsadaloi, A.; Kaewgahya, M.; Intachai, W.; German, R.; Koparal, M.; Leethanakul, C.; Tolun, A.; Cairns, J.R.K. *WNT10B* mutations associated with isolated dental anomalies. *Clin. Genet.* **2018**, *93*, 992–999. [CrossRef]
87. Yu, P.; Yang, W.; Han, D.; Wang, X.; Guo, S.; Li, J.; Li, F.; Zhang, X.; Wong, S.-W.; Bai, B.; et al. Mutations in *WNT10B* Are Identified in Individuals with Oligodontia. *Am. J. Hum. Genet.* **2016**, *99*, 195–201. [CrossRef]
88. Massink, M.P.; Créton, M.A.; Spanevello, F.; Fennis, W.M.; Cune, M.S.; Savelberg, S.M.; Nijman, I.J.; Maurice, M.M.; van den Boogaard, M.-J.H.; van Haaften, G. Loss-of-Function Mutations in the WNT Co-receptor *LRP6* Cause Autosomal-Dominant Oligodontia. *Am. J. Hum. Genet.* **2015**, *97*, 621–626. [CrossRef]
89. Dinckan, N.; Du, R.; Petty, L.E.; Coban-Akdemir, Z.; Jhangiani, S.N.; Paine, I.; Baugh, E.H.; Erdem, A.P.; Kayserili, H.; Doddapaneni, H.; et al. Whole-Exome Sequencing Identifies Novel Variants for Tooth Agenesis. *J. Dent. Res.* **2017**, *97*, 49–59. [CrossRef]
90. Issa, Y.A.; Kamal, L.; Abu Rayyan, A.; Dweik, D.; Pierce, S.; Lee, M.K.; King, M.-C.; Walsh, T.; Kanaan, M. Mutation of *KREMEN1*, a modulator of Wnt signaling, is responsible for ectodermal dysplasia including oligodontia in Palestinian families. *Eur. J. Hum. Genet.* **2016**, *24*, 1430–1435. [CrossRef]
91. Dinckan, N.; Du, R.; Akdemir, Z.C.; Bayram, Y.; Jhangiani, S.N.; Doddapaneni, H.; Hu, J.; Muzny, D.M.; Guven, Y.; Aktoren, O.; et al. A biallelic *ANTXR1* variant expands the anthrax toxin receptor associated phenotype to tooth agenesis. *Am. J. Med. Genet. Part A* **2018**, *176*, 1015–1022. [CrossRef]
92. Kantaputra, P.N.; Kaewgahya, M.; Hatsadaloi, A.; Vogel, P.; Kawasaki, K.; Ohazama, A.; Cairns, J.R.K. *GREMLIN 2* Mutations and Dental Anomalies. *J. Dent. Res.* **2015**, *94*, 1646–1652. [CrossRef]
93. Dabovic, B.; Chen, Y.; Colarossi, C.; Zambuto, L.; Obata, H.; Rifkin, D.B. Bone defects in latent TGF-beta binding protein (Ltbp)-3 null mice; a role for Ltbp in TGF-beta presentation. *J. Endocrinol.* **2002**, *175*, 129–141. [CrossRef]
94. Dabovic, B.; Levasseur, R.; Zambuto, L.; Chen, Y.; Karsenty, G.; Rifkin, D.B. Osteopetrosis-like phenotype in latent TGF-β binding protein 3 deficient mice. *Bone* **2005**, *37*, 25–31. [CrossRef]

95. Bloch-Zupan, A.; Jamet, X.; Etard, C.; Laugel, V.; Muller, J.; Geoffroy, V.; Strauss, J.-P.; Pelletier, V.; Marion, V.; Poch, O.; et al. Homozygosity Mapping and Candidate Prioritization Identify Mutations, Missed by Whole-Exome Sequencing, in SMOC2, Causing Major Dental Developmental Defects. *Am. J. Hum. Genet.* **2011**, *89*, 773–781. [CrossRef]
96. Ye, X.; Attaie, A.B. Genetic Basis of Nonsyndromic and Syndromic Tooth Agenesis. *J. Pediatr. Genet.* **2016**, *5*, 198–208. [CrossRef]
97. AlFawaz, S.; Fong, F.; Plagnol, V.; Wong, F.S.; Fearne, J.; Kelsell, D.P. Recessive oligodontia linked to a homozygous loss-of-function mutation in the *SMOC2* gene. *Arch. Oral Biol.* **2013**, *58*, 462–466. [CrossRef]
98. Botelho, J.; Mascarenhas, P.; Mendes, J.J.; Machado, V. Network Protein Interaction in Parkinson's Disease and Periodontitis Interplay: A Preliminary Bioinformatic Analysis. *Genes* **2020**, *11*, 1385. [CrossRef]
99. Szklarczyk, D.; Kirsch, R.; Koutrouli, M.; Nastou, K.; Mehryary, F.; Hachilif, R.; Gable, A.L.; Fang, T.; Doncheva, N.T.; Pyysalo, S.; et al. The STRING database in 2023: Protein-protein association networks and functional enrichment analyses for any se-quenced genome of interest. *Nucleic Acids Res.* **2023**, *51*, D638–D646. [CrossRef]
100. Leira, Y.; Mascarenhas, P.; Blanco, J.; Sobrino, T.; Mendes, J.J.; Machado, V.; Botelho, J. Network Protein Interaction in the Link between Stroke and Periodontitis Interplay: A Pilot Bioinformatic Analysis. *Genes* **2021**, *12*, 787. [CrossRef]
101. Liu, T.; Wang, S.; Wornow, M.; Altman, R.B. Construction of disease-specific cytokine profiles by associating disease genes with immune responses. *PLOS Comput. Biol.* **2022**, *18*, e1009497. [CrossRef]
102. Szklarczyk, D.; Franceschini, A.; Kuhn, M.; Simonovic, M.; Roth, A.; Minguez, P.; Doerks, T.; Stark, M.; Muller, J.; Bork, P.; et al. The STRING database in 2011: Functional interaction networks of proteins, globally integrated and scored. *Nucleic Acids Res.* **2011**, *39*, D561–D568. [CrossRef]
103. Huang, X.-F.; Chai, Y. Molecular regulatory mechanism of tooth root development. *Int. J. Oral Sci.* **2012**, *4*, 177–181. [CrossRef] [PubMed]
104. Thesleff, I.; Sharpe, P. Signalling networks regulating dental development. *Mech. Dev.* **1997**, *67*, 111–123. [CrossRef]
105. Hardman, M.J.; Liu, K.; Avilion, A.A.; Merritt, A.; Brennan, K.; Garrod, D.R.; Byrne, C. Desmosomal Cadherin Misexpression Alters β-Catenin Stability and Epidermal Differentiation. *Mol. Cell. Biol.* **2005**, *25*, 969–978. [CrossRef]
106. Wang, L.; Liu, T.; Wang, Y.; Cao, L.; Nishioka, M.; Aguirre, R.L.; Ishikawa, A.; Geng, L.; Okada, N. Altered expression of desmocollin 3, desmoglein 3, and β-catenin in oral squamous cell carcinoma: Correlation with lymph node metastasis and cell proliferation. *Virchows Arch.* **2007**, *451*, 959–966. [CrossRef] [PubMed]
107. Cheng, J.; Yang, J.; Xue, K.; Zhao, Y.; Zhao, C.; Li, S.; Wang, Z. Desmoglein 3 Silencing Inhibits Inflammation and Goblet Cell Mucin Secretion in a Mouse Model of Chronic Rhinosinusitis via Disruption of the Wnt/β-Catenin Signaling Pathway. *Inflammation* **2019**, *42*, 1370–1382. [CrossRef]
108. Sawa, Y.; Kuroshima, S.-I.; Yamaoka, Y.; Yoshida, S. Intracellular Distribution of Desmoplakin in Human Odontoblasts. *J. Histochem. Cytochem.* **2005**, *53*, 1099–1108. [CrossRef]
109. Matsuda-Hirose, H.; Yamate, T.; Goto, M.; Katoh, A.; Kouji, H.; Yamamoto, Y.; Sakai, T.; Uemura, N.; Kobayashi, T.; Hatano, Y. Selective Inhibition of β-Catenin/Co-Activator Cyclic AMP Response Element-Binding Protein-Dependent Signaling Prevents the Emergence of Hapten-Induced Atopic Dermatitis-Like Dermatitis. *Ann. Dermatol.* **2019**, *31*, 631–639. [CrossRef] [PubMed]
110. Reuter, S.; Beckert, H.; Taube, C. Take the Wnt out of the inflammatory sails: Modulatory effects of Wnt in airway diseases. *Lab. Investig.* **2016**, *96*, 177–185. [CrossRef] [PubMed]
111. Takahashi-Yanaga, F.; Kahn, M. Targeting Wnt Signaling: Can We Safely Eradicate Cancer Stem Cells? *Clin. Cancer Res.* **2010**, *16*, 3153–3162. [CrossRef]
112. Lee, S.-Y.; Kim, S.; Kang, M.J.; Song, K.-B.; Choi, E.J.; Jung, S.; Yoon, J.-S.; Suh, D.I.; Shin, Y.H.; Kim, K.W.; et al. Phenotype of Atopic Dermatitis with Food Allergy Predicts Development of Childhood Asthma via Gut Wnt Signaling. *Allergy Asthma Immunol. Res.* **2022**, *14*, 674–686. [CrossRef] [PubMed]

International Journal of *Molecular Sciences*

Article

Cross-Reactivity of Intraoral Allergic Contact Mucositis in the Nickel-Sensitized Ear Model of Metal Allergy

Ryota Matsubara [1,2], Kenichi Kumagai [2,3,*], Keisuke Nasu [2,4], Takamasa Yoshizawa [2,4], Kazutaka Kitaura [2,5], Motoaki Suzuki [2,6], Yoshiki Hamada [4] and Ryuji Suzuki [2,5]

1 Department of Oral and Maxillofacial Surgery, Sendai Tokushukai Hospital, Sendai 981-3116, Japan
2 Department of Rheumatology and Clinical Immunology, Clinical Research Center for Rheumatology and Allergy, Sagamihara National Hospital, National Hospital Organization, Sagamihara 252-0392, Japan
3 Department of Oral and Maxillofacial Surgery, Dentistry and Orthodontics, The University of Tokyo Hospital, Tokyo 113-8655, Japan
4 Department of Oral and Maxillofacial Surgery, School of Dental Medicine, Tsurumi University, Yokohama 230-8501, Japan
5 Repertoire Genesis Inc., Osaka 567-0085, Japan
6 Department of Anatomy and Physiology, Faculty of Medicine, Saga University, Saga 849-8501, Japan
* Correspondence: kumagaik-ora@h.u-tokyo.ac.jp; Tel.: +81-3-5800-8669; Fax: +81-3-5800-6832

Abstract: Cross-reactivity of metal allergies can make metal allergy treatment complicated because the background of immune response in cross-reactions remains unknown. In clinical settings, cross-reactivity among several metals has been suspected. However, the precise mechanism of immune response in cross-reactivity is unclear. Two sensitizations with nickel, palladium, and chromium plus lipopolysaccharide solution into the postauricular skin were followed by a single nickel, palladium, and chromium challenge of the oral mucosa to generate the intraoral metal contact allergy mouse model. Results showed that the infiltrating T cells in nickel-sensitized, palladium- or chromium-challenged mice expressed CD8+ cells, cytotoxic granules, and inflammation-related cytokines. Thus, nickel ear sensitization can cause cross-reactive intraoral metal allergy.

Keywords: metal allergy; nickel; palladium; chromium; cross-reactivity; allergic contact mucositis; metal-specific T cells

Citation: Matsubara, R.; Kumagai, K.; Nasu, K.; Yoshizawa, T.; Kitaura, K.; Suzuki, M.; Hamada, Y.; Suzuki, R. Cross-Reactivity of Intraoral Allergic Contact Mucositis in the Nickel-Sensitized Ear Model of Metal Allergy. *Int. J. Mol. Sci.* **2023**, *24*, 3965. https://doi.org/10.3390/ijms24043965

Academic Editor: Peter Proff

Received: 31 December 2022
Revised: 4 February 2023
Accepted: 12 February 2023
Published: 16 February 2023

1. Introduction

Metal allergy is a delayed-type hypersensitivity reaction in which T cell-dependent macrophage activation and inflammation cause tissue injury [1]. Unlike immediate hypersensitivity reactions, cellular hypersensitivity reactions, such as delayed-type hypersensitivity, are mediated by antigen-specific effector T cells. The liquefaction of metal materials causes metal hypersensitivity or allergic reactions that mediate antigen-specific T cell sensitization.

Previous studies aimed to evaluate antigen-specific immune mechanisms by developing mouse models of palladium (Pd), nickel (Ni), chromium (Cr), and titanium (Ti) allergies by sensitizing the skin with chloride and lipopolysaccharide (LPS) solutions and by challenging it with the injection of these metal solutions into the footpads and oral mucosa [2–6].

Previous data showed a heterogeneous group of patients with different manifestations of oral contact allergy to dental metals [7]. Generally, Pd-ACD is almost observed together with Ni-ACD [8,9]. Antibodies against one antigen may bind with other structurally similar antigens. Such binding to similar epitopes is referred to as cross-reaction [1]. This phenomenon may be mainly attributed to cross-reactivity between Pd and Ni. However, the precise mechanism is not completely understood [10–13]. Therefore, a novel mouse model of cross-reactive metal allergy in the footpad skin was established, and immune response was investigated [14]. However, the mucosal immune system has a different immune response than the skin immune response because it encounters antigens more frequently and more extensively [15,16].

Ni is the most common metal causing contact dermatitis [17]. It is found in several personal products, such as ear piercing, jewelry, belt buckles, metal fasteners on clothing, and eyeglass frames [18]. Among them, ear piercing is a sensitizer for developing Ni allergies [19]. However, whether allergies to other metals are involved is unknown.

Adverse reactions to metal ions, such as cheilitis, perioral dermatitis, burning mouth syndrome, lichenoid reaction, orofacial granulomatosis, pustulosis palmaris et plantaris, rheumatoid arthritis, and systemic lupus erythematosus, can cause serious issues due to incompatibility reactions to metal-containing biomaterials [20–23]. Nevertheless, the precise mechanism of cross-reactivity among metal allergens in the mucosa remains unknown. To elucidate the immune response of cross-reactive Ni allergy, the current study aimed to establish a novel mouse model of Ni ear sensitization and characterize intraoral-infiltrating T cells during the elicitation phase in terms of phenotypic T cell markers and cytokine expressions.

2. Results

2.1. *Oral Mucosa Swelling in a Metal Allergy Cross-Reaction Mouse Model*

All experimental protocols are depicted in the Materials and Methods section (Table 1). In all groups, the peak of buccal mucosa swelling was observed at 1 day after challenge. At 7 days after challenge, buccal mucosa swelling was significantly higher in the Ni-Pd and Ni-Cr groups than that in the control group. Meanwhile, swelling did not significantly differ between Pd-Ni and Cr-Ni mice and control mice. Buccal mucosa swelling was significantly higher in sensitization of Ni-induced allergic mice compared with sensitization of Pd-induced allergic mice (Figure 1). Additionally, visually significant swelling was observed in the buccal area of mice with Ni-Pd and Ni-Cr allergic contact mucositis (ACM) compared with control mice at day seven after the first challenge (Figure 2).

Table 1. Experimental groups of the metal allergy cross-reaction mouse model.

Groups ACM	Sensitization Metal Salts	Challenge for Elicitation Metal Salts
Ni *-Pd **	$NiCl_2$	$PdCl_2$
Ni-Cr ***	$NiCl_2$	$CrCl_2$
Pd-Ni	$PdCl_2$	$NiCl_2$
Cr-Ni	$CrCl_2$	$NiCl_2$

* Nickel; ** Palladium; *** Chromium.

2.2. *Histological and Immunohistochemical Analyses of CD3 and F4/80 in the Oral Mucosa of Mice with Cross-Reactive Metal-Induced Allergy*

To validate whether antigen-presenting cells (APCs) and T cells infiltrated into the site of inflamed skin, we analyzed the oral mucosa of metal-induced ACM and control mice at 1 and 7 days after the challenge. Hematoxylin and eosin (H&E) staining showed epithelial acanthosis and epidermal spongiosis and liquefaction degeneration of the epithelial basal layer infiltrated with dense mononuclear cells in the epithelial basal layer and upper dermis of ACM mice (Figure 3 Day1-C, Day7-B,C). Immunohistochemical staining showed that CD3-positive T cells existed in the epithelial basal layer and upper dermis of ACM mice (Figure 3 Day1-F,G, Day7-F,G). Immunohistochemical staining revealed that F4/80-positive cells predominantly existed in the epithelial basal layer and upper dermis of ACM mice (Figure 3 Day1-J,K, Day7-J–L). In the Pd-Ni groups, F4/80 was present only after 7 days. In contrast, inflammatory reactions (H&E, CD3-positive T cells, F4/80-positive cells) were not observed in the oral mucosa of the control mice (Figure 3).

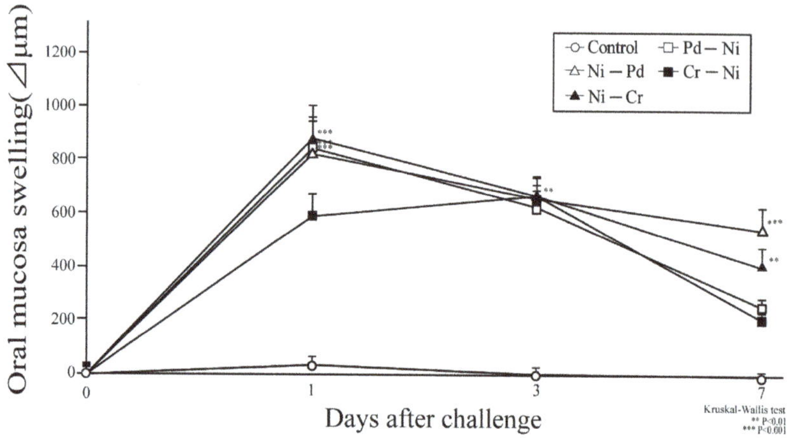

Figure 1. Oral mucosa swelling in mice with cross-reactive metal-induced allergy. In all groups, swelling was measured 1, 3, and 7 days after the first challenge. Furthermore, sensitization to Ni-, Pd-, and Cr-induced allergies was evaluated 7 days after the first challenge. Bars and error bars indicate the mean + standard deviation (SD). Statistical significance was evaluated using the Kruskal–Wallis test, followed by Dunn's multiple comparison tests. ** p value of <0.01 was considered very significant, and *** p value of <0.001 was considered extremely significant.

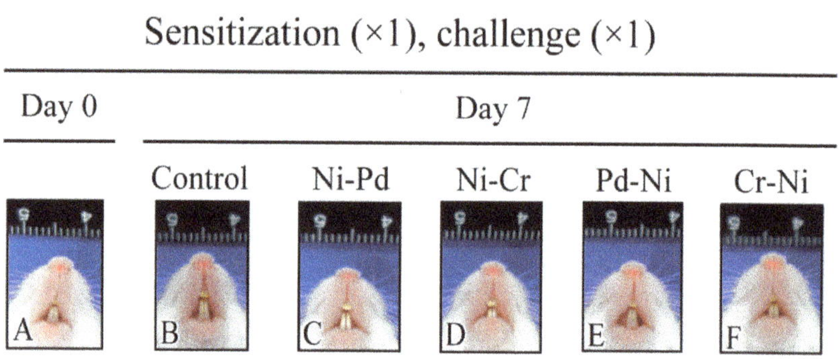

Figure 2. Swelling of the buccal mucosa of mice before challenge (**A**), and 7 days after the first challenge (**B–F**) were shown. Comparison of buccal mucosal swelling 7 days after the first challenge in control mice (**B**) and cross-reactive metal-induced allergic mice (**C–F**).

2.3. Expression Levels of T Cell Markers, Related Cytokines, and APC-Related Markers in the Oral Mucosa of Mice with Cross-Reactive Metal-Induced Allergy

We investigated the expression levels of T cell markers, related cytokines, and APC-related markers via quantitative polymerase chain reaction (qPCR). Messenger RNA (mRNA) expression levels of CD4 and CD8 in the left and right buccal mucosa were assessed at 1 and 7 days after the challenge. To validate whether T cells infiltrated into the inflamed oral mucosa in ACM mice, we performed qPCR analysis of CD4 and CD8 expressions. Metal-induced ACM mice had significantly higher CD8 levels than control mice 1 and 7 days after the challenge. The CD8 levels of Pd-Ni group and Ni-sensitized mice (Ni-Pd, Ni-Cr) were significantly higher than control mice at 1 day after challenge, and Ni mice (Ni-Pd, Ni-Cr) were significantly higher than control mice even at 7 days after challenge (Figure 4).

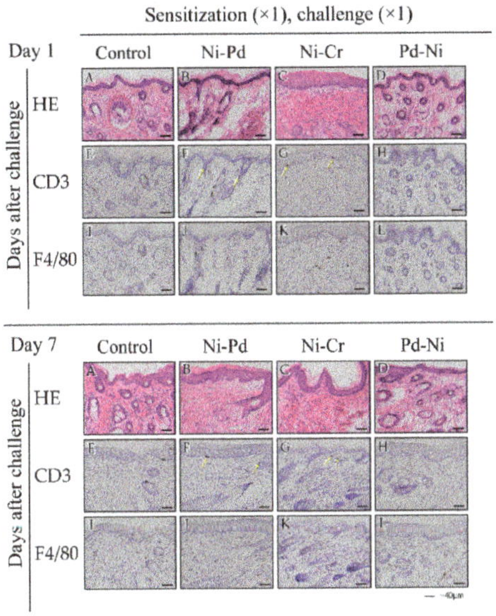

Figure 3. Histopathology and immunohistochemical analyses of accumulated T cells and antigen-presenting cells (APCs) in mice with cross-reactive metal induced allergy. Histopathology and immunohistochemical analyses of monoclonal antibody (mAb) that binds to a surface molecule on CD3-positive T cells and mature macrophages and dendritic cells (F4/80-positive cells) in buccal mucosa tissues. Frozen oral mucosa tissue sections were stained with hematoxylin and eosin (H&E) (**A–D**) and anti-CD3 (**E–H**) and anti F4/80 (**I–L**) antibodies 1 and 7 days after the challenge. Representative examples of CD3-positive T cells are indicated by arrows. Scale bar = 40 μm.

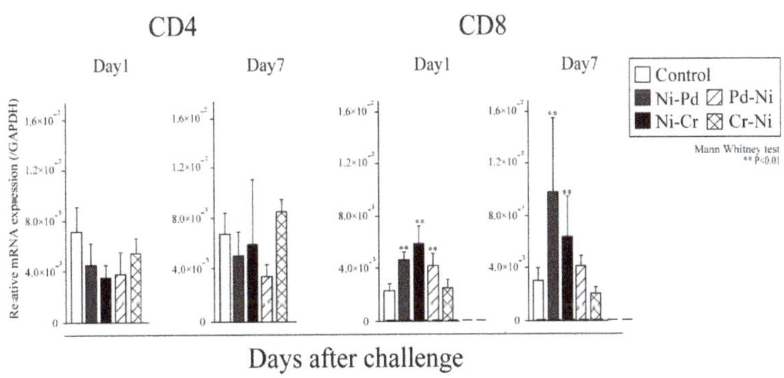

Figure 4. mRNA expression of T cell phenotypes in the oral mucosa of mice with cross-reactive metal-induced allergy. The mRNA expression of CD4 and CD8 in the buccal mucosa tissue was assessed 1 and 7 days after the first challenge. GAPDH gene expression was used as an internal control. Bars and error bars indicate the mean + standard deviation (SD). Statistical significance was tested using the unpaired Mann–Whitney test. ** p value of <0.01 was considered very significant.

Further, we examined the expression levels of Th1 cytokines (tumor necrosis factor [TNF]-α and interferon [IFN]-γ), Th2 cytokines (IL-4 and IL-5), cytotoxic granules (granzyme A and B), transcription factors of regulatory T cells (Foxp3), CD1d-restricted T cells (CD-1d), and MHC-related protein 1 (MR1) in the left and right buccal mucosa at 7 days after challenge (Figure 5). The levels of IFN-γ, IL-4 and granzyme B levels were significantly higher in sensitization of Ni mice (Ni-Pd, Ni-Cr) than control mice at 7 days after challenge (Figure 5A–C). The levels of TNF-α, granzyme A, CD-1d, and MR1 levels were significantly higher in Pd-Ni groups and sensitization of Ni mice (Ni-Pd, Ni-Cr) than control mice at 7 days after challenge (Figure 5A,B,D).

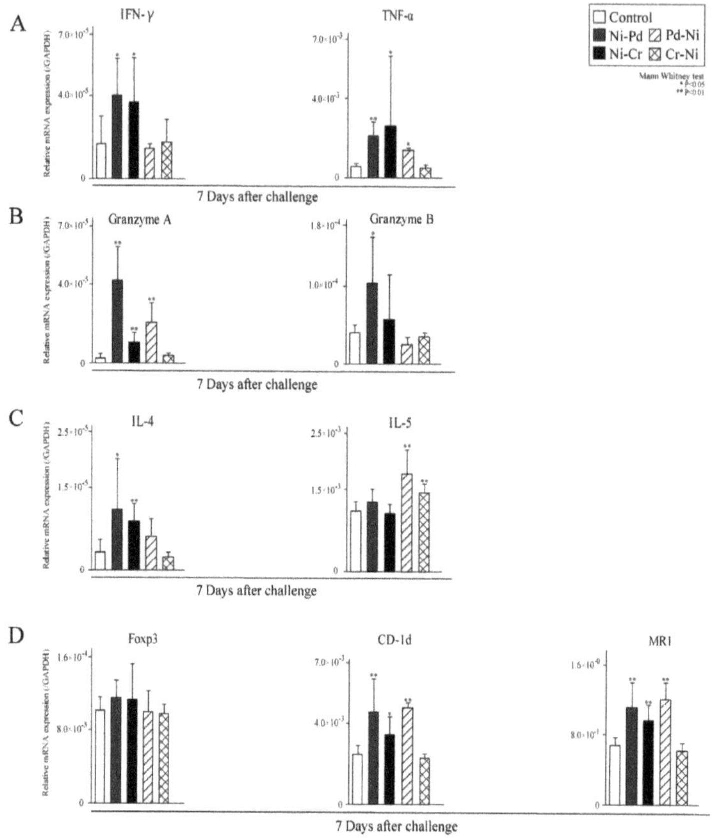

Figure 5. mRNA expression of T cell-related cytokines, cytotoxic granules, and T cell-related markers genes in the oral mucosa of cross-reactive metal-induced allergic mice. mRNA expression of (**A**) T helper type (Th) 1 cytokines (tumor necrosis factor(TNF)-α, Interferon(IFN)-γ), (**B**) cytotoxic granules (granzyme A and B), (**C**) T helper type (Th) 2 cytokines (IL-4 and IL-5), (**D**) transcription factors of regulatory T cells, CD1d-restricted T cells, and MHC-related protein 1 in the buccal mucosa tissue were assessed 7 days after the first challenge. GAPDH gene expression was used as an internal control. Bars and error bars indicate the mean + standard deviation (SD). Statistical significance was tested using the unpaired Mann–Whitney test. * p value of <0.05 was considered significant, and ** p value of <0.01 was considered very significant.

3. Discussion

Delayed type hypersensitivity generally occurs when the swelling takes 24 to 48 h after challenge, and onset of the pathology may take three to five days [24]. Clinically, the patch test is evaluated until 1 week later. In this study, in all groups, the peak of buccal mucosa swelling was observed at 1 day after challenge. Moreover, allergic reactions were most observed at 7 days after challenge. Infiltrating T cells in the Ni-sensitized, Pd- and Cr-challenged mice expressed the most CD8+ cells 7 days after the challenge. However, infiltrating T cells in Pd- and Cr-sensitized, Ni-challenged mice did not express CD8+. Ni-sensitized mice had significantly higher expression levels of almost all immune response-related genes. Cr-sensitized, Ni-challenged mice almost did not show significant differences in buccal mucosal swelling or in the expression levels of immune response-related genes. Thus, we suggested that the sensitization of Ni caused significant cross-reactivity with Pd and Cr. In our previous study, the oral mucosal allergic mouse model sensitized and challenged with nickel had the highest CD8 expression at day one of elicitation for challenge [5]. It suggests that allergic reactions may differ depending on the metal exposed to challenge, even when sensitized to Ni. Cross-reactivity in allergic contact dermatitis occurs due to the similar structure of antigens. Thus, cross-reactivity is found among several foreign substances, such as food [25], antibiotics [26], antifungals [27], anti-inflammatory analgesics [28], and steroids [29]. In metals, cross-reactions between Ni and Pd are the main reactions. A previous study has shown that only Ni and Pd cross-reacted in ear skin [30]. Ni and Pd have physiochemical similarities, and they belong to the same group in the periodic table. This fact considers that cross-reactivity relates near to the periodic table of the chemical elements (PSE) [31]. Identical structures form the same complexes and cause similar modifications, and they may be recognized by the same T cells [31]. Franziska et al. reported up to 80% Pd co-sensitization with Ni [32]. However, in the oral cavity, during Ni sensitization, immune responses were observed even during Cr challenge for elicitation. This fact may be correlated with differences in the immune mechanisms of the skin and oral cavity. The skin and mucous membranes have different immune responses. Contact dermatitis is a disease that can be cured by identifying the causative allergen and discontinuing contact. However, when the cause of the disease is not clear and appropriate protective measures cannot be taken, it is often intractable and difficult to treat. Symptomatic treatment of contact dermatitis without identifying the cause is not a desirable option, with consideration of the risk of side effects due to the continuous use of topical steroids and the unnecessary expenditure of medical expenses. Treatment is mainly recommended with the removal of the causative metal and treatment with steroids or antihistamines.

In clinical settings, the treatments of metal allergy include allergen avoidance and use of antihistamine drugs and corticosteroids. However, it is challenging to determine removal because it is difficult to identify the substances causing cross-reactions [33]. The use of corticosteroids is effective in immune system diseases. However, it also has side effects. A previous study reported that antihistamines inhibited immune responses in metal allergies compared with corticosteroids [34]. However, treatment of the metal allergy intraorally are only the removal of the causative agent and topical application steroid. In addition, in rare cases, the topical steroid ointment itself can cause allergies [35]. Allergic reactions to topical medications can be caused not only by the drug but also by the base or preservative. Whether the allergy is caused by a single metal or a cross-reaction is challenging to determine due to the presence of multiple metals in a small area in the oral cavity. In such cases, it is difficult to investigate cross-reactivity. However, steroids should not be continually used if the cause is unknown. Testing methods for metal allergies include blood tests, lymphocyte transformation tests [36], oral challenge tests [37], hair mineral analyses [38], and principal component analyses [39]. Among these tests, the patch tests are the most generic test [40]. They have long been used to identify allergenic substances. Identifying allergens via patch testing can help treat refractory and recurrent allergic dermatitis. In patients with a history of metal allergy, patch testing is recommended prior to the use of metals with treatment [41]. The problem is that judgments about how to implement, judge, discuss, and guide patients' lives based on patch test results require some practice because of the bias of the users. In addition, whether the patients have a history of immunologically relevant co-exposures is not known, and this is an issue with analyzing cross-reactivity in patch tests. Hence, the timing of patch testing is not similar, and analyzing the causality of cross-reactivity is challenging. Furthermore, some patients, including those on steroids, those who need to shave, and pregnant women, should not undergo the test. Due to the abovementioned reasons, everyone cannot be tested. However, there is still no more reliable and useful test method than the patch test in identifying the cause of the problem. Therefore, it is important to elucidate the background of immune response in metal allergy to obtain a better diagnosis.

Ni causes an allergic reaction correlated with the number of ear piercings [19]. However, the correlation between piercing and mouth allergies in Ni is unclear. It is known that contact hypersensitivity reactions to divalent cations such as Ni have been observed. These divalent cations can alter the conformation or the peptide binding of MHC class II molecules and, thus, provoke a T cell response [24]. Additionally, it is well-known that LPS is required for the development of metal allergy in mice [42]. To determine whether the challenge of metal allergy is hapten-specific, there have been reports of metal solution plus LPS sensitized mice challenged with metal solution or 1-fluoro-2,4-dinitrobenzene (DNFB) solution to induce allergic reactions [43]. Sensitized, metal ion-challenged mice showed allergic reactions, while sensitized, DNFB-challenged mice did not. DNFB-sensitized, DNFB-challenged mice exhibited an ear swelling response. However, DNFB-sensitized, Pd-challenged mice did not. Moreover, no allergic reactions were observed when unsensitized, metal solution alone sensitized, and LPS alone sensitized mice were challenged with injections of the same amount of metal. The results are known that metal allergy is hapten-specific and that a metal-specific immunological response develops in mice. Therefore, they also showed that LPS is essential for the metal allergy response. In our previous study, we used this information as the basis for an ear-sensitized mouse and challenged the oral cavity to generate a Ni allergic mouse model [5]. Results showed that mice presented with Ni allergy in the oral cavity. Therefore, we suggested that metals outside the oral cavity can be the main cause of allergies, and secondary allergic reactions may occur in the oral cavity. Thereby, in this study as well, we used LPS plus metal solution for all experimental groups in sensitization.

In the case of Ni allergy, NK T cells are involved in both the skin and oral cavity and are believed to be involved in allergic reactions [2,5]. Our study has previously suggested that natural killer (NK) T cells may be correlated with cross-reactivity in metal allergy [44]. The ability of invariant natural killer T (iNKT) cells to recognize different glycolipid constituents from microorganisms presented by CD1d molecules places them in an innate category. Meanwhile, their possession of a fully rearranged T cell receptor, despite its relatively limited repertoire, makes them adaptive [24]. Therefore, iNKT cells can be involved in innate and acquired immunity. iNKT cells acquire a defined effector program during their development in the thymus [24]. They exhibit a memory-cell phenotype when they leave the thymus and migrate to peripheral lymphoid tissues and mucosal surfaces [24]. Hence, NK T cells may act as a bridge between innate and adaptive immunity. The T cell population called mucosal-associated invariant T cells (MAIT cells) are recognized vitamin B9 metabolites presented by the MR1 MHC class Ib molecule, suggesting that the MAIT cells also have a 'transitional' role between innate and adaptive immunity [24,45]. Since allergy to Pd alone is rare, cross-reactivity is known to be involved in Pd allergy [46]. Lymphocyte transformation studies showed simultaneous patch test reactivity to palladium and nickel, suggesting cross-reactivity [47]. However, during Pd sensitization and Ni elicitation, there was no allergic reaction. The involvement of T cells was observed. In contrast, there were significant differences in gene expression levels of CD1d and MRI in Pd and Cr induced by Ni sensitization and Ni induced by Pd sensitization compared with controls. This study suggested a possible involvement of NK T cells and MAIT cells in the challenge of Pd and Cr by Ni sensitization and Ni challenge by Pd sensitization in the oral cavity. Therefore, in the case of Pd sensitization, innate immune-derived T cells may be involved in the development of allergy. As described in the Materials and Methods, in this mouse model, sensitization was achieved via injection into the ears and elicitation via injection into the oral cavity. Therefore, cross-reactivity between Ni and Pd in the oral cavity may be caused by Ni sensitization in the skin. Haptens can induce an early response via innate immune mechanisms [48]. Therefore, suppression of products that may cause Ni ionization may inhibit allergic reactions. Successful avoidance of this sensitization can reduce the development of allergies with complex pathologies, including cross-reactivity. In fact, Ni regulation could reduce Ni allergies. However, its use is not regulated in several countries.

This study showed cross-reactivity between Ni and other metals. This suggests that Ni suppression may inhibit allergic reactions to other metals in which Ni is an inflamer.

We consider that oral metal allergies can be controlled by properly regulating the use of metals outside the oral cavity. Therefore, a patch test even before treatment with other metals may be effective in people who are constantly exposed to Ni. In addition, since the degree of sensitization by metals varies depending on individual susceptibility, it may be important to regulate the Ni content of piercings and other products in several countries. Recently, it has been suggested that even artificial joint implants used in orthopedic surgery on human subjects be patch tested beforehand and substituted with zirconia or other alternatives, since nickel can be allergenic [49]. More genetic information on cross-reactive T cells will further elucidate metal allergy and facilitate safety in dermatology and dental treatments. This new mouse model is useful for the diagnosis of intraoral metal contact allergy, and the development of new treatments to metal-specific T cells in the oral mucosa.

These results suggest that metal allergy immune response in the oral mucosa differs from the skin. Ni exposure is believed to cause metal allergy [50]. Thus, immunological information in the cross-reactivity of metal allergy could be important for selecting dental materials to prevent incompatibility reactions.

4. Materials and Methods

4.1. Animals

Four-week-old female BALB/cAJcl mice (n = 56) were purchased from CLEA Japan (CLEA Japan, Tokyo, Japan) and housed under standard conditions. During the study period, all mice remained in good health, and they were assigned randomly to various groups. The mice were acclimated for at least 7 days before experimental use. They were kept in standard conditions (plastic and aluminum cages with a lid made of stainless-steel wire at our conventional animal facility that maintained the temperature at approximately 23 °C ± 1 °C and humidity at 30–70% with a 12 h day/night cycle). Food and water were available ad libitum.

4.2. Reagents

$PdCl_2$ (>99% pure), $NiCl_2$ (>99% pure), and $CrCl_2$(>99% pure) were purchased from Wako Pure Chemical Industries ((FUJIFILM Wako Pure Chemical Co., Ltd., Osaka, Japan). Lipopolysaccharide (LPS) from Escherichia coli (O55:B5) prepared via phenol-water extraction was purchased from Sigma (Sigma-Aldrich, St Louis, MO, USA). $PdCl_2$, $NiCl_2$, $CrCl_2$, and LPS were dissolved in sterile saline (Otsuka Normal Saline, Otsuka Pharmaceutical Factory, Inc., Tokushima, Japan).

4.3. Anesthetic Agents

The following anesthetics were prepared: medetomidine hydrochloride (Nippon Zenyaku Kogyo Co., Ltd., Fukushima, Japan), midazolam (Sandoz, Tokyo, Japan), and butorphanol tartrate (Meiji Seika Pharma Co., Ltd., Tokyo, Japan). These anesthetic agents were kept at room temperature (RT).

Medetomidine hydrochloride was prepared at a dose of 0.3 mg/kg, midazolam at a dose of 4 mg/kg, and butorphanol tartrate at a dose of 5 mg/kg. The concentration ratio of the three types of mixed anesthetic agents was determined based on a previous study [51]. Therefore, 0.75 mL of medetomidine hydrochloride was mixed with 2 mL of midazolam and 2.50 mL of butorphanol tartrate and was adjusted to a volume of 19.75 mL with sterile saline. All agents were diluted in sterile saline and stored at 4 °C in the dark. The three types of mixed anesthetic agents were administered to all mice at a volume of 0.01 mL/g of body weight.

4.4. Experimental Protocol of the Metal Allergy Cross-Reaction Mouse Model

The protocols were used based on previous protocols for the induction of metal allergy in the oral mucosa [5]. Each experimental group of mice was separated into six sets, each comprising randomly chosen mice (Table 1). All experiments were carried out in another room after transfer from the animal holding room.

Sensitization: In total, 125 μL of 10 mm $PdCl_2$, $NiCl_2$, and $CrCl_2$ and 10 μg/mL of LPS in sterile saline were injected twice at an interval of 7 days via the intradermal route into the left and right postauricular skin of mice (250 μL each). Seven days after the second sensitization, the mice were challenged for the first time.

Challenge for elicitation: At day 7 after the second sensitization, the ACM mice were challenged for elicitation with 25 μL of 10 mm $PdCl_2$, $NiCl_2$, and $CrCl_2$ without LPS in sterile saline into the left and right buccal mucosa via submucosal injection under anesthesia with the three types of mixed anesthetic agents. The mice with metal allergy cross-reactions were classified into four groups: sensitization to $NiCl_2$ with LPS and challenged with $PdCl_2$ (Ni-Pd) group (n = 5), $CrCl_2$ (Ni-Cr) group (n = 5), sensitization to $CrCl_2$ with LPS and challenged with $NiCl_2$ (Cr-Ni) group (n = 5), and sensitization to $PdCl_2$ with LPS and challenged with $NiCl_2$ (Pd-Ni) group (n = 5). Sensitization and challenge for elicitation used different metal solutions for each. Mice sensitized with $NiCl_2$ plus LPS and then challenged with sterile saline were used as a control.

4.5. Measurement of Oral Mucosa Swelling

Buccal mucosa swelling was measured before challenge and at 24 h, 72 h, and 1 week after the first challenge using a Peacock dial thickness gage (Ozaki MFG Co., Ltd., Tokyo, Japan). The difference in oral mucosa thickness before and after challenge was recorded. All procedures were performed by the same experimenter.

4.6. Immunohistochemistry

Buccal mucosa specimens were obtained from mice with metal allergy cross-reaction ACM for histology and immunohistochemical analyses. Tissue samples were immersed in 4% paraformaldehyde–lysine–periodate overnight at 4 °C. After washing with phosphate-buffered saline (PBS) for 10 min, fixed tissues were penetrated by soaking in 5% sucrose/PBS for 1 h, 15% sucrose/PBS for 3 h, and then 30% sucrose/PBS overnight at 4 °C. Tissue samples were embedded in Tissue Mount (Chiba Medical, Saitama, Japan) and snap-frozen into a mixture of acetone and dry ice. Frozen sections were sliced into 6-μm-thick cryosections and air dried on poly-L-lysine-coated glass slides. For histological analyses, the cryosections were stained with H&E. Antigen retrieval was performed for immunohistochemical analyses. Cryosections were stained with anti-mouse F4/80 (1:1000; Cl-A3-1, Abcam, Cambridge, UK) and anti-CD3 (1:500; SP7, Abcam, Cambridge, UK) monoclonal antibodies (mAbs). Non-specific binding of mAbs was blocked via the incubation of sections in PBS containing 5% normal goat and rabbit serum, 0.025% Triton X-100 (FUJIFILM Wako Pure Chemical, Osaka, Japan), and 5% bovine serum albumin (Sigma, Aldrich St. Louis, MO, USA) for 30 min at RT. The sections were incubated with primary mAbs for 1 h at RT. After washing three times with PBS for 5 min, intrinsic peroxidase was quenched using 3% hydrogen peroxide (H_2O_2) in methanol. After soaking the sections in distilled water, they were washed twice and then incubated with a secondary antibody (biotinylated goat anti-hamster IgG or biotinylated rabbit anti-rat IgG) for 1 h at RT. After soaking the sections in distilled water, they were washed twice. Then, sections were incubated with a secondary antibody (biotinylated goat anti-hamster IgG antibody or biotinylated rabbit anti-rat IgG antibody) for 1 h at RT. After washing three times, the sections were incubated with Vectastain ABC Reagent (Vector Laboratories, Burlingame, CA, USA) for 30 min at RT, followed by 3,3-diaminobenzidine staining (0.06% diaminobenzidine and 0.03% H_2O_2 in 0.1 M Tris-HCl, pH 7.6; Wako Pure Chemicals Co., Ltd., Osaka, Japan). The tissue sections were counterstained with hematoxylin to visualize the cell of nuclei.

4.7. RNA Extraction and cDNA Synthesis

Fresh buccal mucosa tissue specimens were obtained from each mouse and immediately soaked in RNAlater RNA Stabilization Reagent (Qiagen, Hilden, Germany). Total RNA from the buccal mucosa tissue was extracted using the RNeasy Lipid Tissue Mini Kit (Qiagen) according to the manufacturer's instructions. Complementary DNA (cDNA) was synthesized from DNA-free RNA using the PrimeScript™ RT reagent Kit (Takara Bio, Tokyo, Japan) according to the manufacturer's instructions.

4.8. Quantitative Polymerase Chain Reaction

The expression levels of immune response-related genes, including T cell-related CD antigens, cytokines, cytotoxic granules, transcription factors of regulatory T cells, CD1d-restricted T cells, and MHC-related protein 1 were evaluated via quantitative polymerase chain reaction (qPCR) using the Bio-Rad CFX96 System (Bio-Rad, Hercules, CA, USA). Specific primers for GAPDH, CD4, CD8, IFN-γ, TNF-α, IL-4, IL-5, Foxp3, CD1d, MR1, and granzymes A and B have been described in previous studies [52,53]. Freshly isolated total RNA from the buccal mucosa tissue of mice was converted to cDNA using PrimeScript RT Reagent Kit (Takara Bio) according to the manufacturer's instructions. The PCR comprised 5 μL of SsoFast™ EvaGreen® Supermix (Bio-Rad), 3.5 μL of RNase/DNase-free water, 0.5 μL of 5-μM primer mix, and 1 μL of cDNA, with a final volume of 10 μL. The cycling conditions were as follows: 30 s at 95 °C, followed by 45 cycles of 1 s at 95 °C and 5 s at 60 °C. At the end of each cycle, melting curve analysis was performed from 65 °C to 95 °C to confirm the homogeneity of PCR products. All assays were repeated three times, and the mean values were calculated at the gene expression levels. Five 10-fold serial dilutions of each standard transcript were used to determine the absolute quantification, specification, and amplification efficiency of each primer set. Standard transcripts were generated by the in vitro transcription of the corresponding PCR product in a plasmid. The nucleotide sequences were confirmed via DNA sequencing using the CEQ8000 Genetic Analysis System (Beckman Coulter, Fullerton, CA, USA). Their quality and concentration were validated using the Agilent DNA 7500 Kit in an Agilent 2100 Bioanalyzer (Agilent, Santa Clara, CA, USA). The expression of the GAPDH gene was used as an internal control. The expression levels of each target gene were normalized to GAPDH expression.

4.9. Statistical Analysis

Differences between the mean values of each experimental group were analyzed using the Kruskal–Wallis test, followed by Dunn's multiple comparison tests and the Mann–Whitney U-test using GraphPad Prism 7 software for Windows (GraphPad Software, Inc., San Diego, CA, USA). p value of <0.05 was considered significant; p value of <0.01, highly significant; and p value of <0.001, extremely significant.

5. Conclusions

The Ni-sensitized group showed significant differences in buccal mucosal swelling, and the expression of CD8, cytotoxic granules, and inflammation-related cytokines compared with the control and Pd-sensitized groups. Ni sensitization and Pd and Cr challenge can cause cross-reactivity in intraoral metal allergy.

Author Contributions: R.M. and R.S. conceived and designed the experiments; R.M., K.N., T.Y., M.S. and R.S. performed the experiments; R.M., K.N., T.Y. and K.K. (Kazutaka Kitaura) analyzed the data; R.M., K.K. (Kenichi Kumagai) and Y.H. contributed reagents/materials/analysis tools; R.M. wrote the paper. All authors have read and agreed to the published version of the manuscript.

Funding: This work was supported by the Japan Society for the Promotion of Science KAKENHI Grant-in-Aid for Scientific Research C Grant No. 19K10371.

Institutional Review Board Statement: All animal experiments in this study were performed in strict accordance with the recommendations in the Guidelines for Care and Use of Laboratory Animals of the Clinical Research Center of Sagamihara National Hospital, Japan. All procedures were performed according to the relevant ethical requirements with approval from the committees for animal experiments at the Clinical Research Center for Rheumatology and Allergy, Sagamihara National Hospital (approval number: H22-2010-1).

Informed Consent Statement: Not applicable.

Data Availability Statement: This data presented in this study are available on request from the corresponding author.

Conflicts of Interest: The authors declare no conflict of interest.

References

1. Abbas, A.K.; Lichtman, A.H.; Pillai, S. *Basic Immunology: Functions and Disorders of the Immune System*, 6th ed.; Elsevier: Amsterdam, The Netherlands, 2020; pp. 78, 267.
2. Eguchi, T.; Kumagai, K.; Kobayashi, H.; Shigematsu, H.; Kitaura, K.; Suzuki, S.; Horikawa, T.; Hamada, Y.; Ogasawara, K.; Suzuki, R. Accumulation of invariant NKT cells into inflamed skin in a novel murine model of nickel allergy. *Cell. Immunol.* **2013**, *284*, 163–171. [CrossRef] [PubMed]
3. Kobayashi, H.; Kumagai, K.; Eguchi, T.; Shigematsu, H.; Kitaura, K.; Kawano, M.; Horikawa, T.; Suzuki, S.; Matsutani, T.; Ogasawara, K.; et al. Characterization of T Cell Receptors of Th1 Cells Infiltrating Inflamed Skin of a Novel Murine Model of Palladium-Induced Metal Allergy. *PLoS ONE* **2013**, *8*, e76385. [CrossRef] [PubMed]
4. Shigematsu, H.; Kumagai, K.; Kobayashi, H.; Eguchi, T.; Kitaura, K.; Suzuki, S.; Horikawa, T.; Matsutani, T.; Ogasawara, K.; Hamada, Y.; et al. Accumulation of Metal-Specific T Cells in Inflamed Skin in a Novel Murine Model of Chromium-Induced Allergic Contact Dermatitis. *PLoS ONE* **2014**, *9*, e85983. [CrossRef] [PubMed]
5. Nakasone, Y.; Kumagai, K.; Matsubara, R.; Shigematsu, H.; Kitaura, K.; Suzuki, S.; Satoh, M.; Hamada, Y.; Suzuki, R. Characterization of T cell receptors in a novel murine model of nickel-induced intraoral metal contact allergy. *PLoS ONE* **2018**, *13*, e0209248. [CrossRef]
6. Kumagai, K.; Matsubara, R.; Nakasone, Y.; Shigematsu, H.; Kitaura, K.; Suzuki, S.; Haneji, K.; Hamada, Y.; Suzuki, R. Possible involvement of invariant natural killer T cells and mucosal-associated invariant T cells in a murine model of titanium allergy. *J. Oral Maxillofac. Surg. Med. Pathol.* **2018**, *30*, 1–9. [CrossRef]
7. Raap, U.; Stiesch, M.; Reh, H.; Kapp, A.; Werfel, T. Investigation of contact allergy to dental metals in 206 patients. *Contact Dermat.* **2009**, *60*, 339–343. [CrossRef]
8. Faurschou, A.; Menné, T.; Johansen, J.D.; Thyssen, J.P. Metal allergen of the 21st century-a review on exposure, epidemiology and clinical manifestations of palladium allergy. *Contact Dermat.* **2011**, *64*, 185–195. [CrossRef]
9. Watsky, K.L. Occupational allergic contact dermatitis to platinum, palladium, and gold. *Contact Dermat.* **2007**, *57*, 382–383. [CrossRef]
10. Gawkrodger, D.J.; Lewis, F.M.; Shah, M. Contact sensitivity to nickel and other metals in jewelry reactors. *J. Am. Acad. Dermatol.* **2000**, *43*, 31–36. [CrossRef]
11. Larese, F.F.; Uderzo, D.; Bagnato, E. Sensitization to palladium chloride: A 10-year evaluation. *Am. J. Contact Dermat.* **2003**, *14*, 78–81.
12. Wahlberg, J.E.; Boman, A.S. Cross-reactivity to palladium and nickel studied in the guinea pig. *Acta Dermato-Venereologica* **1992**, *72*, 95–97. [PubMed]
13. Hindsén, M.; Spiren, A.; Bruze, M. Cross-reactivity between nickel and palladium demonstrated by systemic administration of nickel. *Contact Dermat.* **2005**, *53*, 2–8. [CrossRef] [PubMed]
14. Shigematsu, H.; Kumagai, K.; Suzuki, M.; Eguchi, T.; Matsubara, R.; Nakasone, Y.; Nasu, K.; Yoshizawa, T.; Ichikawa, H.; Mori, T.; et al. Cross-Reactivity of Palladium in a Murine Model of Metal-induced Allergic Contact Dermatitis. *Int. J. Mol. Sci.* **2020**, *21*, 4061. [CrossRef] [PubMed]
15. Ahlfors, E.; Czerkinsky, C. Contact sensitivity in the murine oral mucosa. I. An experimental model of delayed-type hypersensitivity reactions at mucosal surfaces. *Clin. Exp. Immunol.* **1991**, *86*, 449–456. [CrossRef]
16. Hirunwidchayarat, W.; Furusawa, E.; Kang, S.; Ohno, T.; Takeuchi, S.; Rungsiyanont, S.; Azuma, M. Site-specific regulation of oral mucosa-recruiting CD8+ T cells in a mouse contact allergy model. *Biochem. Biophys. Res. Commun.* **2017**, *490*, 1294–1300. [CrossRef]
17. Garner, L.A. Contact dermatitis to metals. *Dermatol. Ther.* **2004**, *17*, 321–327. [CrossRef]
18. Mortz, C.G.; Lauritsen, J.M.; Bindslev-Jensen, C.; Andersen, K.E. Nickel sensitization in adolescents and association with ear piercing, use of dental braces and hand eczema. The Odense Adolescence Cohort Study on Atopic Diseases and Dermatitis (TOACS). *Acta Dermat. Venereol.* **2002**, *82*, 359–364. [CrossRef]
19. Larsson-Stymne, B.; Widstrom, L. Ear piercing–a cause of nickel allergy in schoolgirls? *Contact Dermat.* **1985**, *13*, 289–293. [CrossRef]

20. Stejskal, V.; Reynolds, T.; Bjørklund, G. Increased frequency of delayed type hypersensitivity to metals in patients with connective tissue disease. *J. Trace Elements Med. Biol.* **2015**, *31*, 230–236. [CrossRef]
21. Nakamura, K.; Imakado, S.; Takizawa, M.; Adachi, M.; Sugaya, M.; Wakugawa, M.; Asahina, A.; Tamaki, K. Exacerbation of pustulosis palmaris et plantaris after topical application of metals accompanied by elevated levels of leukotriene B4 in pustules. *J. Am. Acad. Dermatol.* **2000**, *42*, 1021–1025. [CrossRef]
22. Hanafusa, T.; Yoshioka, E.; Azukizawa, H.; Itoi, S.; Tani, M.; Kira, M.; Katayama, I. Systemic allergic contact dermatitis to palladium inlay manifesting as annular erythema. *Eur. J. Dermatol.* **2012**, *22*, 697–698. [CrossRef] [PubMed]
23. Khamaysi, Z.; Bergman, R.; Weltfriend, S. Positive patch test reactions to allergens of the dental series and the relation to the clinical presentations. *Contact Dermat.* **2006**, *55*, 216–218. [CrossRef] [PubMed]
24. Kenneth, M.; Casey, W. *Janeway's Immune Biology*, 9th ed.; Garland Science: New York, NY, USA, 2017; pp. 247, 250, 335, 336, 631, 633.
25. Oka, K.; Saito, F.; Yasuhara, T.; Sugimoto, A. A study of cross-reactions between mango contact allergens and urushiol. *Contact Dermat.* **2004**, *51*, 292–296. [CrossRef] [PubMed]
26. Saxon, A.; Adelman, D.C.; Patel, A.; Hajdu, R.; Calandra, G.B. Imipene631m cross-reactivity with penicillin in humans. *J. Allergy Clin. Immunol.* **1988**, *82*, 213–217. [CrossRef]
27. Fujimoto, K.; Yamaguchi, H.; Otsuka, Y.; Mayumi, N.; Saeki, H. Contact Dermatitis Caused by Efinaconazole and Luliconazole. *J. Nippon. Med Sch.* **2021**, *88*, 253–257. [CrossRef]
28. Li, L.; Laidlaw, T. Cross-reactivity and tolerability of celecoxib in adult patients with NSAID hypersensitivity. *J. Allergy Clin. Immunol. Pr.* **2019**, *7*, 2891–2893.e4. [CrossRef]
29. Stokes, F.; Bailey, L.; Ganguli, A.; Davison, A.S. Assessment of endogenous, oral and inhaled steroid cross-reactivity in the Roche cortisol immunoassay. *Ann. Clin. Biochem. Int. J. Biochem. Lab. Med.* **2014**, *51*, 503–506. [CrossRef]
30. Kinbara, M.; Nagai, Y.; Takano-Yamamoto, T.; Sugawara, S.; Endo, Y. Cross-reactivity among some metals in a murine metal allergy model. *Br. J. Dermatol.* **2011**, *165*, 1022–1029. [CrossRef]
31. Santucci, B.; Cristaudo, A.; Cannistraci, C.; Picardo, M. Interaction of palladium ions with the skin. *Exp. Dermatol.* **1995**, *4*, 207–210. [CrossRef]
32. Riedel, F.; Aparicio-Soto, M.; Curato, C.; Thierse, H.; Siewert, K.J.; Luch, A. Immunological Mechanisms of Metal Allergies and the Nickel-Specific TCR-pMHC Interface. *Int. J. Environ. Res. Public Health* **2021**, *18*, 10867. [CrossRef]
33. Roach, K.A.; Stefaniak, A.B.; Roberts, J.R. Metal nanomaterials: Immune effects and implications of physicochemical properties on sensitization, elicitation, and exacerbation of allergic disease. *J. Immunotoxicol.* **2019**, *16*, 87–124. [CrossRef] [PubMed]
34. Matsubara, R.; Kumagai, K.; Shigematsu, H.; Kitaura, K.; Nakasone, Y.; Suzuki, S.; Hamada, Y.; Suzuki, R. Fexofenadine Suppresses Delayed-Type Hypersensitivity in the Murine Model of Palladium Allergy. *Int. J. Mol. Sci.* **2017**, *18*, 1357. [CrossRef] [PubMed]
35. Zirwas, M. Allergy to topical steroids. *J. Drugs Dermatol.* **2012**, *11*, s9–s11. [PubMed]
36. Richards, L.J.; Streifel, A.; Rodrigues, J.M. Utility of Patch Testing and Lymphocyte Transformation Testing in the Evaluation of Metal Allergy in Patients with Orthopedic Implants. *Cureus* **2019**, *11*, e5761. [CrossRef]
37. Ricciardi, L.; Carnì, A.; Loschiavo, G.; Gangemi, S.; Tigano, V.; Arena, E.; Mannucci, C.; Calapai, G. Systemic Nickel Allergy: Oral Desensitization and Possible Role of Cytokines Interleukins 2 and 10. *Int. J. Immunopathol. Pharmacol.* **2013**, *26*, 251–257. [CrossRef] [PubMed]
38. Michalak, I.; Mikulewicz, M.; Chojnacka, K.; Wołowiec, P.; Saeid, A.; Górecki, H. Exposure to nickel by hair mineral analysis. *Environ. Toxicol. Pharmacol.* **2012**, *34*, 727–734. [CrossRef]
39. Bezerra, M.A.; Bruns, R.E.; Ferreira, S.L. Statistical design principal component analysis optimization of a multiple response procedure using cloud point extraction and simultaneous determination of metals by ICP OES. *Anal. Chim. Acta* **2006**, *580*, 251–257. [CrossRef]
40. Wahlberg, J.E.; Lindberg, M. Patch testing. In *Contact Dermatitis*, 4th ed.; Frosch, P.J., Menne, T., Lepoittevin, J.P., Eds.; Springer: Berlin/Heidelberg, Germany, 2005; pp. 366–386.
41. Schalock, P.C.; Crawford, G.; Nedorost, S.; Scheinman, P.L.; Atwater, A.R.; Mowad, C.; Brod, B.; Ehrlich, A.; Watsky, K.L.; Sasseville, D.; et al. Patch Testing for Evaluation of Hypersensitivity to Implanted Metal Devices: A Perspective From the American Contact Dermatitis Society. *Dermatitis* **2016**, *27*, 241–247. [CrossRef]
42. Schmidt, M.; Raghavan, B.; Müller, V.; Vogl, T.; Fejer, G.; Tchaptchet, S.; Keck, S.; Kalis, C.; Nielsen, P.J.; Galanos, C.; et al. Crucial role for human Toll-like receptor 4 in the development of contact allergy to nickel. *Nat. Immunol.* **2010**, *11*, 814–819. [CrossRef]
43. Kawano, M.; Nakayama, M.; Aoshima, Y.; Nakamura, K.; Ono, M.; Nishiya, T.; Nakamura, S.; Takeda, Y.; Dobashi, A.; Takahashi, A.; et al. NKG2D+ IFN-γ+ CD8+ T Cells Are Responsible for Palladium Allergy. *PLoS ONE* **2014**, *9*, e86810. [CrossRef]
44. Kumagai, K.; Horikawa, T.; Shigematsu, H.; Matsubara, R.; Kitaura, K.; Eguchi, T.; Kobayashi, H.; Nakasone, Y.; Sato, K.; Yamada, H.; et al. Possible Immune Regulation of Natural Killer T Cells in a Murine Model of Metal Ion-Induced Allergic Contact Dermatitis. *Int. J. Mol. Sci.* **2016**, *17*, 87. [CrossRef] [PubMed]
45. Keller, A.N.; Corbett, A.J.; Wubben, J.M.; McCluskey, J.; Rossjohn, J. MAIT cells and MR1-antigen recognition. *Curr. Opin. Immunol.* **2017**, *46*, 66–74. [CrossRef] [PubMed]
46. González-Ruiz, L.; De Caso, E.V.; Peña-Sánchez, R.; Silvestre-Salvador, J.F.; Peña-Sáncez, R. Delayed hypersensitivity to palladium dichloride: 15-year retrospective study in a skin allergy unit. *Contact Dermat.* **2019**, *81*, 249–253. [CrossRef] [PubMed]

47. Kapp, F.; Summer, B.; Thomas, P. Usefulness of lymphocyte transformation test and in vitro cytokine release in differentiating between independent and cross-reacting nickel/palladium allergy. *Immunity, Inflamm. Dis.* **2020**, *8*, 483–492. [CrossRef]
48. Kaplan, D.H.; Igyártó, B.Z.; Gaspari, A.A. Early immune events in the induction of allergic contact dermatitis. *Nat. Rev. Immunol.* **2012**, *12*, 114–124. [CrossRef]
49. Desai, M.M.; Shah, K.A.; Mohapatra, A.; Patel, D.C. Prevalence of metal hypersensitivity in total knee replacement. *J. Orthop.* **2019**, *16*, 468–472. [CrossRef]
50. Thyssen, J.P.; Skare, L.; Lundgren, L.; Menné, T.; Johansen, J.D.; Maibach, H.I.; Lidén, C. Sensitivity and specificity of the nickel spot (dimethylglyoxime) test. *Contact Dermat.* **2010**, *62*, 279–288. [CrossRef]
51. Kawai, S.; Takagi, Y.; Kaneko, S.; Kurosawa, T. Effect of Three Types of Mixed Anesthetic Agents Alternate to Ketamine in Mice. *Exp. Anim.* **2011**, *60*, 481–487. [CrossRef]
52. Fujii, Y.; Kitaura, K.; Nakamichi, K.; Takasaki, T.; Suzuki, R.; Kurane, I. Accumulation of T-cells with selected T-cell receptors in the brains of Japanese encephalitis virus-infected mice. *Jpn. J. Infect. Dis.* **2008**, *61*, 40–48.
53. Kitaura, K.; Fujii, Y.; Hayasaka, D.; Matsutani, T.; Shirai, K.; Nagata, N.; Lim, C.-K.; Suzuki, S.; Takasaki, T.; Suzuki, R.; et al. High Clonality of Virus-Specific T Lymphocytes Defined by TCR Usage in the Brains of Mice Infected with West Nile Virus. *J. Immunol.* **2011**, *187*, 3919–3930. [CrossRef]

International Journal of *Molecular Sciences*

MDPI

Article

Dual Role of Interleukin-20 in Different Stages of Osteoclast Differentiation and Its Osteoimmune Regulation during Alveolar Bone Remodeling

Bowen Meng [1,2,3,†], Benyi Yang [1,2,3,†], Yan Qu [1,2,3], Yuanbo Liu [1,2,3], Dongle Wu [1,2,3], Chaoran Fu [1,2,3], Yifan He [1,2,3], Xi Chen [1,2,3], Chufeng Liu [4], Xiaoxing Kou [1,2,3,*] and Yang Cao [1,2,3,*]

1 Hospital of Stomatology, Sun Yat-sen University, Guangzhou 510055, China
2 Guangdong Provincial Key Laboratory of Stomatology, Guangzhou 510055, China
3 South China Center of Craniofacial Stem Cell Research, Guanghua School of Stomatology, Sun Yat-sen University, Guangzhou 510055, China
4 Department of Orthodontics, Stomatological Hospital, Southern Medical University, Guangzhou 510260, China
* Correspondence: kouxiaoxing@mail.sysu.edu.cn (X.K.); caoyang@mail.sysu.edu.cn (Y.C.)
† These authors contributed equally to this work.

Abstract: Osteoimmunology mediators are critical to balance osteoblastogenesis and osteoclastogenesis to maintain bone homeostasis. A lot of the osteoimmunology mediators are regulated by interleukin-20 (IL-20). However, little is known about the role of IL-20 in bone remodeling. Here, we showed that IL-20 expression was correlated with osteoclast (OC) activity in remodeled alveolar bone during orthodontic tooth movement (OTM). Ovariectomize (OVX) in rats promoted OC activity and enhanced IL-20 expression, while blocking OC inhibited IL-20 expression in osteoclasts. In vitro, IL-20 treatment promoted survival, inhibited apoptosis of the preosteoclast at the early stages of osteoclast differentiation, and boosted the formation of osteoclasts and their bone resorption function at the late stages. More importantly, anti-IL-20 antibody treatment blocked IL-20-induced osteoclastogenesis and the subsequent bone resorption function. Mechanistically, we showed that IL-20 synergistically acts with RANKL to activate the NF-κB signaling pathway to promote the expression of c-Fos and NFATc1 to promote osteoclastogenesis. Moreover, we found that local injection of IL-20 or anti-IL-20 antibody enhanced osteoclast activity and accelerated OTM in rats, while blocking IL-20 reversed this phenomenon. This study revealed a previously unknown role of IL-20 in regulating alveolar bone remodeling and implies the application of IL-20 to accelerated OTM.

Keywords: interleukin-20; osteoimmunology; osteoclast differentiation; signaling pathways; orthodontic tooth movement

Citation: Meng, B.; Yang, B.; Qu, Y.; Liu, Y.; Wu, D.; Fu, C.; He, Y.; Chen, X.; Liu, C.; Kou, X.; et al. Dual Role of Interleukin-20 in Different Stages of Osteoclast Differentiation and Its Osteoimmune Regulation during Alveolar Bone Remodeling. *Int. J. Mol. Sci.* **2023**, *24*, 3810. https://doi.org/10.3390/ijms24043810

Academic Editor: Kenichi Kumagai

Received: 30 November 2022
Revised: 6 February 2023
Accepted: 9 February 2023
Published: 14 February 2023

1. Introduction

Osteoimmunology is an emerging and interdisciplinary concept encompassing the interplay between the skeletal and immune systems in the bone turnover mechanism under physiology and pathology conditions [1]. The balance of bone remodeling is maintained with various cells, including osteocytes, osteoblasts, osteoclasts macrophages, T cells, and B cells [2–4], which derive from the same bone marrow micromilieu and share plenty of cytokines, receptors, transcription factors, and signaling pathways. An imbalance of bone and the immune system contributes to plenty of bone loss diseases, such as osteoporosis, periodontitis, and bone cracking in orthodontic tooth movement (OTM) [5–7]. Therefore, there is a need to find new alternative osteoclast-targeting agents for the treatment of bone loss diseases.

Several bone-affecting inflammatory cytokines such as TNF, IL-1, IL-6, IL-17, IL-22, IL-23 and IL-33, secreted by immune cells, synergistically act with RANKL and disrupt the balance of bone resorption and bone formation, leading to bone homeostasis disorders in bone loss diseases [7–10]. Interestingly, most of the above-mentioned cytokines are regulated by IL-20, one of the IL-20 subfamily cytokines that belongs to the IL-10 large family [11–21]. IL-20 is mainly produced by activated macrophages and skin cells [22,23]. Previous researchers have focused on the ability of IL-20 to act as a proinflammatory, chemotactic, and angiogenic cytokine in skin inflammation diseases [24], RA [16], liver fibrosis [14] and ischemic diseases [25] by activating its heterodimeric receptors (either IL-20RA/IL-20RB or IL-22RA1/IL-20RB). Recently, clinical trials and animal experiments have found elevated serum IL-20 levels in rheumatoid arthritis [16,26–28], osteoporosis [29], and cancer-related osteolytic diseases [30]. However, whether IL-20 contributes to bone loss diseases and whether IL-20 can serve as a potential target for bone remodeling is unclear.

OTM is involved in the complex biomechanical responses of periodontal tissue [31]. Mechanical force acts on the periodontal cells to produce a series of cytokines, including prostaglandins, IL-1, IL-6, and IL-17 [32,33], which initiate osteoclast procedures. Meanwhile, the receptor activator of nuclear factor (RANK)/RANK ligand (RANKL)/osteoprotegerin (OPG) and tumor necrosis factor (TNF)-α superfamily are increased in the periodontium. Thus, there is predominant clinical value and paramount scientific importance in identifying alternative osteoimmunology mediators targeting osteoclasts and treatments for bone-loss diseases. Our previous studies found that IL-20 was involved in osteoclastogenesis through the OPG/RANKL/RANK axis and the Notch pathway in vitro [34,35]. However, little is known about the role of IL-20 in orthodontic tooth movement and the specific downstream molecular mechanism of IL-20 in RANKL-induced osteoclastogenesis.

In this study, we explored the roles of IL-20 in regulating osteoclast fate in vitro and in vivo. We found that IL-20 promoted osteoclastogenesis through NF-κB-mediated signaling pathways. Moreover, blocking IL-20 decreased alveolar bone remodeling and orthodontic tooth movement, which indicated that IL-20 is a promising direction for the targeted regulation of osteoclastogenesis.

2. Results

2.1. Osteoclasts and IL-20 Were Synchronously Activated in OTM

To explore the relationship between IL-20 and osteoclasts during alveolar bone remodeling, we constructed the rat OTM model (Figures 1A and S1A). After mechanical force application for 16 days, the tooth movement distance increased to 0.18 mm. As shown by TRAP staining and immunohistology staining, the number of IL-20-positive cells and TRAP-positive osteoclasts simultaneously increased (Figures 1A–D and S1D). To confirm whether IL-20 was expressed by the osteoclast, we used double immunofluorescent staining to show that IL-20 was colocalized with osteoclast markers RANK and TRAP (Figures 1C,E,F and S1B,D,E). To further confirm whether IL-20 was synchronously activated with the osteoclast, we constructed the rat ovariectomize (OVX) model and applicated mechanical force. Compared to the Force group, the OVX + Force group increased tooth movement distance to 0.33 mm, accompanying an increased number of IL-20-positive cells colocalized with TRAP and RANK-positive osteoclasts (Figures 1A–F and S1B,D,E). Additionally, immunofluorescence staining showed that IL-20^{+} and CD11b^{+} double-positive cells were observed in the alveolar bone of OTM mice, and OVX + Force treatment enhanced the numbers of IL-20 and CD11b-positive cells, suggesting that macrophages expressed IL-20 and participated in osteoclast activation (Figure S1C). In addition, compared to the OVX group, mechanical force enhanced the distance of tooth movement and simultaneously increased the number of IL-20-positive cells and TRAP-positive osteoclasts (Figure 1G–J). Moreover, risedronate, an anti-osteoporosis drug, decreased the distance of tooth movement in OVX rats and simultaneously decreased the number of IL-20-positive cells and TRAP-positive osteoclasts (Figure 1G–J). Synchronously, immunofluorescence showed that IL-20 colocalized with osteoclast markers TRAP and RANK was inhibited (Figures 1K and S1B,F,G). Taken together, these results suggested that IL-20 was synchronously activated with osteoclasts in OTM, and may play a key role in osteoclastogenesis.

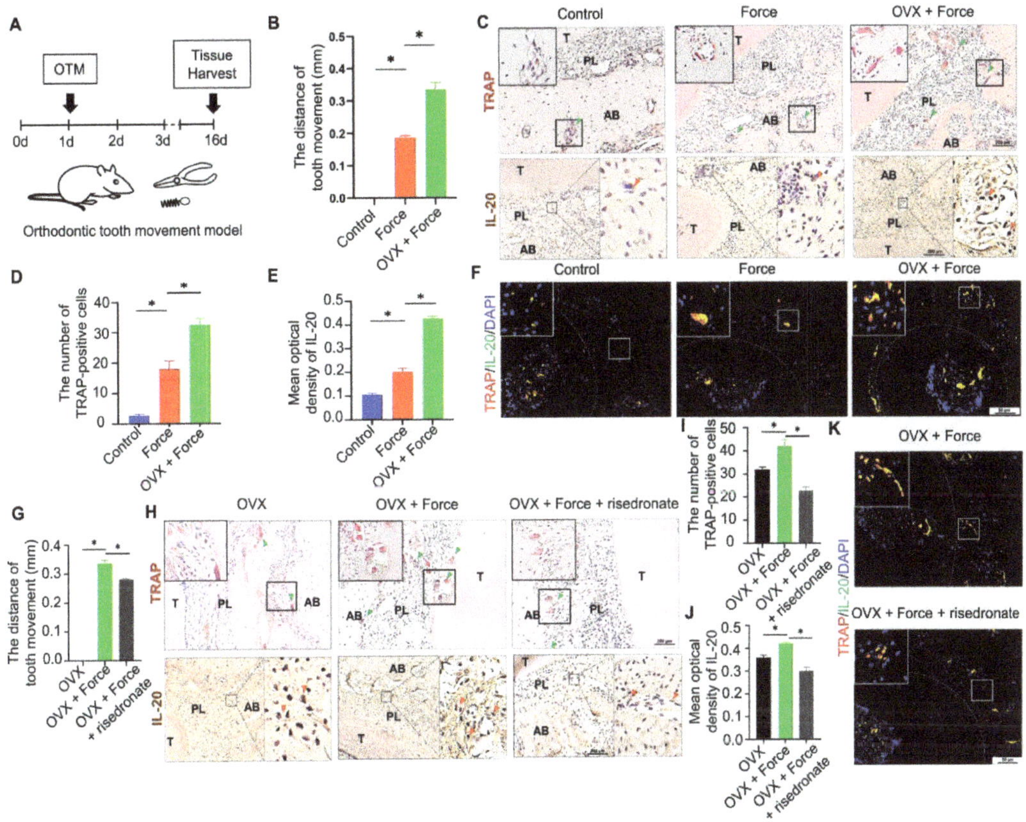

Figure 1. IL-20 accelerated OVX-induced bone loss in orthodontic tooth movement by promoting osteoclastogenesis. (**A**) Scheme illustrating the establishment of orthodontic tooth movement. (**B**) The distance of orthodontic tooth movement in the Control group, Force group, and OVX + Force group on day 16. (**C**) TRAP and Immunohistochemical staining showed that TRAP and IL-20 -positive cells and in the Control group, Force group, and OVX + Force group. Green triangles showed TRAP-positive osteoclasts. Red triangles showed IL-20-positive cells. T: tooth, PL: periodontal ligament, AB: alveolar bone. (**D**) The quantification of TRAP-positive osteoclasts in the Control group, Force group, and OVX + Force group. (**E**) Immunohistochemical staining and semiquantification of IL-20 in the Control group, Force group, and OVX + Force group. (**F**) Double-labelled immunofluorescence staining showed that, in the context of orthodontic force, the expression levels of IL-20 and osteoclast marker protein TRAP increased in the first molar periodontal ligament. (**G**) The distance of orthodontic tooth movement in the OVX group, OVX + Force group, and OVX + Force + risedronate group on day 16. (**H**) TRAP and immunohistochemical staining showed that TRAP and IL-20 -positive cells and in the OVX group, OVX + Force group, and OVX + Force + risedronate group. (**I**) The quantification of TRAP-positive osteoclasts in the OVX group, ovariectomy + Force group, and Ovariectomy + Force + risedronate group. (**J**) Immunohistochemical staining and semiquantification of IL-20 in the OVX group, OVX + Force group, and OVX + Force + risedronate group. (**K**) Immunofluorescence staining showed that the expression levels of IL-20 and osteoclast marker protein TRAP increased in the first molar periodontal ligament. * $p < 0.05$ vs. the control group. $n = 6$.

2.2. *IL-20 Promoted Preosteoclast Proliferation by MAPK Pathways*

To identify whether IL-20 was expressed and to identify its function in osteoclast differentiation, the BMMs were induced into preosteoclasts and stained using cellular immunochemistry and immunofluorescence staining. Here, the immunofluorescence staining results revealed that different concentrations of M-CSF promoted IL-20 expression in preosteoclasts, and the cellular fluorescence intensity of IL-20 was highest after treatment with 30 ng/mL M-CSF (Figure 2A,B). Moreover, the qRT-PCR results showed that M-CSF differentially regulated the expression of IL-20 and its receptors IL-20RA, IL-20RB, and IL-22RA1 in preosteoclasts (Figure 2C–F). Additionally, we found that the expression of IL-20 in BMMs time-dependently increased with the treatment of 30 ng/mL M-CSF (Figure S2F). What is more, the cellular immunochemistry further confirmed that IL-20 and its receptors were expressed in preosteoclasts (Figure S2A). To explore the effect of IL-20 on preosteoclasts, we added different concentrations of IL-20 and detected preosteoclast function with CCK8 and flow cytometry. We found that IL-20 promoted the proliferation of preosteoclasts in a dose-dependent manner, and simultaneously suppressed preosteoclasts apoptosis (Figures 2G and S2B–E). Western blotting suggested that IL-20 could perform a function by activating MAPK pathways, including ERK, p38, and JNK pathways. Moreover, blocking IL-20 with an anti-IL-20 antibody could partly inhibit the above-mentioned pathways' activation (Figure 2H). In conclusion, IL-20 was expressed in MCF-induced preosteoclasts and promoted its function by the MAPK pathway.

2.3. *IL-20 Had No Effect on Osteoclasts Differentiation and Functions at the Early Stage of Osteoclast Differentiation*

To prove the possibility of an interaction between IL-20 and osteoclastogenesis at the early stage of osteoclast differentiation from BMMs to preosteoclasts, we cultured BMMs with M-CSF (30 ng/mL) and IL-20 or anti-IL-20 antibody for 3 days and then changed to osteoclast medium without IL-20 for 6 days (Figure 3A). The TRAP staining and bone resorption experiments results revealed that there was no significant difference in TRAP-positive osteoclasts and bone resorption function compared to that of the respective control groups (Figure 3B,C). To further explore whether RANKL was essential for IL-20 regulation of osteoclastogenesis, we treated preosteoclasts with IL-20 without RANKL. The results of TRAP staining showed that IL-20 alone could not stimulate the formation of TRAP-positive cells (Figure 3D), which suggested that IL-20 was incapable of independently inducing osteoclastogenesis. Therefore, IL-20 had no effect on RANKL-induced osteoclast formation at the early stage of osteoclast differentiation.

2.4. *IL-20 Promoted Osteoclasts Differentiation and Functions at the Late Stage of Osteoclast Differentiation*

To identify the role of IL-20 in osteoclastogenesis at the late stage of differentiation, preosteoclasts were cultured in an osteoclast medium containing IL-20. After 6 days, the number and size of mature osteoclasts were determined by TRAP staining (Figure 4A). Compared with the control group, IL-20 promoted osteoclastogenesis in both numbers of TRAP+ cells and size with the increase in RANKL concentration, while IL-20-block using anti-IL-20 antibody could eliminate the promotion effects, even in a high concentration of RANKL (Figure 4B). Resorption assays showed the same pattern as TRAP staining and exhibited an increased bone resorption area and pits which were blocked with anti-IL-20 antibody (Figure 4C). Therefore, IL-20 promoted RANKL-induced osteoclast differentiation and bone resorption function.

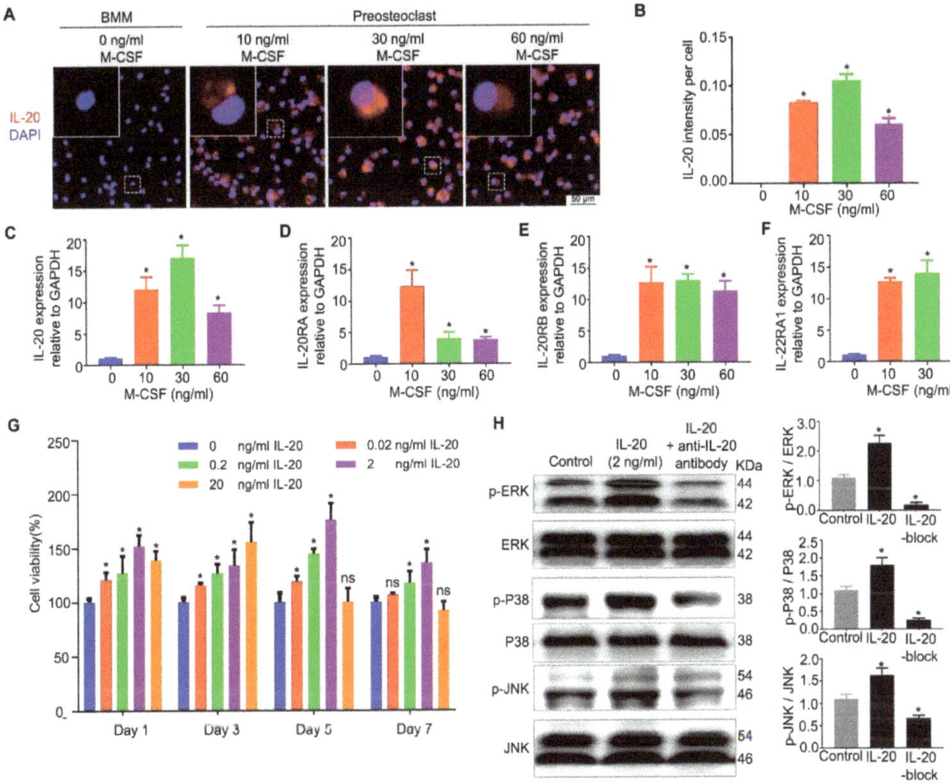

Figure 2. IL-20 promoted preosteoclast viability via MAPK pathway at the early stage of osteoclast differentiation. Cell viability was examined in M-CSF-induced preosteoclasts by a CCK8 assay. (**A**,**B**) The expression of IL-20 in M-CSF-induced preosteoclasts was determined by cellular immunofluorescence staining on day 2. (**C**–**F**) The mRNA expression levels of IL-20, IL-20RA, IL-20RB, and IL-22RA1 in preosteoclasts were evaluated by qRT-PCR after 2 days of M-CSF treatment. (**G**) Cell viability detection in preosteoclasts treated with a gradient of concentrations of IL-20 on days 1, 3, 5, and 7. (**H**) Preosteoclasts were treated with 2 ng/mL IL-20 or anti-IL-20 antibody. The levels of phosphorylation of proteins in the IL-20-mediated signaling pathway, including p38, phos-pho-p38, ERK, phospho-ERK, JNK, and phospho-JNK, were analyzed using Western blotting. IL-20-block group meant that cells were treated with IL-20 and anti-IL-20 antibody. * $p < 0.05$ vs. the 0 ng/mL IL-20 group. ns $p > 0.05$ vs. the 0 ng/mL IL-20 group. $n = 6$.

2.5. IL-20 Promoted Osteoclast Differentiation through the NF-κB Pathway

To determine the molecular mechanism of IL-20 in osteoclastogenesis, immunofluorescence staining, and Western blotting, assays were performed. We found that IL-20 or RANKL alone promoted the expression of p-P65 and its translocation to the nucleus (Figure 5A,B). Furthermore, a combination treatment of IL-20 and RANKL could further enhance RANKL-induced activation of p-P65 (Figure 5A,B). Therefore, the results suggested that IL-20 was involved in osteoclast differentiation by a synergistic reaction with RANKL. In addition, TPCA-1, a NF-κB pathway inhibitor, notably suppressed the activation of p-NF-κB in preosteoclasts (Figure 5A,B). As the time of IL-20 treatment went by, Western blotting demonstrated that IL-20 promoted the phosphorylation of NF-κB cascade pathway, including IKKα/β, IκBα and NF-κB pathway (Figure 5C). To test the activation of the upstream and downstream pathways of the OPG/RANKL/RANK axis, Western blotting was performed and showed that downstream pathways TRAF6, c-FOS, and NFATc1 were activated at different times of IL-20 treatment. However, in upstream pathways, RANK showed no significant change (Figure 5D). However, TPCA-1 could partly block the activation of these pathways (Figure S3A,B). Similarly, after 6 days of IL-20 treatment, osteoclast-related proteins such as TRAP, Cathepsin K, TRAF6, c-FOS, and NFATc1 in RANKL-induced osteoclasts were verified and all upregulated by IL-20, and suppressed by TPCA-1 (Figures 5E and S3C). To further confirm the results, qPCR has been used to examine the expression of osteoclastic genes with a gradient concentration of RANKL, and results indicated that IL-20 promoted osteoclastic gene expression via the NF-κB signaling pathway, including TRAP, Cathepsin K, MMP9, MT1-MMP, c-Fos, and NFATc1 (Figure 5F–K). Taken together, IL-20 may promote osteoclast differentiation and function by the NF-κB pathway.

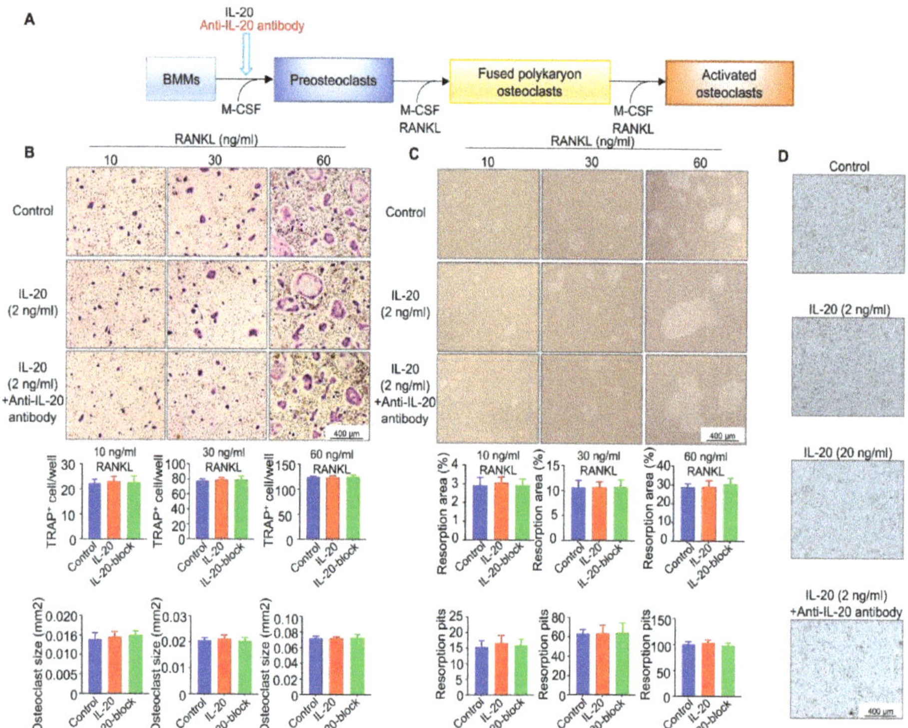

Figure 3. IL-20 had no effect on osteoclast formation at the early stage of osteoclast differentiation. (**A**) Scheme illustrating bone marrow-derived macrophages were cultured in osteoclast medium containing M-CSF and IL-20 or anti-IL-20 antibody at the early stage of osteoclast differentiation, and then induced with the presence of 10, 30, or 60 ng/mL RANKL. The control group included M-CSF-induced preosteoclasts induced with 10, 30, or 60 ng/mL RANKL. (**B**) TRAP staining was performed, and the number and size of TRAP-positive osteoclasts with more than three nuclei were quantified on day 6. IL-20-block group meant that cells were treated with IL-20 or anti-IL-20 antibody. (**C**) A bone resorption pit assay was performed to detect osteoclast function, and bone resorption pits were counted, and the area and number of bone resorption were quantified on day 6. IL-20-block group meant that cells were treated with IL-20 and anti-IL-20 antibody. (**D**) M-CSF-induced preosteoclasts were cultured in osteoclast medium containing different concentrations of IL-20 or anti-IL-20 antibody without RANKL. $n = 6$.

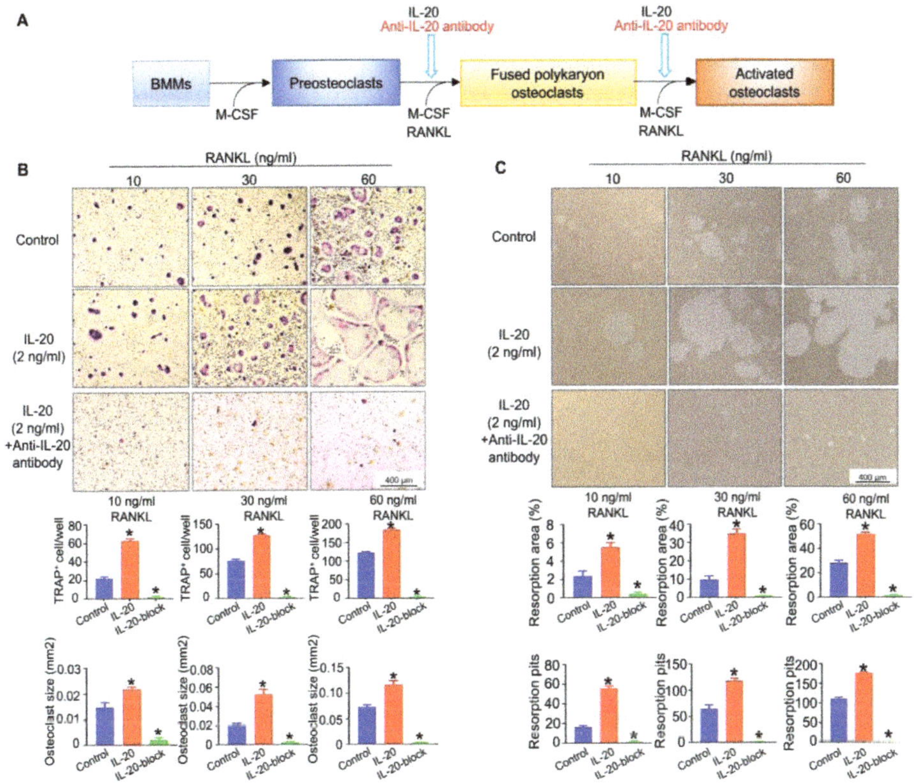

Figure 4. IL-20 promoted RANKL-induced osteoclast differentiation and bone resorption function. (**A**) Scheme illustrating M-CSF-induced preosteoclasts were cultured in osteoclast medium containing IL-20 or anti-IL-20 antibody at the late stage of osteoclast differentiation in the presence of 10, 30, or 60 ng/mL RANKL. (**B**) TRAP staining was performed, and the number and size of TRAP-positive osteoclasts with more than three nuclei were quantified at the late stage of differentiation on day 6. The control group included M-CSF-induced preosteoclasts induced with 10, 30, or 60 ng/mL RANKL. IL-20-block group meant that cells were treated with IL-20 and anti-IL-20 antibody. (**C**) A resorption pit assay was performed to detect osteoclast function, and resorption pits were counted at the late stage of differentiation on day 6, and the area and number of bone resorption were quantified at the stage of differentiation. IL-20-block group meant that cells were treated with IL-20 and anti-IL-20 antibody. * $p < 0.05$ vs. the control group. $n = 6$.

2.6. Exogenous Injection of IL-20 Accelerated Tooth Movement in OTM

To demonstrate the potential clinical effects of IL-20, we locally infused IL-20 or anti-IL-20 antibody after the application of orthodontic force. HE staining verified the success of the rat model (Figure S4A). Micro CT manifested that local injection of IL-20 could increase the distance of OTM to 0.21 mm compared to the OTM group (Figure 6A,B). As shown by TRAP staining, local injection of IL-20 increased the number of TRAP-positive osteoclasts (Figure 6C,D). Additionally, immunofluorescent staining showed that IL-20 treatment increased the number of RANK-positive osteoclasts (Figure 6E,F). Moreover, local injection of anti-IL-20 antibody decreased the distance of OTM, TRAP-positive osteoclasts, and RANK-positive osteoclasts (Figure 6A–F). Monocyte chemoattractant protein-1 (MCP-1) is one of the key chemokines that regulates infiltration of monocytes/macrophages [36]. A previous study showed that IL-20 controlled T cell infiltration by regulating the expression of MCP-1 in mouse psoriasis models [19]. Thus, we investigated whether IL-20 affects

MCP-1 expression and CD11b-positive macrophage infiltration in mice alveolar bone during OTM. Immunofluorescence staining showed that both MCP-1 staining, and CD11b-positive macrophages were observed in the alveolar bone of OTM mice, and IL-20 treatment enhanced the numbers of MCP-1 and CD11b-positive cells (Figure 6G,H). Moreover, local injection of anti-IL-20 antibody repressed IL-20 induced MCP-1 and CD11b expression in OTM mice (Figures 6G,H and S4B). These results indicated that IL-20 may promote macrophage infiltration via upregulation of MCP-1. The results suggested that IL-20 could accelerate orthodontic tooth movement by activating osteoclasts (Figure S5).

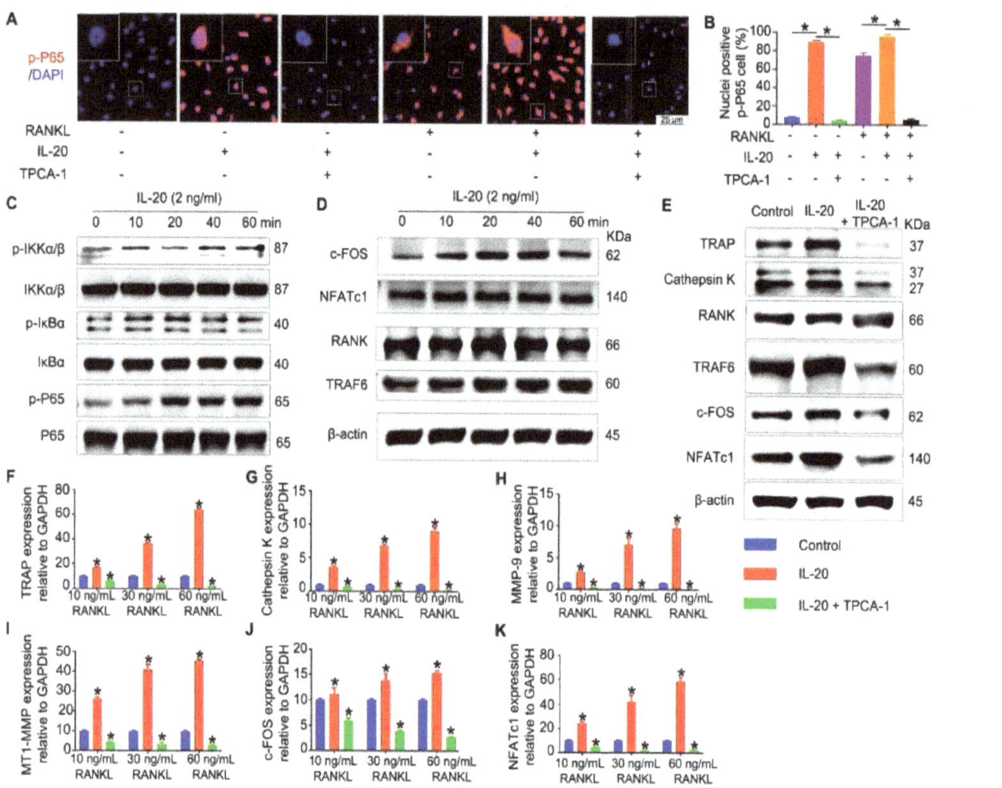

Figure 5. IL-20 promoted the expression of osteoclast-specific genes and proteins via NF-κB pathway during RANKL-induced osteoclast differentiation. Preosteoclasts were stimulated with IL-20 or NF-κB pathway inhibitor TPCA-1. (**A**,**B**) IL-20 promoted phospho-P65 nuclear translocation and was blocked by TPCA-1 in preosteoclasts after 1 h treatment, with or without RANKL. (**C**) The levels of phosphorylation for proteins in the NF-κB pathway, including the IKKα/β, IκB-α, and P65 proteins in preosteoclasts were detected using Western blotting. (**D**) The levels of activated proteins in preosteoclasts, including the RANK, TRAF6, c-Fos, and NFATc1 proteins without RANKL were detected using Western blotting. (**E**) The protein expression levels of TRAP, Cathepsin K, RANK, TRAF6, c-Fos and NFATc1 in osteoclasts were examined using Western blotting after 6 days of IL-20 treatment. (**F–K**) The mRNA expression levels of TRAP, Cathepsin K, MMP9, MT1-MMP, c-Fos, and NFATc1 in osteoclasts were detected by qRT-PCR after 6 days of IL-20 treatment. * $p < 0.05$ vs. the 0 ng/mL IL-20 group. $n = 6$.

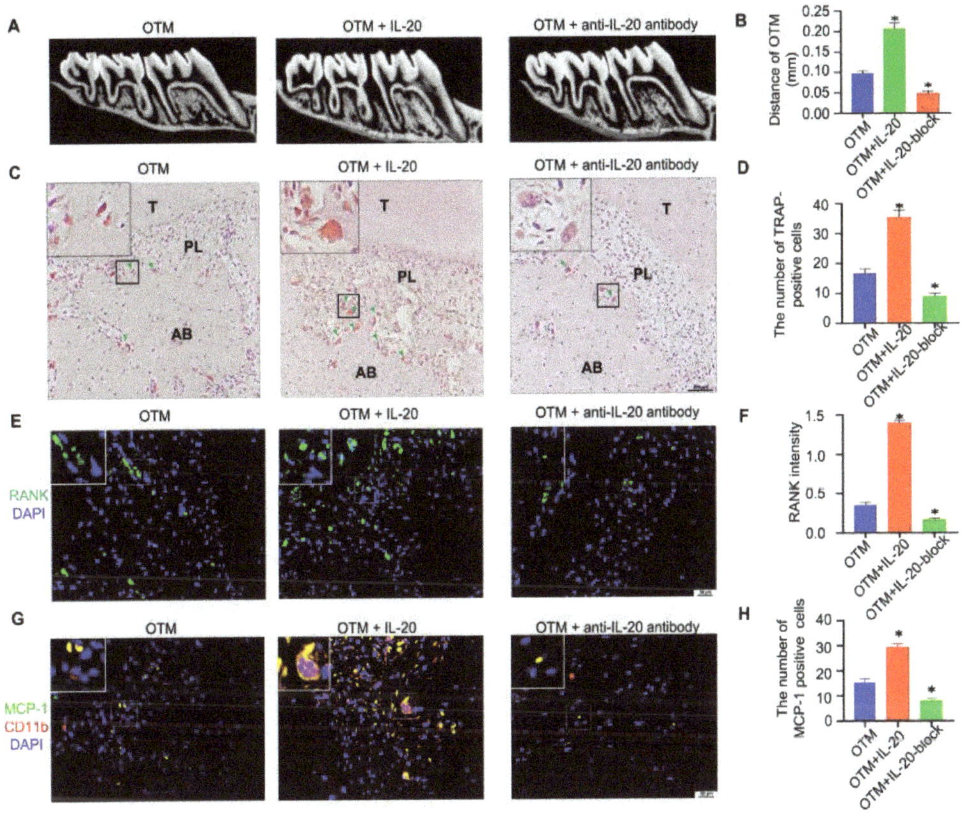

Figure 6. Therapeutic effect of IL-20 on orthodontic tooth movement. (**A**,**B**) Micro-CT showed the distance of orthodontic tooth movement between first and second molars treated with local infusion of IL-20 or an-ti-IL-20 antibody, seven days after the application of orthodontic force. OTM + IL-20-block group meant that rats were locally infused with anti-IL-20 antibody. (**C**,**D**) TRAP staining and the quantification of TRAP-positive osteoclasts in the OTM group, OTM + IL-20 group, and OTM + anti-IL-20 antibody group. Green triangles showed TRAP-positive osteoclasts. OTM + IL-20 block group meant that rats were locally infused with anti-IL-20 antibody. (**E**,**F**) Immunofluorescence staining showed that the expression levels of osteoclast marker protein RANK in the first molar periodontal ligament after the application of orthodontic force. OTM + IL-20-block group meant that rats were locally infused with anti-IL-20 antibody. (**G**,**H**) Immunofluorescence staining showed that the expression levels of MCP-1 and CD11b in the first molar periodontal ligament after the application of orthodontic force. * $p < 0.05$ vs. the control group. $n = 6$.

3. Discussion

In this study, we discovered for the first time that IL-20 is involved in orthodontic tooth movement and may promote osteoclastogenesis, leading to accelerated orthodontic tooth movement in ovariectomy-induced rats. What is more, IL-20 and its receptors targeted M-CSF-induced preosteoclasts, and IL-20 promoted preosteoclast survival and inhibited apoptosis by MAPK pathways. Moreover, IL-20 promoted the osteoclast formation and bone resorption function induced by different concentrations of RANKL at the late stages of osteogenic differentiation, but had no effect at the early stages. Anti-IL-20 antibody notably blocked IL-20-induced osteoclastogenesis at the late stages. With respect to the molecular mechanisms, IL-20 activated the NF-κB signaling pathways and p-P65 translocation to the nucleus in the presence or absence of RANKL, and subsequently activated c-Fos and

NFATc1. Finally, IL-20 promoted the expression of osteoclast-specific genes and proteins, including TRAP, CK, MMP9, MT1-MMP, c-Fos, NFATc1, and RANK.

It is well known that the survival factor M-CSF regulates the proliferation and apoptosis of BMMs and induces their differentiation into preosteoclasts through its receptor c-Fms [37,38]. In the bone marrow, osteoclasts are derived from granulocyte–macrophage progenitors (CFU-GM) of hematopoietic origin, and gene mutation of M-CSF completely inhibits both macrophage and osteoclast differentiation and formation [39,40]. However, Liu et al. reported that IL-20 specifically enhanced the proliferation, cell cycling, and colony formation of multipotential progenitors (CFU-GEMM) [41]. In addition, recent research has shown that IL-20, a potent angiogenic, chemotactic, and proinflammatory molecule, targets endothelial cells and epithelial cells and regulates their proliferation and apoptosis through related signaling pathways and genes, such as MAPK, JAK/STAT3, caspase 9 [42,43]. In our experiment, we found that IL-20 and its receptors were secreted by M-CSF-induced preosteoclasts in an autocrine manner, and that IL-20 promoted the proliferation and inhibited apoptosis of M-CSF-induced preosteoclasts in a dose-dependent manner. These findings suggested that IL-20 promoted the proliferation and inhibited apoptosis of M-CSF-induced preosteoclasts at the early stages of osteoclast differentiation.

In addition to M-CSF, RANKL has been shown to be at the forefront of osteoimmunology as another factor that is both necessary and sufficient for mature osteoclast formation, and it controls the level of bone resorption by binding to its specific receptor RANK [7,44–47]. RANKL is secreted not only by bone mesenchymal stem cells (BMSCs) or osteoblasts to sustain osteoclastogenesis by direct cell contact in vitro [48,49], but also by activated immune T cells that are involved in pathological bone loss [50,51]. RANKL–RANK binding subsequently activates the key transcription factors c-Fos and NFATc1, and finally leads to the expression of osteoclast marker molecules, such as TRAP, CK, MMP9, and MT1-MMP, that degrade the bone collagen and mineral matrix [52]. Previous studies showed that certain proinflammatory cytokines, such as TNF, IL-1, IL-6, and IL-22, were also involved in pathologic bone loss via synergistic action with RANKL [8,10,53]. In this study, we showed that IL-20 promoted the expression of RANK in response to RANKL and further contributed to the differential expression of c-Fos, NFATc1, TRAP, CK, MMP9, and MT1-MMP, which differentially regulated RANKL-induced osteoclast formation and bone resorption pits in vitro. Thus, we speculated that IL-20 was a downstream regulator of the OPG/RANKL/RANK axis, which mediated osteoclastogenesis by targeting preosteoclasts.

RANKL binding to its receptor RANK leads to the recruitment of the main adaptor molecule TRAF6, which activates downstream signaling pathways, including MAPK (p38, ERK, JNK), AKT and NF-κB. These pathways control osteoclast survival, apoptosis, and differentiation, and the functions of bone degradation and resorption [52]. Activation of p38 leads the transcriptional regulator mi/Mitf to enter the nucleus, where it regulates the gene expression of TRAP and CK [54]. The JNK pathway promotes formation of the activator protein-1 (AP1), which includes c-Fos and c-Jun, by activation of c-Jun phosphorylation [55]. In addition to M-CSF, RANKL can also activate the ERK and AKT pathways, which regulate the survival and proliferation of preosteoclasts and cytoskeletal rearrangement and motility in mature osteoclasts [52,56]. The NF-κB pathway is activated by the degradation of IκB proteins, which is induced by IκB kinase (IKK). Both the P65 and c-Fos transcription factors of the activator protein-1 (AP1) component are crucial for the activation of NFATc1, which controls the expression of osteoclast-specific genes, as mentioned above. In this study, we further found that IL-20 activated the phosphorylation of IKKα/β, IκB-α, P65, p38, ERK, and JNK by activating TRAF6 without RANKL, which ultimately resulted in the activation of c-Fos and NFATc1. Collectively, these results indicated that IL-20 activated TRAF6-mediated downstream NF-κB signaling pathways, and differentially induced preosteoclast differentiation into mature osteoclasts to perform bone resorption by synergistic action with RANKL, showing that NF-κB activation was not sufficient but indispensable for osteoclastogenesis.

Several proinflammatory cytokines are involved in osteoporosis. Neutralizing them with specific blocking molecules, including soluble receptors and antibodies, represents a significant therapeutic strategy for preventing bone loss in osteoporotic conditions [57–60]. Previous studies showed that the level of IL-20 was increased in osteoporotic patients and OVX-induced osteoporotic patients [29]. Moreover, soluble IL-20RA receptor blocked IL-20-induced osteoclast formation and inhibited the pathophysiology of bone loss in collagen-induced arthritis (CIA) [16]. In this study, we found that anti-IL-20 antibody inhibited IL-20-induced osteoclastogenesis at the late stages. Moreover, we found that IL-20 promoted osteoclastogenesis, leading to accelerated orthodontic tooth movement in OVX-induced rats. Risedronate, as a clinical drug to target osteoporosis, partly decreased the level of IL-20 and inhibited IL-20-induced osteoclast formation. Additionally, in our previous study, we found that risedronate inhibited orthodontic tooth movement by regulating the OPG/RANKL/RANK axis in ovariectomized rats [61]. Fortunately, specific IL-20 antibody can inhibit osteoclastogenesis and promote osteoblastogenesis. Moreover, IL-20 affected bone formation and downregulated osteoblastogenesis on osteoblasts; whereas, IL-20 antibody increased bone formation during fracture healing [62]. Our previous studies found that IL-20 differentially regulated the expression of OPG and RANKL in BMSCs and inhibited the osteogenic differentiation of MC3T3-E1 cells via the GSK3β/β-catenin signaling pathway [34,63]. It has the therapeutic potential to decrease osteoporotic bone loss and increase bone mineral density [29,62]. These findings suggested that blocking IL-20 might be a promising direction for targeted regulation in bone loss diseases.

Overall, this study elucidates the downstream mechanism of IL-20 in different stages of osteoclast differentiation and function in vitro, and the detailed function of how IL-20 participates in bone remodeling by regulating osteoclast differentiation and the osteoimmune microenvironment in vivo. Therefore, targeting IL-20 maybe a promising direction for the treatment of bone remolding-related diseases. Inhibition of IL-20 may repress osteoclast activity to ameliorate bone loss, such as in osteoporosis. On the other hand, IL-20 local injection may activate osteoclasts to accelerate the tooth movement process for orthodontic patients.

4. Materials and Methods

4.1. Animals

All Sprague-Dawley rats were purchased from the Animal Experimental Center of Guangzhou University of Chinese Medicine. The diet, housing conditions, and maintenance of rats were in accordance with the Institutional Animal Care and Use Committee (IACUC) of Sun Yat-sen University. All conducted experiments were approved by the Animal Ethical and Welfare Committee of Sun Yat-sen University (SYSU-IACUC-2018-000099, Guangzhou, China), which was approved on 10/25/2019. The rats were randomly divided into 7 groups (n = 6 in each group): group 1 (sham), group 2 (sham + Force), group 3 (ovariectomy), group 4 (ovariectomy + Force), group 5 (ovariectomy + risedronate + Force), group 6 (IL-20 + Force), and group 7 (anti-IL-20 + Force).

4.2. The Rat Model of Orthodontic Tooth Movement and Ovariectomy

The female rats were anesthetized for the bilateral ovariectomy operation. Briefly, we anesthetized the rats, removed the hair, disinfected the site, made 1 cm back incisions to remove the ovaries, and closed the incisions. After 1 week of postoperative recovery, 15 μg/kg risedronate was injected intraperitoneally in group 4 every 3 days for 3 weeks, while the other groups were given an equal volume of saline or drug.

Four weeks after ovariectomy, the rat models of orthodontic tooth movement were established as described in our published articles [61]. Under general anesthesia, we placed an orthodontic device consisting of nickel–titanium coiled springs and stainless steel (Xinya, Hang Zhou, China) between the incisors and the first molars on the maxilla, which caused the first molars to move medially. After self-etching using a dental etching agent (Heraeus Kulzer GmbH, Hanau, Germany), orthodontic devices were bonded with a chemically cured resin (3M Unitek, Sao Paulo, MN, USA) to provide 50 g orthodontic forces. On day 16, we measured the distance of orthodontic tooth movement and sacrificed the rats to collect the bilateral maxillary for the subsequent analysis.

4.3. Immunohistochemistry and TRAP Staining In Vivo

For immunohistochemistry, the maxillaries containing the molars were treated as described. Briefly, the decalcified samples were embedded for cutting, deparaffinized, and rehydrated, and endogenous peroxidase activity was blocked. Then, the slides were blocked, incubated with the primary and secondary antibodies, and stained with the chromogenic agent DAB and Meyer's hematoxylin. For TRAP staining, we stained TRAP+ multinucleated osteoclasts with a TRAP staining kit (#G1050-50T, Servicebio, Wuhan, China), and counterstained the nucleus with hematoxylin.

4.4. Osteoclastogenesis Assays In Vitro

Osteoclasts were generated by the methods previously described [64,65]. Primary rat bone marrow macrophages (BMMs) from the whole bone marrow were isolated from the femur and tibia cavities of 4-week-old Sprague-Dawley rats. Rats were sacrificed after anesthesia; then, we flushed the bone marrow cavity with α-MEM, and purified it with RBC lysis buffer (CWBIO, Beijing, China) to collect the cell pellet. The cells were resuspended in 3 mL of α-MEM culture medium and incubated at 37 °C in a 5% CO_2 incubator overnight. Then, we transferred the supernatant to a 15 mL centrifuge tube to collect the unattached cells (BMMs) the next morning. At the early stage of osteoclastogenesis, we treated BMMs with 30 ng/mL M-CSF (#400-28, Peprotech, Rocky Hill, NJ, USA) at 37 °C in a 5% CO_2 incubator for 2–3 days to generate preosteoclasts, with or without IL-20 (#80422-RNAE, Sino Biological Inc., Beijing, China) or anti-IL-20 antibody (#80187/80453-R08H, Sino Biological Inc., Beijing, China). At the late stage of osteoclastogenesis, RANKL (#9366-TN-025, R&D Systems, Minneapolis, MN, USA) was added to the osteoclast medium containing IL-20 or anti-IL-20 antibody. The medium was replaced every 2 days until the formation of mature multinuclear osteoclasts was observed.

4.5. TRAP Staining In Vitro

According to the manufacturer's protocol, TRAP staining was used to assess mature osteoclasts with an Acid Phosphatase Leukocyte (TRAP) Kit (#CS0740, Sigma, MO, USA) in vitro. After 6 days of osteoclastogenic differentiation, we counted TRAP-positive multinucleated cells with three or more nuclei as mature osteoclasts under an inverted fluorescence microscope (Zeiss, Jena, Germany). Six microscopic images were randomly taken at 50× magnification from three samples, and the average number was counted.

4.6. Resorption Pit Assay

A resorption pit assay was performed to detect the resorption function of osteoclasts. We cultured M-CSF-induced preosteoclasts in an Osteo Assay Surface multiple-well plate (Corning, New York, NY, USA), which was coated with an inorganic crystalline material on the surface. Briefly, mature osteoclasts were stripped with sodium hypochlorite, washed with distilled water and air-dried so that resorption pits could be easily observed with an inverted fluorescence microscope (Zeiss, Jena, Germany).

4.7. Cellular Immunochemistry and Immunofluorescence Staining

For cellular immunochemistry and immunofluorescence staining, preosteoclasts were cultured with $1 \times 10^{\hat{}6}$ cells/well in 6-well plate containing glass slips and $5 \times 10^{\hat{}5}$ cells/well in a laser confocal dish in an aseptic environment, respectively. When the cells covered 70–80% of the surface of the coverslips, we fixed the coverslips with 4% paraformaldehyde, permeabilized the cells with 0.2–0.25% Triton X-100, and blocked the cells with 3% BSA. Then, we incubated the cells with primary and secondary antibodies (anti-rabbit). Subsequently, we stained preosteoclasts with the chromogenic agent DAB and counterstained the cell nucleus with Meyer's hematoxylin for immunocytochemistry or DAPI for immunocytofluorescence. Finally, glass coverslips were mounted with neutral balsam or antifluorescence quenching agent. An LSM780 confocal microscope (Zeiss, Germany) or an inverted fluorescence micro-scope (Zeiss, Germany) was used to capture cell images.

4.8. Cell Viability and Cytotoxicity Assay

Cell viability and proliferation were investigated using the Cell Counting Kit-8 (#CK04, CCK-8, Kumamoto, Dojindo, Japan). Briefly, we treated M-CSF-induced preosteoclasts with a gradient of concentrations of IL-20 (0–20 ng/mL) in 96-well plates (4000 cells/well). The medium was changed every 2 days. Then, we washed cells with PBS and added 100 μL of FBS-free DMEM containing 10 μL of WST-8 solution per well, and the cells were incubated for 1 h without light for the indicated times. Finally, we detected the absorbance at 450 nm using a microplate reader (Tecan SUNRISE microplate reader, Shanghai, Tecan, Männedorf, Switzerland).

4.9. Flow Cytometry Detection of Cell Apoptosis

Cell apoptosis was detected with an Annexin V-APC/7-AAD apoptosis kit (#AT105, MultiSciences Bio-tech, Hangzhou, China) by flow cytometry. Preosteoclasts were treated with various concentrations of IL-20 (0–20 ng/mL). According to the manufacturer's protocol, we obtained the cells via trypsinization, washed the cells with PBS, and stained the cells with an apoptosis kit. After careful analysis with a Cytoflex flow cytometer (Beckman, Brea, CA, USA), four cell populations with apoptotic staining were distinguished as early apoptotic cells (APC+ 7-AAD-), late apoptotic cells (APC+ 7-AAD+), and viable cells (APC- 7-AAD-).

4.10. qRT-PCR Analysis

Briefly, we extracted total mRNA from 2×10^6 cells/well in 6-well plate using an RNA-Quick Purification Kit (#ES-RN001, ES Science, Shanghai, China). PrimeScriptTM RT Master Mix (#RR047, Perfect Real Time) (Takara, Osaka, Japan) was used to obtain cDNA. The primers used for this study were commercially synthesized by Takara, and their sequences are listed in Table 1. Then, we quantified gene transcript levels with SYBR® Premix Ex Taq™ (Tli RNaseH Plus) (Takara, Japan) in a QuantStudio 5 or 7 Flex Real-Time PCR System (Applied Biosystems™, Foster City, CA, USA) and analyzed the results using the $2^{-\triangle\triangle CT}$ method.

4.11. Western Blotting Analysis

After the treatment of IL-20, anti-IL-20 antibody, or TPCA-1 (MCE, Monmouth Junction, NJ, USA), we obtained protein from the samples with cold RIPA containing 1% protease inhibitor cocktail and 1% phosphatase inhibitors. Then, we detected the total concentration of sample proteins with a BCA Protein Assay Kit (CWBIO, Beijing, China). For Western blotting, we added 50 μg of sample protein per lane and then transferred the sample protein to a PVDF membrane. Subsequently, we blocked the membranes with TBST containing 5% BSA and incubated them with primary and secondary antibodies. Primary antibodies were specific for IKKα/β (#11930/8943, Cell Signaling Technology Inc., Beverly, MA, USA), phospho-IKKα/β (#2697, Cell Signaling Technology Inc., Beverly, MA, USA), IκBα (#4814, Cell Signaling Technology Inc., Beverly, MA, USA), phospho-IκBα (#2859, Cell Signaling Technology Inc., Beverly, MA, USA), P65 (#8242, Cell Signaling Technology Inc., Beverly, MA, USA), phospho-P65 (#3033, Cell Signaling Technology Inc., Beverly, MA, USA), p38 (#8690, Cell Signaling Technology Inc., Beverly, MA, USA), phos-pho-p38 (#4511, Cell Signaling Technology Inc., Beverly, MA, USA), JNK (#9252, Cell Signaling Technology Inc., Beverly, MA, USA), phospho-JNK, ERK (#4695, Cell Signaling Technology Inc., Beverly, MA, USA), phospho-ERK (#4370, Cell Signaling Technology Inc., Beverly, MA, USA), and c-Fos (#2250, Cell Signaling Technology Inc., Beverly, MA, USA); TRAP (#ab191406,

Abcam, Cambridge, MA, USA), Cathepsin K (CK, #ab300569, Abcam, Cambridge, MA, USA), NFATc1 (#ab253477, Abcam, Cambridge, MA, USA), and TRAF6 (#ab33915, Abcam, Cambridge, MA, USA); and RANK (#sc-374360, Santa Cruz Biotechnology, Santa Cruz, CA, USA). The secondary antibodies were HRP Affinipure goat anti-rabbit IgG and HRP Affinipure goat anti-mouse IgG (#EM35111-01, EMAR, Beijing, China).

Table 1. Primers sequences used for real-time PCR.

Gene	Forward Primer Sequence (5′-3′) (Tm) Reverse Primer Sequence (5′-3′) (Tm)	Product Size
IL-20	ACTGCAAACCTACAGGCGATACAA (64.1 °C) AGAACCTCACTAGATGGCGGAGA (63.7 °C)	163 bp
IL-20RA	GGGTCTACACGGAGTCGAAGTCA (64.9 °C) ACGCTCATAGTCCGAGGTCTCAA (64.5 °C)	139 bp
IL-20RB	AGCACTTGATGGGTTAACAGCC (61.1 °C) AAAACAGAGACACAGCCCTCC (60.2 °C)	72 bp
IL-22RA1	CCTACACGTGCCGAGTGAAGA (63.6 °C) AAAGCTCAGGACACGCTGGA (63.4 °C)	176 bp
Cathepsin K	CGGCTATATGACCACTGCCTTC (63.0 °C) TTTGCCGTGGCGTTATACATACA (64.3 °C)	114 bp
TRAP	TGGCAATGTCTCGGCACAA (64.9 °C) AGCATCACGGTGTCCAGCATAA (65.0 °C)	138 bp
MT1-MMP	GAGAACTTCGTGTTGCCTGATGAC (64.5 °C) TTTCTGGGCTTATCTGGGACAGAG (64.9 °C)	134 bp
MMP9	CATGCGCTGGGCTTAGATCA (64.6 °C) GAGGCCTTGGGTCAGGTTTAGAG (64.5 °C)	148 bp
NFATc1	CAAGTCTCACCACAGGGCTCACTA (64.0 °C) TCAGCCGTCCCAATGAACAG (62.2 °C)	144 bp
c-Fos	CGTCTTCCTTTGTCTTCACCTACC (64.8 °C) TTGCTGCTGCTGCCCTTT (63.7 °C)	81 bp
GAPDH	GGCACAGTCAAGGCTGAGAATG (64.4 °C) ATGGTGGTGAAGACGCCAGTA (62.8 °C)	143 bp

4.12. Statistical Analysis

All results in this study are presented as the mean ± standard deviation, with values from at least 3 or 6 independent experiments in vitro or in vivo, respectively. Using Student's t-test or one-way ANOVA with Tukey's post hoc analysis, statistically significant differences with values of $p < 0.05$ were determined by GraphPad Prism 7.04 software.

Supplementary Materials: The following supporting information can be downloaded at https://www.mdpi.com/article/10.3390/ijms24043810/s1.

Author Contributions: Authors' roles: Study design: Y.C., X.K. and C.L. Conducting Study: B.M. and B.Y. Data collection: Y.Q. and Y.L. Data analysis: D.W. and X.C. Data interpretation: C.F. and Y.H. Drafting manuscript: B.M., B.Y. and Y.Q. Revising manuscript content: B.M., B.Y. and Y.Q. All authors take responsibility for the integrity of the data analysis. All authors have read and agreed to the published version of the manuscript.

Funding: This work was supported by grants from the National Natural Science Foundation of China (82170987; 81970963) and the Natural Science Foundation of Guangdong Province, China (2022A1515010871; 2021A1515012535).

Institutional Review Board Statement: All conducted experiments were approved by the Animal Ethical and Welfare Committee of Sun Yat-sen University (SYSU-IACUC-2018-000099, Guangzhou, China).

Informed Consent Statement: Not applicable.

Data Availability Statement: Not applicable.

Conflicts of Interest: The authors declare no conflict of interest.

References

1. Okamoto, K.; Nakashima, T.; Shinohara, M.; Negishi-Koga, T.; Komatsu, N.; Terashima, A.; Sawa, S.; Nitta, T.; Takayanagi, H. Osteoimmunology: The Conceptual Framework Unifying the Immune and Skeletal Systems. *Physiol. Rev.* **2017**, *97*, 1295–1349. [CrossRef] [PubMed]
2. Manolagas, S.C.; Jilka, R.L. Bone marrow, cytokines, and bone remodeling. Emerging insights into the pathophysiology of osteoporosis. *N. Engl. J. Med.* **1995**, *332*, 305–311. [CrossRef] [PubMed]
3. Redlich, K.; Smolen, J.S. Inflammatory bone loss: Pathogenesis and therapeutic intervention. *Nat. Rev. Drug Discov.* **2012**, *11*, 234–250. [CrossRef]
4. Weitzmann, M.N.; Ofotokun, I. Physiological and pathophysiological bone turnover—Role of the immune system. *Nat. Rev. Endocrinol.* **2016**, *12*, 518–532. [CrossRef] [PubMed]
5. Arron, J.R.; Choi, Y. Bone versus immune system. *Nature* **2000**, *408*, 535–536. [CrossRef] [PubMed]
6. Takayanagi, H.; Ogasawara, K.; Hida, S.; Chiba, T.; Murata, S.; Sato, K.; Takaoka, A.; Yokochi, T.; Oda, H.; Tanaka, K.; et al. T-cell-mediated regulation of osteoclastogenesis by signalling cross-talk between RANKL and IFN-gamma. *Nature* **2000**, *408*, 600–605. [CrossRef] [PubMed]
7. Tsukasaki, M.; Takayanagi, H. Osteoimmunology: Evolving concepts in bone-immune interactions in health and disease. *Nat. Rev. Immunol.* **2019**, *19*, 626–642. [CrossRef]
8. Kim, K.-W.; Kim, H.-R.; Park, J.-Y.; Oh, H.-J.; Woo, Y.-J.; Park, M.-K.; Cho, M.-L.; Lee, S.-H. Interleukin-22 promotes osteoclastogenesis in rheumatoid arthritis through induction of RANKL in human synovial fibroblasts. *Arthritis Rheum.* **2011**, *64*, 1015–1023. [CrossRef]
9. Schulze, J.; Bickert, T.; Beil, F.T.; Zaiss, M.M.; Albers, J.; Wintges, K.; Streichert, T.; Klaetschke, K.; Keller, J., Hissnauer, T.-N.; et al. Interleukin-33 is expressed in differentiated osteoblasts and blocks osteoclast formation from bone marrow precursor cells. *J. Bone Miner. Res.* **2010**, *26*, 704–717. [CrossRef]
10. O'Gradaigh, D.; Ireland, D.; Bord, S.; Compston, J.E. Joint erosion in rheumatoid arthritis: Interactions between tumour necrosis factor alpha, interleukin 1, and receptor activator of nuclear factor kappaB ligand (RANKL) regulate osteoclasts. *Ann. Rheum. Dis.* **2004**, *63*, 354–359. [CrossRef]
11. Rich, B.E.; Kupper, T.S. Cytokines: IL-20—A new effector in skin inflammation. *Curr. Biol.* **2001**, *11*, R531–R534. [CrossRef]
12. Li, H.-H.; Hsu, Y.-H.; Wei, C.-C.; Lee, P.-T.; Chen, W.-C.; Chang, M.-S. Interleukin-20 induced cell death in renal epithelial cells and was associated with acute renal failure. *Genes Immun.* **2008**, *9*, 395–404. [CrossRef] [PubMed]
13. Hsu, Y.-H.; Wei, C.-C.; Shieh, D.-B.; Chan, C.-H.; Chang, M.-S. Anti-IL-20 Monoclonal Antibody Alleviates Inflammation in Oral Cancer and Suppresses Tumor Growth. *Mol. Cancer Res.* **2012**, *10*, 1430–1439. [CrossRef] [PubMed]
14. Chiu, Y.-S.; Wei, C.-C.; Lin, Y.-J.; Hsu, Y.-H.; Chang, M.-S. IL-20 and IL-20R1 antibodies protect against liver fibrosis. *Hepatology* **2014**, *60*, 1003–1014. [CrossRef] [PubMed]
15. Huang, K.-Y.; Lin, R.-M.; Chen, W.-Y.; Lee, C.-L.; Yan, J.-J.; Chang, M.-S. IL-20 may contribute to the pathogenesis of human intervertebral disc herniation. *Spine* **2008**, *33*, 2034–2040. [CrossRef]
16. Hsu, Y.-H.; Li, H.-H.; Hsieh, M.-Y.; Liu, M.-F.; Huang, K.-Y., Chin, L.-S.; Chen, P.-C.; Cheng, H.-H.; Chang, M.-S. Function of interleukin-20 as a proinflammatory molecule in rheumatoid and experimental arthritis. *Arthritis Rheum.* **2006**, *54*, 2722–2733. [CrossRef]
17. Hsieh, M.-Y.; Chen, W.-Y.; Jiang, M.-J.; Cheng, B.-C.; Huang, T.-Y.; Chang, M.-S. Interleukin-20 promotes angiogenesis in a direct and indirect manner. *Genes Immun.* **2006**, *7*, 234–242. [CrossRef]
18. Li, A.; Dubey, S.; Varney, M.L.; Dave, B.J.; Singh, R.K. IL-8 directly enhanced endothelial cell survival, proliferation, and matrix metalloproteinases production and regulated angiogenesis. *J. Immunol.* **2003**, *170*, 3369–3376. [CrossRef]
19. Ha, H.-L.; Wang, H.; Claudio, E.; Tang, W.; Siebenlist, U. IL-20-Receptor Signaling Delimits IL-17 Production in Psoriatic Inflammation. *J. Investig. Dermatol.* **2019**, *140*, 143–151.e3. [CrossRef]
20. Sabat, R.; Wolk, K. Research in practice: IL-22 and IL-20: Significance for epithelial homeostasis and psoriasis pathogenesis. *J. der Dtsch. Dermatol. Ges.* **2011**, *9*, 518–523. [CrossRef] [PubMed]
21. Chan, J.R.; Blumenschein, W.; Murphy, E.; Diveu, C.; Wiekowski, M.; Abbondanzo, S.; Lucian, L.; Geissler, R.; Brodie, S.; Kimball, A.B.; et al. IL-23 stimulates epidermal hyperplasia via TNF and IL-20R2-dependent mechanisms with implications for psoriasis pathogenesis. *J. Exp. Med.* **2006**, *203*, 2577–2587. [CrossRef]
22. Blumberg, H.; Conklin, D.; Xu, W.; Grossmann, A.; Brender, T.; Carollo, S.; Eagan, M.; Foster, D.; Haldeman, B.A.; Hammond, A.; et al. Interleukin 20: Discovery, receptor identification, and role in epidermal function. *Cell* **2001**, *104*, 9–19. [CrossRef]
23. Commins, S.; Steinke, J.; Borish, L. The extended IL-10 superfamily: IL-10, IL-19, IL-20, IL-22, IL-24, IL-26, IL-28, and IL-29. *J. Allergy Clin. Immunol.* **2008**, *121*, 1108–1111. [CrossRef] [PubMed]
24. Myles, I.; Fontecilla, N.M.; Valdez, P.A.; Vithayathil, P.J.; Naik, S.; Belkaid, Y.; Ouyang, W.; Datta, S. Signaling via the IL-20 receptor inhibits cutaneous production of IL-1β and IL-17A to promote infection with methicillin-resistant Staphylococcus aureus. *Nat. Immunol.* **2013**, *14*, 804–811. [CrossRef]
25. Tritsaris, K.; Myren, M.; Ditlev, S.B.; Hübschmann, M.V.; van der Blom, I.; Hansen, A.J.; Olsen, U.B.; Cao, R.; Zhang, J.; Jia, T.; et al. IL-20 is an arteriogenic cytokine that remodels collateral networks and improves functions of ischemic hind limbs. *Proc. Natl. Acad. Sci. USA* **2007**, *104*, 15364–15369. [CrossRef]

26. Waszczykowski, M.; Fabiś-Strobin, A.; Bednarski, I.; Narbutt, J.; Fabiś, J. Serum and synovial fluid concentrations of interleukin-18 and interleukin-20 in patients with osteoarthritis of the knee and their correlation with other markers of inflammation and turnover of joint cartilage. *AMS* **2022**, *18*, 448–458. [CrossRef]
27. Šenolt, L.; Prajzlerová, K.; Hulejová, H.; Šumová, B.; Filková, M.; Veigl, D.; Pavelka, K.; Vencovský, J. Interleukin-20 is triggered by TLR ligands and associates with disease activity in patients with rheumatoid arthritis. *Cytokine* **2017**, *97*, 187–192. [CrossRef] [PubMed]
28. Valentina, M.; Jan, F.; Peder, N.L.; Bo, Z.; Hongjie, D.; Pernille, K. Cytokine detection and simultaneous assessment of rheumatoid factor interference in human serum and synovial fluid using high-sensitivity protein arrays on plasmonic gold chips. *BMC Biotechnol.* **2015**, *15*, 73. [CrossRef]
29. Hsu, Y.-H.; Chen, W.-Y.; Chan, C.-H.; Wu, C.-H.; Sun, Z.-J.; Chang, M.-S. Anti-IL-20 monoclonal antibody inhibits the differentiation of osteoclasts and protects against osteoporotic bone loss. *J. Exp. Med.* **2011**, *208*, 1849–1861. [CrossRef]
30. Hsu, Y.-H.; Hsing, C.-H.; Li, C.-F.; Chan, C.-H.; Chang, M.-C.; Yan, J.-J. Anti-IL-20 monoclonal antibody suppresses breast cancer progression and bone osteolysis in murine models. *J. Immunol.* **2012**, *188*, 1981–1991. [CrossRef]
31. Yamaguchi, M.; Fukasawa, S. Is Inflammation a Friend or Foe for Orthodontic Treatment?: Inflammation in Orthodontically Induced Inflammatory Root Resorption and Accelerating Tooth Movement. *Int. J. Mol. Sci.* **2021**, *22*, 2388. [CrossRef]
32. Antoun, J.S.; Mei, L.; Gibbs, K.; Farella, M. Effect of orthodontic treatment on the periodontal tissues. *Periodontology 2000* **2017**, *74*, 140–157. [CrossRef]
33. Li, Y.; Zhan, Q.; Bao, M.; Yi, J.; Li, Y. Biomechanical and biological responses of periodontium in orthodontic tooth movement: Up-date in a new decade. *Int. J. Oral Sci.* **2021**, *13*, 20. [CrossRef]
34. Meng, B.; Wu, D.; Cheng, Y.; Huang, P.; Liu, Y.; Gan, L.; Liu, C.; Cao, Y. Interleukin-20 differentially regulates bone mesenchymal stem cell activities in RANKL-induced osteoclastogenesis through the OPG/RANKL/RANK axis and the NF-κB, MAPK and AKT signalling pathways. *Scand. J. Immunol.* **2020**, *91*, e12874. [CrossRef]
35. Yang, B.; Fu, C.; Wu, Y.; Liu, Y.; Zhang, Z.; Chen, X.; Wu, D.; Gan, Z.; Chen, Z.; Cao, Y. γ-Secretase inhibitors suppress IL-20-mediated osteoclastogenesis via Notch signalling and are affected by Notch2 in vitro. *Scand. J. Immunol.* **2022**, *96*, e13169. [CrossRef]
36. Singh, S.; Anshita, D.; Ravichandiran, V. MCP-1: Function, regulation, and involvement in disease. *Int. Immunopharmacol.* **2021**, *101*, 107598. [CrossRef]
37. Otero, K.; Turnbull, I.; Poliani, P.; Vermi, W.; Cerutti, E.; Aoshi, T.; Tassi, I.; Takai, T.; Stanley, S.; Miller, M.; et al. Macrophage colony-stimulating factor induces the proliferation and survival of macrophages via a pathway involving DAP12 and beta-catenin. *Nat. Immunol.* **2009**, *10*, 734–743. [CrossRef]
38. Yoshida, H.; Hayashi, S.-I.; Kunisada, T.; Ogawa, M.; Nishikawa, S.; Okamura, H.; Sudo, T.; Shultz, L.D.; Nishikawa, S.-I. The murine mutation osteopetrosis is in the coding region of the macrophage colony stimulating factor gene. *Nature* **1990**, *345*, 442–444. [CrossRef]
39. Zur, Y.; Rosenfeld, L.; Keshelman, C.; Dalal, N.; Guterman-Ram, G.; Orenbuch, A.; Einav, Y.; Levaot, N.; Papo, N. A dual-specific macrophage colony-stimulating factor antagonist of c-FMS and alphavbeta3 integrin for osteoporosis therapy. *PLoS Biol.* **2018**, *16*, e2002979. [CrossRef]
40. Wiktor-Jedrzejczak, W.; Gordon, S. Cytokine regulation of the macrophage (M phi) system studied using the colony stimulating factor-1-deficient op/op mouse. *Physiol. Rev.* **1996**, *76*, 927–947. [CrossRef]
41. Liu, L.; Ding, C.; Zeng, W.; Heuer, J.G.; Tetreault, J.W.; Noblitt, T.W.; Hangoc, G.; Cooper, S.; Brune, K.A.; Sharma, G.; et al. Selective enhancement of multipotential hematopoietic progenitors in vitro and in vivo by IL-20. *Blood* **2003**, *102*, 3206–3209. [CrossRef] [PubMed]
42. Rutz, S.; Wang, X.; Ouyang, W. The IL-20 subfamily of cytokines–from host defence to tissue homeostasis. *Nat. Rev. Immunol.* **2014**, *14*, 783–795. [CrossRef] [PubMed]
43. Zhang, W.; Magadi, S.; Li, Z.; Smith, C.W.; Burns, A.R. IL-20 promotes epithelial healing of the injured mouse cornea. *Exp. Eye Res.* **2017**, *154*, 22–29. [CrossRef] [PubMed]
44. Ikebuchi, Y.; Aoki, S.; Honma, M.; Hayashi, M.; Sugamori, Y.; Khan, M.; Kariya, Y.; Kato, G.; Tabata, Y.; Penninger, J.M.; et al. Coupling of bone resorption and formation by RANKL reverse signalling. *Nature* **2018**, *561*, 195–200. [CrossRef]
45. Coury, F.; Peyruchaud, O.; Machuca-Gayet, I. Osteoimmunology of Bone Loss in Inflammatory Rheumatic Diseases. *Front. Immunol.* **2019**, *10*, 679. [CrossRef]
46. Ralston, S.; Schett, G. Osteoimmunology. *Calcif. Tissue Int.* **2018**, *102*, 501–502. [CrossRef]
47. Okamoto, K.; Takayanagi, H. Osteoimmunology. *Cold Spring Harb. Perspect. Med.* **2019**, *9*, a031245. [CrossRef]
48. Xiong, J.; Onal, M.; Jilka, R.L.; Weinstein, R.S.; Manolagas, S.C.; O'Brien, C.A. Matrix-embedded cells control osteoclast formation. *Nat. Med.* **2011**, *17*, 1235–1241. [CrossRef]
49. Cao, X. Targeting osteoclast-osteoblast communication. *Nat. Med.* **2011**, *17*, 1344–1346. [CrossRef]
50. Theill, L.E.; Boyle, W.J.; Penninger, J.M. RANK-L and RANK: T cells, bone loss, and mammalian evolution. *Annu. Rev. Immunol.* **2022**, *20*, 795–823. [CrossRef]
51. Danks, L.; Komatsu, N.; Guerrini, M.M.; Sawa, S.; Armaka, M.; Kollias, G.; Nakashima, T.; Takayanagi, H. RANKL expressed on synovial fibroblasts is primarily responsible for bone erosions during joint inflammation. *Ann. Rheum. Dis.* **2015**, *75*, 1187–1195. [CrossRef] [PubMed]

52. Boyle, W.J.; Simonet, W.S.; Lacey, D.L. Osteoclast differentiation and activation. *Nature* **2003**, *423*, 337–342. [CrossRef] [PubMed]
53. Lechner, J.; Rudi, T.; von Baehr, V. Osteoimmunology of tumor necrosis factor-alpha, IL-6, and RANTES/CCL5: A review of known and poorly understood inflammatory patterns in osteonecrosis. *Clin. Cosmet. Investig. Dent.* **2018**, *10*, 251–262. [CrossRef] [PubMed]
54. Mansky, K.C.; Sankar, U.; Han, J.; Ostrowski, M.C. Microphthalmia transcription factor is a target of the p38 MAPK pathway in response to receptor activator of NF-kappa B ligand signaling. *J. Biol. Chem.* **2002**, *277*, 11077–11083. [CrossRef]
55. Kim, T.; Yoon, J.; Cho, H.; Lee, W.B.; Kim, J.; Song, Y.H.; Kim, S.N.; Yoon, J.H.; Kim-Ha, J.; Kim, Y.J. Downregulation of lipopolysaccharide response in Drosophila by negative crosstalk between the AP1 and NF-kappaB signaling modules. *Nat. Immunol.* **2005**, *6*, 211–218. [CrossRef] [PubMed]
56. Seeman, E.; Delmas, P.D. Bone quality–the material and structural basis of bone strength and fragility. *N. Engl. J. Med.* **2006**, *354*, 2250–2261. [CrossRef]
57. Lacey, D.L.; Boyle, W.J.; Simonet, W.S.; Kostenuik, P.J.; Dougall, W.C.; Sullivan, J.K.; Martin, J.S.; Dansey, R. Bench to bedside: Elucidation of the OPG-RANK-RANKL pathway and the development of denosumab. *Nat. Rev. Drug Discov.* **2012**, *11*, 401–419. [CrossRef]
58. Nishida, H.; Suzuki, H.; Madokoro, H.; Hayashi, M.; Morimoto, C.; Sakamoto, M.; Yamada, T. Blockade of CD26 signaling inhibits human osteoclast development. *J. Bone Miner. Res. Off. J. Am. Soc. Bone Miner. Res.* **2014**, *29*, 2439–2455. [CrossRef]
59. Zhu, M.; Sun, B.H.; Saar, K.; Simpson, C.; Troiano, N.; Dallas, S.L.; Tiede-Lewis, L.M.; Nevius, E.; Pereira, J.P.; Weinstein, R.S.; et al. Deletion of Rac in Mature Osteoclasts Causes Osteopetrosis, an Age-Dependent Change in Osteoclast Number, and a Reduced Number of Osteoblasts In Vivo. *J. Bone Miner. Res. Off. J. Am. Soc. Bone Miner. Res.* **2016**, *31*, 864–873. [CrossRef]
60. Su, N.; Li, X.; Tang, Y.; Yang, J.; Wen, X.; Guo, J.; Tang, J.; Du, X.; Chen, L. Deletion of FGFR3 in Osteoclast Lineage Cells Results in Increased Bone Mass in Mice by Inhibiting Osteoclastic Bone Resorption. *J. Bone Miner. Res. Off. J. Am. Soc. Bone Miner. Res.* **2016**, *31*, 1676–1687. [CrossRef]
61. Wu, D.; Meng, B.; Cheng, Y.; Gan, L.; Huang, P.; Cao, Y. The effect of risedronate on orthodontic tooth movement in ovariectomized rats. *Arch. Oral Biol.* **2019**, *105*, 59–64. [CrossRef] [PubMed]
62. Hsu, Y.-H.; Chiu, Y.-S.; Chen, W.-Y.; Huang, K.-Y.; Jou, I.-M.; Wu, P.-T.; Wu, C.-H.; Chang, M.-S. Anti-IL-20 monoclonal antibody promotes bone fracture healing through regulating IL-20-mediated osteoblastogenesis. *Sci. Rep.* **2016**, *6*, 24339. [CrossRef] [PubMed]
63. Chen, X.; Liu, Y.; Meng, B.; Wu, D.; Wu, Y.; Cao, Y. Interleukin-20 inhibits the osteogenic differentiation of MC3T3-E1 cells via the GSK3β/β-catenin signalling pathway. *Arch. Oral Biol.* **2021**, *125*, 105111. [CrossRef] [PubMed]
64. Dai, Q.; Han, Y.; Xie, F.; Ma, X.; Xu, Z.; Liu, X.; Zou, W.; Wang, J. A RANKL-based Osteoclast Culture Assay of Mouse Bone Marrow to Investigate the Role of mTORC1 in Osteoclast Formation. *J. Vis. Exp.* **2018**, *133*, e56468.
65. Marino, S.; Logan, J.G.; Mellis, D.; Capulli, M. Generation and culture of osteoclasts. *BoneKEy Rep.* **2014**, *3*, 570. [CrossRef]

International Journal of
Molecular Sciences

MDPI

Article

Type IVb Hypersensitivity Reaction in the Novel Murine Model of Palladium–Induced Intraoral Allergic Contact Mucositis

Keisuke Nasu [1,2,†], Kenichi Kumagai [2,3,†], Takamasa Yoshizawa [1,2], Kazutaka Kitaura [2,4], Ryota Matsubara [2,5], Motoaki Suzuki [2,6], Ryuji Suzuki [2,4] and Yoshiki Hamada [1,*]

1 Department of Oral and Maxillofacial Surgery, School of Dental Medicine, Tsurumi University, Yokohama 230-8501, Japan
2 Department of Rheumatology and Clinical Immunology, Clinical Research Center for Rheumatology and Allergy, Sagamihara National Hospital, National Hospital Organization, Sagamihara 252-0392, Japan
3 Department of Oral and Maxillofacial Surgery, Dentistry and Orthodontics, The University of Tokyo Hospital, Tokyo 113-8655, Japan
4 Repertoire Genesis Inc., Osaka 567-0085, Japan
5 Department of Oral and Maxillofacial Surgery, Sendai Tokushukai Hospital, Sendai 981-3116, Japan
6 Department of Anatomy and Physiology, Faculty of Medicine, Saga University, Saga 849-8501, Japan
* Correspondence: hamada-y@tsurumi-u.ac.jp; Tel./Fax: +81-45-580-8327
† These authors contributed equally to this work.

Citation: Nasu, K.; Kumagai, K.; Yoshizawa, T.; Kitaura, K.; Matsubara, R.; Suzuki, M.; Suzuki, R.; Hamada, Y. Type IVb Hypersensitivity Reaction in the Novel Murine Model of Palladium–Induced Intraoral Allergic Contact Mucositis. *Int. J. Mol. Sci.* **2023**, *24*, 3137. https://doi.org/10.3390/ijms24043137

Academic Editor: Kenneth Michael Pollard

Received: 29 December 2022
Revised: 30 January 2023
Accepted: 1 February 2023
Published: 5 February 2023

Abstract: Palladium (Pd) is a component of several alloy types that are widely used in our environment, including several dental alloy types that cause adverse reactions such as hypersensitivity in the oral mucosa. However, the pathological mechanism of intraoral Pd allergies remains unclear because its animal model in the oral mucosa has not been established. In this study, we established a novel murine model of Pd–induced allergies in the oral mucosa, and explored the immune response of cytokine profiles and T cell diversity in terms of the T cell receptor. The Pd–induced allergy mouse was generated by two sensitizations with $PdCl_2$, plus a lipopolysaccharide solution into the postauricular skin followed by a single Pd challenge of the buccal mucosa. Significant swelling and pathological features were histologically evident at five days after the challenge, and CD4–positive T cells producing high levels of T helper 2 type cytokines had accumulated in the allergic oral mucosa. Characterization of the T cell receptor repertoire in Palladium allergic mice indicated that Pd–specific T cell populations were limited in V and J genes but were diverse at the clonal level. Our model demonstrated that a Pd–specific T cell population with Th2 type response tendencies may be involved in the Pd–induced intraoral metal contact allergy.

Keywords: palladium allergy; contact dermatitis; allergic contact mucositis; metal allergy

1. Introduction

Metals are ubiquitous in our environment, and often cause allergic contact dermatitis (ACD), which is an inflammatory disease categorized as a delayed–type hypersensitivity (DTH) reaction [1]. Epidemiological studies reported palladium (Pd) as the hapten with the highest prevalence in ACD pathogenesis after nickel (Ni) [2,3]. Pd has a close chemical resemblance with Ni and platinum (Pt), and its main applications include the crown and bridgework in dentistry and is sometimes used in jewelry. Pd allergy incidents have recently increased in patients with dermatitis and dental disorders [4].

Most dental appliances to restore or replace decayed teeth are partially composed of alloys that may contain a large variety of metals. Corrosion is an inevitable chemical reaction between the oral environment and dental alloys that may lead to substantial and clinically relevant ion release and in turn, result in adverse reactions, such as hypersensitivity and ACD. A strong relationship was found between the exposure to dental alloys and ACD; thus, dental crowns seem to play a key role [5,6]. However, the relationship between hypersensitivity to metals and objective oral abnormalities is unclear. Most likely, the ability of various metals to trigger the innate immune system and the tolerant character of the oral mucosa (OM) play a key role.

The DTH immune response in the OM differs from that in the skin mainly due to differences in the accumulation of local antigen–presenting cells and T cells [7,8]. Previous studies reported that DTH reactions in the oral mucosa have the hallmarks of skin DTH reactions with T cells and macrophages [9]. The involvement of pathogenic T cells in the development of a metal allergy in the oral environment has not yet been explored using animal models, although metal allergies have long been known to be T cell–dependent and the metal ions caused by the corrosion from dental alloys are known to function as haptens. We recently generated the novel murine model of Ni–induced allergic contact mucosa (ACM) and have analyzed antigen–specific immune responses [10]. However, the conformation and toxicity of Ni and Pd were different, and the mechanism of the specific immune response to Pd in the OM remained unknown.

Recent advances in next–generation sequencing (NGS) have enabled massive sequencing and long–read sequencing, and the development of NGS–compatible bioinformatics software has enabled NGS–based T cell receptor (TCR) repertoire analysis, which is a quantitative and comprehensive analysis of a large–scale repertoire [11–13].

In the present study, we generated a novel murine model of Pd–induced allergies in the OM to explore how accumulated antigen–specific T cells at the site of allergic inflammation contribute to Pd allergy development in the OM.

2. Results

2.1. Pd–Induced Allergic BALB/cAJcl Mice Develop Swelling in the OM

The swelling of the buccal area of the OM of ACM mice reached a maximum on day 2 after the challenge (Figure 1A). Significant redness and swelling are evident in the buccal area of ACM mice compared with the control and ICM mice on day 2 after the challenge (Figure 1B). Conversely, OM buccal area swelling in irritant contact mucositis (ICM) and control mice reached a maximum on day 1 after the challenge. The difference in buccal mucosal swelling was significantly higher in ACM mice compared with ICM mice on days 2, 3, and 5 after the challenge. The most differences in OM buccal area swelling between ACM and ICM were observed in day 2 after the challenge. We focused our analysis on day 2, when ACM had the largest OM swelling and the most significant difference between ACM and ICM, and day 5, when ACM had significantly larger OM swelling than ICM and the largest CD3 mRNA expression levels in OM.

2.2. Histopathological and Immunohistochemical (IHC) Analyses of F4/80 and CD3 in the OM of Pd–Induced Allergic BALB/cAJcl Mice

We examined the histopathological and IHC analyses in the OM of control and Pd–induced ACM and ICM mice on days 2 and 5 after the challenge to verify whether macrophages and T cells infiltrated into the inflamed OM. Hematoxylin and eosin (HE) staining showed the dense infiltration of inflammatory cells in the epithelial basal layer and upper dermis, as well as swelling of the OM epithelium and epidermal spongiosis in ACM mice on day 5 after the challenge, but not in ICM and control mice (Figure 2A,C,E).

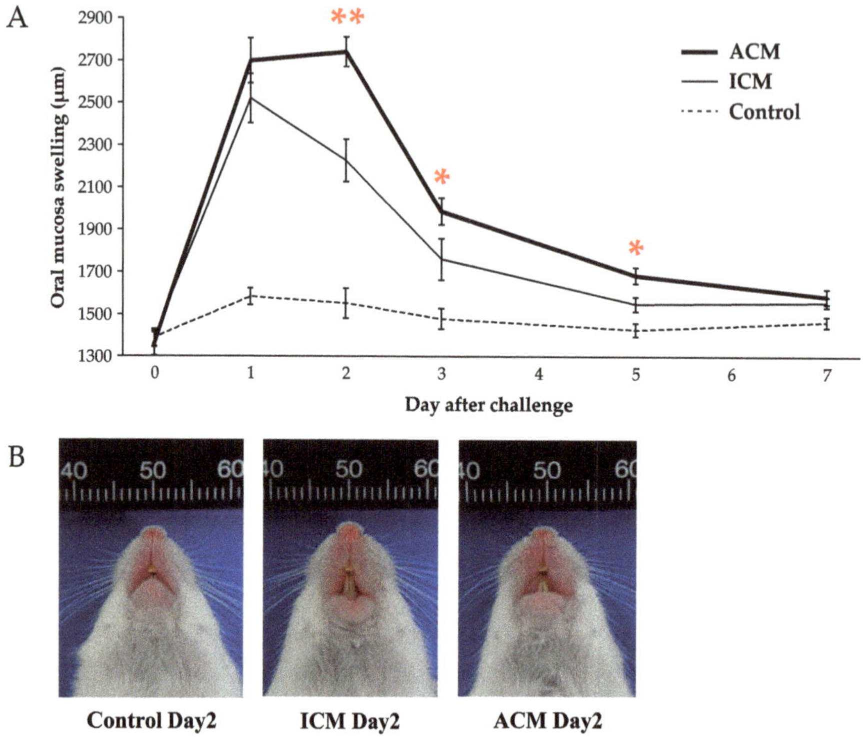

Figure 1. Swelling of the OM in Pd–induced allergic mice. (**A**) OM buccal area swelling from day 1 to day 7 after the challenge was measured in Pd–induced ACM and ICM and control mice ($n = 10$). Each point and error bars indicate the mean values and standard deviations. Statistical significance was tested through the Kruskal–Wallis test followed by Steel–Dwass's multiple comparison tests (* $p < 0.05$, ** $p < 0.01$). (**B**) Macroscopic findings of the oral mucosa in Pd–induced allergic mice.

Next, we performed an IHC analysis of F4/80 and CD3 in the OM of control and Pd–induced ICM and ACM mice to verify whether macrophages and T cells had infiltrated into the inflamed OM of ACM mice (Figure 2F–O). IHC staining showed considerable F4/80+ macrophage infiltration into the dermis of ICM and ACM mice on day 2 after the challenge compared with control mice (Figure 2F,G,I). Moreover, the F4/80 staining intensity was stronger in ACM than in ICM mice. In contrast, F4/80+ cells were lost on day 5 after the challenge, both in ACM and ICM mice. IHC analyses of CD3+ T cells were observed in the epithelial basal layer and upper dermis in the ACM mice, and they were continued between day 2 and day 5 after the challenge (Figure 2N,O). Conversely, CD3+ T cells were slightly observed in ICM mice (Figure 2L,M).

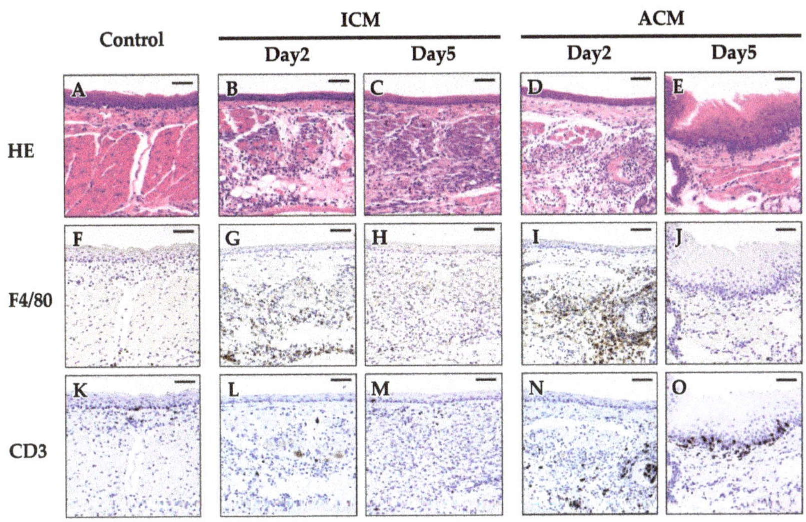

Figure 2. Histopathological and IHC analyses of F4/80 and CD3 in the OM of Pd–induced allergic mice. Representative photomicrographs of OM sections from control and Pd–induced ICM and ACM mice from days 2 and 5 after the challenge, stained with HE (**A–E**), F4/80 (**F–J**), and CD3 (**K–O**). Scale bar = 5 μm.

2.3. mRNA Expression Levels of T Cell Markers in the OM of Pd–Induced Allergic BALB/cAJcl Mice

We performed a quantitative polymerase chain reaction (qPCR) analysis in CD3, CD4, and CD8 expressions to verify whether mRNA expression levels of T cells infiltrated into the inflamed OM. The CD3 expression ratio was significantly increased in control ACM mice compared to ICM mice on days 2 and 5 after the challenge (Figure 3). Furthermore, the ratio of CD3 expression in ACM tended to increase from day 2 to day 5. A similar expression trend to CD3 was observed for CD4, which was significantly increased in ACM mice compared to ICM mice. Conversely, no differences were found in the CD8 expression ratio between ACM and ICM mice.

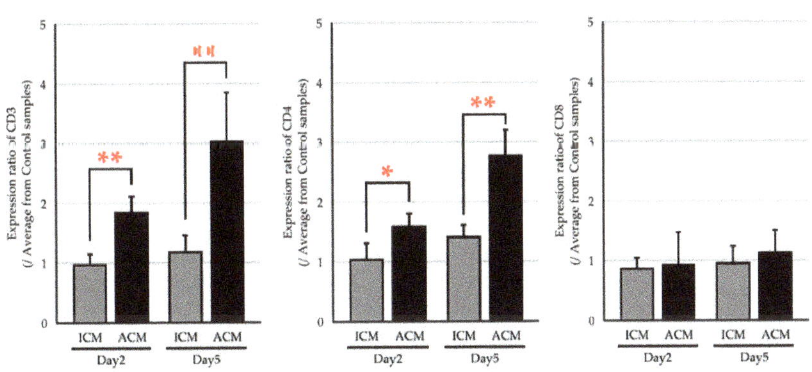

Figure 3. mRNA expression levels of T cell phenotypes in the OM of Pd–induced allergic mice. The mRNA expression levels of CD3, CD4, and CD8 were assessed in the OM of ICM (gray bars) and ACM (black bars) mice (n = 5) on days 2 and 5 after the challenge. GAPDH gene expression was used as an internal control. Each sample was divided by the average of the control sample to calculate the expression ratio (n = 5). Bars and error bars indicate the mean plus standard deviation. Statistical significance was tested by the Mann–Whitney U test (* $p < 0.05$, ** $p < 0.01$).

2.4. mRNA Expression Levels of T Cell–Related Cytokines in the OM of Pd–Induced Allergic BALB/cAJcl Mice

We compared the expression levels of Th1–related cytokines (interleukin (IL)–2, interferons (IFN)–γ, and tumor necrosis factors (TNF)–α), and Th2–related cytokines (IL–4 and IL–10), in the OM of Pd–induced ICM and ACM mice to examine inflammation in allergic OM (Figure 4). The expression level of Th2–type cytokines was higher in ACM mice compared with ICM mice from day 2 to day 5 after the challenge. A large difference in the expression levels of ACM and ICM mice was observed, especially in IL–4, and the difference in IL–10 between ACM and ICM mice appears to be larger on day 5 than on day 2 after the challenge. Conversely, the expression levels of TH1–related cytokines were closely similar between ICM and ACM mice, except for IL–2 on day 2 after the challenge.

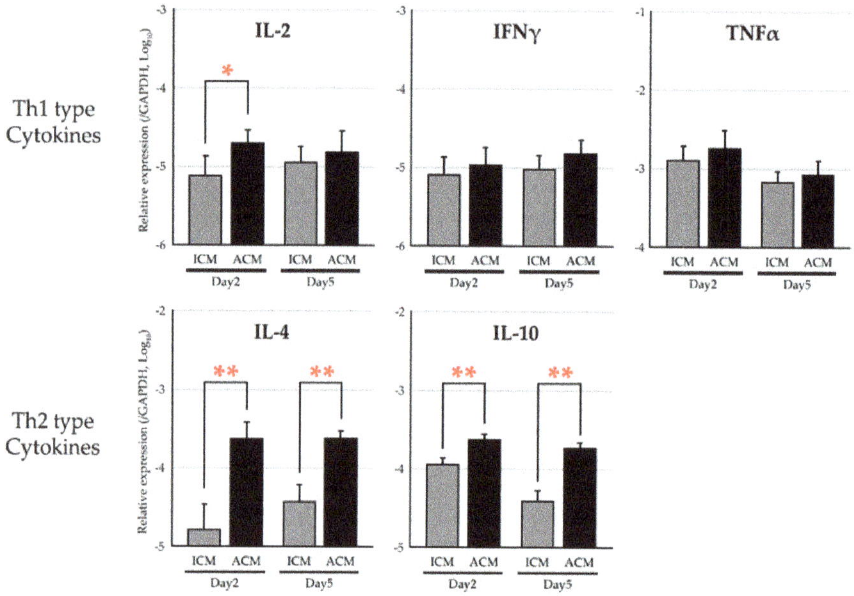

Figure 4. mRNA expression levels of Th1– and Th2–type cytokines in the mucosa of Pd–induced allergic mice. The mRNA expression levels of Th1–type cytokines (IL–2, IFNγ, and TNFα) and Th2 type cytokines (IL–4 and IL–10) were assessed in the OM of ICM (gray bars) and ACM (black bars) mice (n = 5) on days 2 and 5 after the challenge. Each sample was divided by GAPDH gene expression as an internal control. Bars and error bars indicate the mean plus standard deviation. Statistical significance was tested by the Mann–Whitney U test (* $p < 0.05$, ** $p < 0.01$).

2.5. TRV–TRJ Combination and Diversity of TCR Repertoire in the OM and Cervical Lymph Nodes (Ly) of Pd–Induced ICM and ACM Mice on Day 5 after the Challenge

We analyzed the TRV and TRJ expression levels in the inflamed OM and Ly by a NGS–based TCR repertoire analysis to determine the TCR repertoire of T cells in a Pd allergy that had infiltrated into the OM and Ly of ICM and ACM mice on day 5 after the challenge. A representative TRV–TRJ combination three–dimensional graph of the TRA and TRB repertoire in OM revealed the dominance of certain TRV and TRJ gene combinations as well as the extent of the diversity of TCR usage (Figure 5A). Hence, more TRV–TRJ combinations of TRA and TRB repertoire were detected in ACM OM than in ICM OM mice. Next, we investigated the diversity of repertoire in TRA and TRB using the Shannon index in ACM and ICM samples (Figure 5B). The Shannon index was significantly higher in ACM OM than in ICM OM mice. Furthermore, ACM Ly showed a trend toward a higher Shannon index than ICM Ly mice.

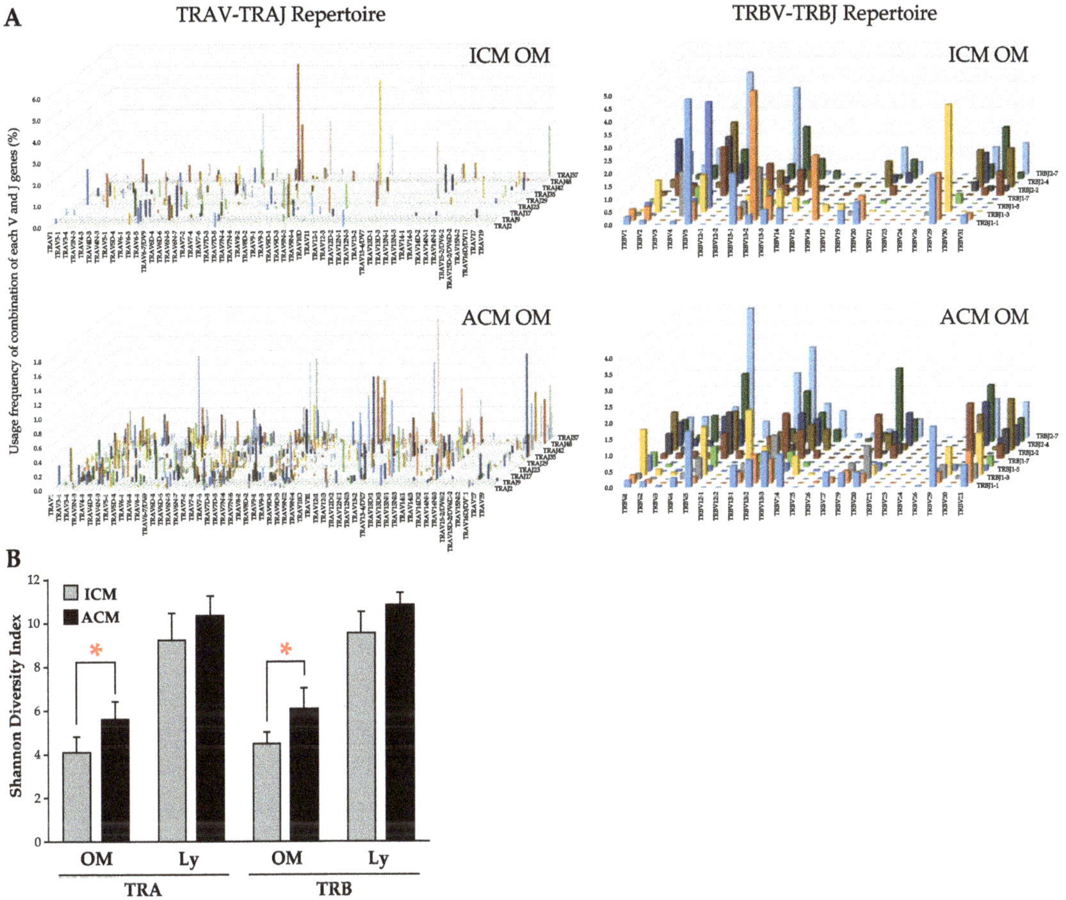

Figure 5. TCR repertoire analysis of the TRA and TRB in ICM and ACM mice. The NGS–based TCR repertoire analysis was performed on the OM and Ly of ICM and ACM mice on day 5 following the challenge. (**A**) Combining TRAV or TRBV on the *X*–axis and TRAJ or TRBJ on the *Z*–axis, with the frequency (percentage) of each clone on the *Y*–axis, 3D images depict the TCR repertoire ($n = 3$, average). (**B**) Shannon Diversity Index is shown in the ACM and ICM, OM and Ly, and TRA and TRB ($n = 3$). Statistical significance was tested by the Mann–Whitney U test (* $p < 0.05$).

2.6. Commonality of TCR Repertoire in the OM and Ly of Pd–Induced ICM and ACM Mice on Day 5 after the Challenge

We compared the top 20 in the ACM OM mice to other samples to assess the commonality of unique reads obtained from the repertoire analysis in TRA and TRB (Figure 6A,B). The top reads comprising the ACM OM mice shared many with ACM Ly (green shading) and some with ICM Ly, but not with ICM OM. The TRA sequences (TRAV11D, TRAJ18, and CVVGDRGSALGRLHF) of iNKT cell (orange shading) were detected in all samples, and the frequency was lower in ACM OM (0.02%) than in ICM OM (1.12%) mice (Figure 6B).

A

Rank	TRAV	TRAJ	CDR3 AA	ACM		ICM	
				OM	Ly	OM	Ly
1	TRAV13D-1	TRAJ58	CAMQQGTGSKLSF	1.5012	0.0054	-	-
2	TRAV21/DV12	TRAJ35	CILRWGFASALTF	1.3412	-	-	-
3	TRAV6-5	TRAJ53	CALGNSGGSNYKLTF	1.1121	0.0005	-	0.0004
4	TRAV13D-1	TRAJ53	CARNSGGSNYKLTF	1.0062	0.0021	-	-
5	TRAV8D-2	TRAJ57	CAQGGSAKLIF	1.0023	-	-	-
6	TRAV12D-1	TRAJ40	CALRSTGNYKYVF	0.9948	0.0075	-	-
7	TRAV13D-2	TRAJ17	CAIEDSAGNKLTF	0.9874	-	-	-
8	TRAV9-1	TRAJ56	CAVSDMATGGNNKLTF	0.8279	-	-	-
9	TRAV12N-2	TRAJ17	CANNSAGNKLTF	0.7628	-	-	-
10	TRAV14-3	TRAJ43	CAASNNNNAPRF	0.7445	0.0008	-	0.0004
11	TRAV12-2	TRAJ42	CALSDRSGGSNAKLTF	0.7252	0.0012	-	-
12	TRAV17	TRAJ52	CALEGPGTGANTGKLTF	0.6209	-	-	-
13	TRAV19	TRAJ42	CAAGSNSGGSNAKLTF	0.6088	-	-	0.0012
14	TRAV12-3	TRAJ49	CALILHTGYQNFYF	0.5966	-	-	-
15	TRAV13D-2	TRAJ22	CAISSGSWQLIF	0.5785	0.0012	-	0.0004
16	TRAV7-1	TRAJ37	CAVRRTGNTRKLIF	0.5613	-	-	-
17	TRAV13N-4	TRAJ18	CAMERRGSALGRLHF	0.5591	-	-	-
18	TRAV14D-3/DV8	TRAJ18	CAARGSALGRLHF	0.5316	0.0104	-	0.0019
19	TRAV5-4	TRAJ18	CAASAGSALGRLHF	0.5218	0.0403	-	-
20	TRAV13D-1	TRAJ37	CAMEKPGNTRKLIF	0.5056	-	-	-
iNKT	TRAV11D	TRAJ18	CVVGDRGSALGRLHF	0.0207	0.2359	1.1174	0.2044

B

Rank	TRBV	TRBJ	CDR3 AA	ACM		ICM	
				OM	Ly	OM	Ly
1	TRBV19	TRBJ2-5	CASSIRTGEDTQYF	1.2700	0.0002	-	-
2	TRBV13-3	TRBJ1-3	CASRTGAGNTLYF	0.9841	0.0005	-	-
3	TRBV5	TRBJ2-7	CASSGTGYEQYF	0.9350	0.0117	-	-
4	TRBV29	TRBJ2-1	CASSLNWDYAEQFF	0.8619	-	-	-
5	TRBV13-2	TRBJ2-7	CASGDWGREEQYF	0.8240	-	-	-
6	TRBV13-1	TRBJ1-4	CASRDNNERLFF	0.7594	0.0002	-	0.0082
7	TRBV14	TRBJ1-1	CASSFPTGRNTEVFF	0.7534	0.0056	-	-
8	TRBV31	TRBJ2-3	CAWSFQLGGLSAETLYF	0.6526	-	-	-
9	TRBV29	TRBJ1-1	CASSIQDTEVFF	0.6147	-	-	-
10	TRBV5	TRBJ1-1	CASSQETNTEVFF	0.6122	0.0037	-	-
11	TRBV5	TRBJ2-7	CASSQVPGGANEQYF	0.5820	-	-	-
12	TRBV19	TRBJ2-1	CASSLDSNYAEQFF	0.5751	0.0005	-	-
13	TRBV13-1	TRBJ2-7	CASRRDWGYEQYF	0.5355	-	-	-
14	TRBV13-1	TRBJ1-4	CASSEQGFSNERLFF	0.4977	0.0004	-	-
15	TRBV31	TRBJ1-2	CAWSLGQGVSDYTF	0.4856	0.0103	-	-
16	TRBV29	TRBJ2-3	CASSLSWGSAETLYF	0.4830	-	-	-
17	TRBV13-3	TRBJ1-1	CASSDQGTNTEVFF	0.4632	-	-	-
18	TRBV13-2	TRBJ2-7	CASGEGTGAEQYF	0.4598	-	-	-
19	TRBV12-1	TRBJ2-7	CASSLGTGGYEQYF	0.4365	0.0004	-	-
20	TRBV17	TRBJ2-4	CASSRTGQNTLYF	0.4322	0.0039	-	-

Figure 6. Commonality of unique reads obtained from the repertoire analysis in TRA and TRB. (**A**,**B**) The %frequency of unique reads in the top 20 in ACM OM mice and their commonality in other samples. The %frequency is the average of the frequency of presence for in–frame reads for each of the three independent samples. The green shading is unique reads shared by ACM OM and ACM Ly mice. (**A**,**B**) Orange shading included iNKT cell sequences as detected in all samples. (**A**) iNKT cell sequence and %frequency of presence in each sample. The %frequency is the average of the frequency of presence for in–frame reads for each of the three independent samples.

2.7. Evaluation of TRV and TRJ Gene Skew in Common Reads of ACM OM and ACM Ly Repertoire

The commonality was assessed for all unique reads in the ACM OM mice for further investigation (Figure 7A,B). Of the 12,608 TRA unique reads of ACM OM mice, 829 reads with ACM Ly, 494 reads with ICM Ly, and 19 reads with ICM OM mice were shared. Of the 15,917 TRB unique reads of ACM OM mice, 630 reads with ACM Ly, 186 reads with ICM Ly, and 16 reads with ICM OM mice were shared.

The number of unique reads present in ACM OM to those that were increased in ACM Ly mice over ICM were 683 and 550 for TRA and TRB, respectively. Furthermore, the reads were limited to 524 and 529 in TRA and TRB, respectively, of the unique reads held by ACM Oms mice, when narrowed down to those that were present only in ACM Ly mice (blue box line).

The frequency counts of TRA and TRB, and TRV and TRJ gene usage were tabulated for the number of unique reads in the two narrowed–down patterns (ACM Ly over ICM, or only in ACM Ly) (Figure 7C–F). The genes in TRAV were distributed in 63 of 112 genes with a mean of 10.8 or 8.3 read counts, and the genes in TRAJ were distributed in 41 of 51 genes with a mean of 16.7 or 12.8 read counts. The genes in TRBV were distributed in 20 of 23 genes with a mean of 27.0 or 26.5 read counts, and the genes in TRBJ were distributed in 12 of 13 genes with a mean of 45.8 or 44.1 read counts. Genes detected at high frequency counts (mean + 2SD, beyond the green or blue line) were TRAV7D–2, TRAV14–1, TRAJ18, TRBV5, and TRBJ2–7.

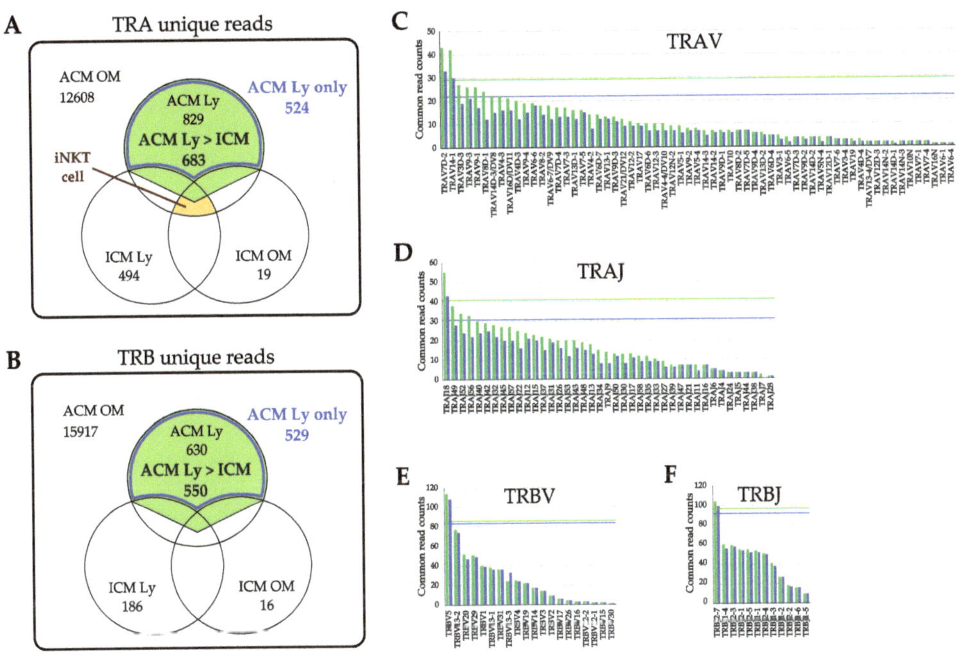

Figure 7. Common read counts of ACM OM and ACM Ly repertoire. (**A**,**B**) The number of unique reads common to the ACM OM mice and other samples is shown in the Venn diagram (n = 3, total). The green shading is unique reads present in ACM OM to those that were increased in ACM Ly mice over ICM. Blue box lines are unique reads present in ACM OM to those that were present only in ACM Ly mice. (**C**–**F**) The common unique reads between ACM OM and Ly mice were counted for TRAV, TRAJ, TRBV, and TRBJ genes, respectively. The green bars are unique reads present in ACM OM to those that were increased in ACM Ly mice over ICM. The blue bars are unique reads present in ACM OM to those that were present only in ACM Ly mice. The green line is the mean + 2SD of green bars. The blue line is the mean + 2SD of blue bars.

3. Discussion

In the present study, we established a novel murine model for Pd–induced allergies in the OM in which CD3+ T cells infiltrated the inflamed area. Previous studies have reported a Pd–allergic model mouse in the footpad skin or auricle regions [14,15]. However, the pathological mechanism of the intraoral Pd allergy remains unclear because an animal model of Pd allergies in the OM has not been established. This is the first report to clarify Pd–specific immune responses in the OM using a murine model of intraoral Pd allergies.

OM swelling of ACM and ICM mice revealed that OM swelling could be caused by nonspecific stimulation to external stimuli only, as shown by the swelling on day 1 without significant difference. However, on day 2, the OM buccal swelling in ICM mice significantly decreased, with the largest significant difference between ACM and ICM mice. Thereafter, significant differences between ACM and ICM mice were observed until day 5, but not on day 7. These findings suggest that inflammation other than nonspecific stimulation to external stimuli is induced in ACM mice [16].

Therefore, we confirmed the histopathological and IHC analyses in ACM and ICM mice from days 2 and 5 after the challenge and control mice. HE staining revealed the dense infiltration of inflammatory cells in the epithelial basal layer and upper dermis, as well as swelling of the OM epithelium and epidermal spongiosis in ACM mice on day 5 after the challenge that was more markedly than others. Spongiform edema in the epithelium and swelling of the epithelium are characteristics of delayed allergy [17,18]. Marked infiltration of F4/80+ macrophages and CD3+ T cells were found in ACM mice on day 2 after the challenge, and CD3+ T cells increased on day 5 from day 2 after the challenge. This innate works in both ACM and ICM mice on day 2 by invading the $PdCl_2$ solution in vivo. Additionally, on day 2, ACM mice admitted CD3 and F4/80 macrophages more remarkably than ICM mice, suggesting that adaptive immunity works in ACM mice. CD3+ T cell infiltration, mucosal epithelial thickening, and supratentorial edema increased from day 2 to day 5 in ACM mice. Our results suggested that allergic inflammation in the OM was initiated by the response of macrophages to Pd, followed by T cell infiltration into the inflamed oral mucosa after antigen presentation by macrophages. This suggested that the ACM mice induced a metal allergy, which was stronger on day 5.

Four categories exist within type IV hypersensitivities, including type IVa, which is a CD4+ T helper (Th) 1 lymphocyte–mediated reaction with macrophage activation; type IVb, which is CD4+ Th2 lymphocyte–mediated with eosinophilic involvement; type IVc, which is cytotoxic CD8+ T lymphocyte–mediated with perforin–granzyme B involvement in apoptosis; and type IVd, which is T cell–driven neutrophilic inflammation [13,19]. A previous study suggested that a metal allergy is associated with either CD4 or CD8 T cell activation depending on the antigen processing pathway involvement [20,21]. The established Pd–induced allergic mouse model in the present study indicated predominant CD4+ T cell infiltration and Th2–type cytokine production in the OM. These results suggest that the established mouse model recapitulates the pathology corresponding to a type IVb delayed allergy.

Diversity was increased in the repertoire analysis of ACM mice compared to ICM mice, suggesting that allergic responses locally induce a variety of T cells and that the affiliated lymph nodes may be the source of these cells. The presence of common unique reads in ACM OM and Ly mice provides evidence for the involvement of the affiliated lymph nodes in the induction of locally generated specific T cells. Conversely, the reason for the partial sharing of ICM Ly unique reads in ACM OM mice may be due to the residual acute–phase clones observed in the early inflammation and dragged by the allergic reaction. Additionally, these unique reads are not detected in ICM OM mice, suggesting that they are lost locally in non–allergic reactions at the time of day 5 after the challenge. The unique reads shared by ACM OM and ACM Ly mice were diverse but biased toward specific TRVs or TRJs, suggesting that specific T cell induction in Pd allergies exists in a wide range.

iNKT cells are characterized by the expression of an invariant TRA encoded by TRAV11 (Vα14)–TRAJ18 (Jα18) in mice and TRAV10 (Vα24)–TRAJ18 (Jα18) in humans [22,23]. Previously, we identified iNKT cells in the lymphocytic infiltrates at a high frequency during the elicitation phase in Ni, Cr, and Ti allergies [24–26]. Additionally, iNKT cells and TRAV6–6–TRAJ57 bearing T cells have been implicated in the immune response to a Ni–induced OM metal contact allergy [10]. Interestingly, the reduced TRA sequence of iNKT cells in ACM OM mice indicates that they were pressured by these Pd allergy–specific clone populations and may not have a major role, unlike Ni–induced intraoral metal contact allergies.

In conclusion, we have established a model of an intraoral Pd–induced metal allergy, and revealed that Pd–specific T cell populations are limited to V or J gene usage. The application of the direct cloning of TCR genes from local sites of inflammation in this model will be a powerful tool for advancing our understanding of T cell–mediated immune disease in metal allergies, as well as providing new insights into antigen recognition by Pd–specific TCR in the OM.

4. Materials and Methods

4.1. Animals

BALB/cAJcl mice (4–week–old females) were purchased from CLEA Japan (CLEA Japan, Tokyo, Japan). All mice were in good health throughout the study and were given 1 week to acclimate to their surroundings before the study began. At the beginning of the experiment, 5–week–old BALB/cAJcl mice were used. All mice were kept in plastic cages (with a lid made of stainless–steel wire) with food and water available ad libitum. They were kept in our conventional animal facility with a temperature of 19–23 °C, a humidity of 30–70%, and a 12–h day/night cycle. All surgeries were performed using three different types of mixed anesthetics, and every effort was made to minimize animal suffering. Three different types of mixed anesthetic agents were used to sacrifice all mice by cervical dislocation to ensure death and prevent pain caused by tissue harvesting.

4.2. Reagents

$PdCl_2$ (purity of >95%) was purchased from FUJIFILM Wako Pure Chemical Co., Ltd. (FUJIFILM Wako Pure Chemical Co., Ltd., Osaka, Japan). Lipopolysaccharide (LPS) from Escherichia coli (O55:B5) prepared by phenol–water extraction was purchased from Sigma–Aldrich (Sigma–Aldrich, St. Louis, MO, USA). The dissolution used $PdCl_2$ plus LPS in sterile saline (Otsuka Normal Saline, Otsuka Pharmaceutical Factory, Inc., Tokushima, Japan).

4.3. Anesthetic Agents

The anesthetic was created by combining three medications. Medetomidine hydrochloride was purchased from Nippon Zenyaku Kogyo Co., Ltd. (Nippon Zenyaku Kogyo Co., Ltd., Fukushima, Japan), midazolam from Sandoz (Sandoz, Tokyo, Japan), and butorphanol tartrate from Meiji Seika Pharma Co., Ltd. (Meiji Seika Pharma Co., Ltd., Tokyo, Japan). These drugs were kept at room temperature (RT). We combined doses of 0.3 mg/kg of medetomidine hydrochloride, 4 mg/kg of midazolam, and 5 mg/kg of butorphanol tartrate. A previous study determined the concentration ratio of the three types of mixed anesthetic agents [27]. Typically, 25 mL of anesthetic agent is prepared by combining 0.75 mL of medetomidine hydrochloride, 2 mL of midazolam, 2.50 mL of butorphanol tartrate, and 19.75 mL of sterile saline. All agents were diluted in sterile saline and stored in the dark at 4 °C and sterile saline at 4 °C. The mice were administered a volume of 0.01 mL/g of body weight of the anesthetic mixture. Every mouse was intraperitoneally injected with the mixture of the three types of anesthetic agents.

4.4. Experimental Protocol of the Mouse Model of Pd–Induced Intraoral Metal Contact Allergy

An experimental protocol for the induction of the metal allergy in the OM was developed based on a previous protocol for metal allergy induction in footpad skin [10,15]. Mice were separated into three groups: ACM mice (n = 15), ICM mice (n = 15), and control mice (n = 15), with each group consisting of randomly chosen mice. All experiments were conducted in another room upon transfer from the animal holding area.

Sensitization. Mice were intradermally injected with 125 µL of 10 mM of $PdCl_2$ and 10 µg/mL of LPS in sterile saline twice with a 7–day interval between injections. ACM was applied to the left and right postauricular skin of mice. Mice were first challenged 7 days after the second sensitization.

Challenge for elicitation. In the preliminary experiments, the concentrations of $PdCl_2$ in 5 µM, 10 µM, and 30 µM were tested. In the 30 µM, almost all mice died. Based on the difference in cheek swelling and CD3 expression levels in the concentration of the 5 µM and 10 µM, the 10 µM were selected as the indicated concentration. The immune response was elicited using 25 µL of 10 mM of $PdCl_2$ without LPS in sterile saline. ACM mice were challenged by submucosal injection in the left and right buccal regions of the OM. Submucosal injections were used to challenge non–sensitized ICM mice in the left and right buccal areas of the OM. Mice sensitized with $PdCl_2$ plus LPS and then challenged with sterile saline were used as a control.

4.5. Measurement of OM Swelling

Swelling of the buccal area was measured before the challenge and at 1, 2, 3, 5, and 7 days after the first challenge using a Peacock Dial Thickness Gauge (Ozaki MFG Co., Ltd., Tokyo, Japan). All procedures on mice under anesthesia were performed by the same person.

4.6. Histological and IHC Analysis

Buccal OM samples were obtained from control mice and Pd–induced ICM and ACM mice for histological and IHC analyses. Furthermore, tissue samples were immersed in 4% paraformaldehyde–lysine–periodate for 48 h at 4 °C. Following a 10–min PBS wash, fixed tissues were soaked in 5% sucrose/PBS for 1 h at 4 °C, 15% sucrose/PBS for 3 h at 4 °C, and finally, 30% sucrose/PBS overnight at 4 °C. Tissue samples were snap–frozen by immersion in a mixture of acetone and dry ice in Tissue Mount (Chiba Medical, Saitama, Japan). Frozen sections were cut into 6–μm thick cryosections and air–dried on poly–L–lysine–coated glass slides. The HE stain was applied to cryosections for histological analysis. Cryosections were stained with anti–mouse F4/80 (1:1000; Cl–A3–1, Abcam, Cambridge, UK) and anti–mouse CD3 (1:500; SP7, Abcam, Cambridge, UK) monoclonal antibodies for IHC analysis. Mouse macrophage populations were detected in many buccal OM tissues using the F4/80 monoclonal antibody. The sections were incubated at RT for 30 min in PBS containing 5% normal goat rabbit serum, 0.025% Triton X–100 (FUJIFILM Wako Pure Chemical, Osaka, Japan), and 5% bovine serum albumin (Sigma–Aldrich St. Louis, MO, USA). Sections were incubated with primary mAbs for 1 h at RT. Intrinsic peroxidase was inhibited with 3% H_2O_2 in methanol after three 5 min washes with PBS. Tissue sections were washed twice and incubated for 1 h at RT with a secondary antibody (biotinylated goat anti–hamster immunoglobulin G or biotinylated rabbit anti–rat immunoglobulin G) after soaking in distilled water. The sections were treated with Vectastain ABC Reagent (Vector Laboratories, Burlingame, CA, USA) for 30 min at RT, followed by 3,3–diaminobenzidine staining (0.06% diaminobenzidine and 0.03% H_2O_2 in 0.1 M Tris–HCl, pH 7.6; FUJIFILM Wako Pure Chemical Co., Ltd., Osaka, Japan). Tissue sections were counterstained with hematoxylin to visualize cell nuclei.

4.7. RNA Extraction and cDNA Synthesis

Fresh OM specimens were obtained from each mouse and immediately soaked in RNAlater RNA Stabilization Reagent (Invitrogen, Carlsbad, CA, USA). Total RNA from the OM was extracted using the RNeasy Lipid Tissue Mini Kit (Qiagen) following the manufacturer's instructions. Complementary DNA (cDNA) was synthesized from DNA–free RNA using the PrimeScript RT reagent Kit (Takara Bio, Tokyo, Japan) following the manufacturer's instructions.

4.8. qPCR

The expression levels of immune response–related genes, including T cell–related CD antigens, cytokines, cytotoxic granule, and transcription factors of regulatory T cells, were measured by qPCR using the Bio–Rad CFX96 system (Bio–Rad, Hercules, CA, USA). Specific primers for GAPDH, CD3, CD4, CD8, IFNγ, TNFα, IL–2, IL–4, and IL–10 were described [11,28]. Freshly isolated total RNA from the OM and submandibular lymph node specimens were converted to cDNA. The PCR consisted of 5 μL of SsoFast EvaGreen Supermix (Bio–Rad), 3.5 μL of RNase/DNase–free water, 0.5 μL of 5 μM of primer mix, and 1 μL of cDNA in a final volume of 10 μL. Cycling conditions were as follows: 30 s at 95 °C followed by 50 cycles of 1 s at 95 °C and 5 s at 60 °C. A melting curve analysis was performed from 70 °C to 90 °C at the end of each program to confirm the homogeneity of the PCR products. All assays were repeated three times, and mean values were used to calculate gene expression levels. Five 10–fold serial dilutions of each standard transcript were used to determine the absolute quantification, specification, and amplification efficiency of each primer set. Standard transcripts were generated by the in vitro transcription of the corresponding PCR product in a plasmid. The nucleotide sequences were confirmed by DNA sequencing using the CEQ8000 Genetic Analysis System (Beckman Coulter, Fullerton, CA, USA). An Agilent DNA 7500 Kit in an Agilent 2100 Bioanalyzer (Agilent, Santa Clara, CA, USA) was used to validate their quality and concentration. GAPDH gene expression was used as an internal control.

4.9. TCR Repertoire Analysis

Total RNA was extracted from the OM and Ly of ICM and ACM mice on day 5 after the challenge and the OM of ACM mice on days 2 and 7 after the challenge. NGS was used to perform a TCR repertoire analysis developed by Repertoire Genesis Inc. (Repertoire Genesis Inc., Osaka, Japan [13]). An unbiased adapter–ligation PCR was executed according to the previous report [29]. Superscript III reverse transcriptase was utilized to convert total RNA to cDNA (Invitrogen). Afterward, double–stranded (ds) cDNA was synthesized, and an adapter was ligated to the 5′ end of the ds–cDNA before it was cut with the SphI restriction enzyme. PCR was performed with a P20EA adapter primer and a TCR α–chain constant region–specific primer (mCA1) for TCRα. The second PCR was conducted using the same PCR conditions and primers, including mCA2 and P20EA. Primers were utilized for TCRβ, mCB1, and mCB2 in the first and second PCR, respectively. Index (barcode) sequences were amplified using a Nextera XT Index Kit v2 setA (Illumina, San Diego, CA, USA) after Tag PCR amplification. The sequencing was performed using the paired–end Illumina MiSeq platform (2 $\times$ 300 base pairs [bp]). Subsequently, data processing, data assignment, and data aggregation were performed automatically using the originally created repertoire analysis software by Repertoire Genesis Inc. TCR (TRA and TRB) sequences were mapped to a reference sequence dataset from the international ImMunoGeneTics information system (IMGT) database (http://www.imgt.org, accessed on 1 November 2022) [30]. Nucleotide sequences of CDR3 regions ranged from a conserved cysteine at position 104 (Cys104) to a conserved phenylalanine at position 118 (Phe118), and the following glycine (Gly119) was translated into an amino acid sequence. A unique sequence read was defined as a sequence read with no identity in TRAV, TRAJ, and the deduced amino acid sequence of CDR3. The copy number of identical unique sequence reads in each sample was automatically counted and then ranked by copy number using software for repertoire analysis. The percentage occurrence frequencies of sequence reads containing TRAV and TRAJ, as well as genes, were calculated.

4.10. Statistical Analysis

Statistically significant differences between the mean values of each experimental group were analyzed using the Kruskal–Wallis test followed by Steel–Dwass' multiple comparison tests and Mann–Whitney U test. All analyses were performed with EZR (Saitama Medical Center, Jichi Medical University, Saitama, Japan) [31], which is a graphical user interface for R (The R Foundation for Statistical Computing, Vienna, Austria). A p–value of <0.05 was considered significant, and a p–value of <0.01 was considered highly significant.

Author Contributions: K.N., K.K. (Kenichi Kumagai), Y.H. and R.S. conceived and designed the experiments; K.N., K.K. (Kenichi Kumagai), R.M., T.Y. and R.S. performed the experiments; K.N., T.Y., K.K. (Kazutaka Kitaura) and M.S. analyzed the data; K.N., K.K. (Kazutaka Kitaura) and T.Y. contributed reagents/materials/analysis tools; K.N., K.K. (Kenichi Kumagai), K.K. (Kazutaka Kitaura), M.S. and R.S. wrote the paper. All authors have read and agreed to the published version of the manuscript.

Funding: This work was supported by the Japan Society for the Promotion of Science KAKENHI Grant–in–Aid for Scientific Research C Grant No. 19K10371.

Institutional Review Board Statement: All animal experiments in this study were carried out according to the relevant ethical requirements with approval from the committees for animal experiments at Tsurumi University (approval number: 21A025).

Informed Consent Statement: Not applicable.

Data Availability Statement: This data presented in this study are available on request from the corresponding author.

Conflicts of Interest: The authors declare no conflict of interest.

Abbreviations

TCR	T cell receptor
TRAV	TCRα–chain variable region
TRAJ	TCRα–chain joining region
TRBV	TCRβ–chain variable region
TRBJ	TCRβ–chain joining region
CDR3	complementarity–determining region 3
CD	cluster of differentiation

References

1. Raap, U.; Stiesch, M.; Reh, H.; Kapp, A.; Werfel, T. Investigation of contact allergy to dental metals in 206 patients. *Contact Dermat.* **2009**, *60*, 339–343. [CrossRef] [PubMed]
2. Kitagawa, M.; Murakami, S.; Akashi, Y.; Oka, H.; Shintani, T.; Ogawa, I.; Inoue, T.; Kurihara, H. Current status of dental metal allergy in Japan. *J. Prosthodont. Res.* **2019**, *63*, 309–312. [CrossRef] [PubMed]
3. Muris, J.; Goossens, A.; Gonçalo, M.; Bircher, A.J.; Giménez-Arnau, A.; Foti, C.; Bruze, M.; Andersen, K.E.; Rustemeyer, T.; Feilzer, A.J.; et al. Sensitization to palladium in Europe. *Contact Dermat.* **2015**, *72*, 11–19. [CrossRef]
4. Faurschou, A.; Menné, T.; Johansen, J.D.; Thyssen, J.P. Metal allergen of the 21st century—A review on exposure, epidemiology and clinical manifestations of palladium allergy. *Contact Dermat.* **2011**, *64*, 185–195. [CrossRef]
5. Kielhorn, J.; Melber, C.; Keller, D.; Mangelsdorf, I. Palladium—A review of exposure and effects to human health. *Int. J. Hyg. Environ. Health* **2002**, *205*, 417–432. [CrossRef] [PubMed]
6. Riedel, F.; Aparicio-Soto, M.; Curato, C.; Thierse, H.J.; Siewert, K.; Luch, A. Immunological Mechanisms of Metal Allergies and the Nickel-Specific TCR-pMHC Interface. *Int. J. Environ. Res. Public Health* **2021**, *18*, 10867. [CrossRef]
7. Ahlfors, E.; Czerkinsky, C. Contact sensitivity in the murine oral mucosa. I. An experimental model of delayed-type hypersensitivity reactions at mucosal surfaces. *Clin. Exp. Immunol.* **1991**, *86*, 449–456. [CrossRef]
8. Larese Filon, F.; Uderzo, D.; Bagnato, E. Sensitization to palladium chloride: A 10-year evaluation. *Am. J. Contact Dermat.* **2003**, *14*, 78–81. [CrossRef]
9. Ahlfors, E.; Jonsson, R.; Czerkinsky, C. Experimental T cell-mediated inflammatory reactions in the murine oral mucosa. II. Immunohistochemical characterization of resident and infiltrating cells. *Clin. Exp. Immunol.* **1996**, *104*, 297–305. [CrossRef]
10. Nakasone, Y.; Kumagai, K.; Matsubara, R.; Shigematsu, H.; Kitaura, K.; Suzuki, S.; Satoh, M.; Hamada, Y.; Suzuki, R. Characterization of T cell receptors in a novel murine model of nickel-induced intraoral metal contact allergy. *PLoS ONE* **2018**, *13*, e0209248. [CrossRef]
11. Kitaura, K.; Fujii, Y.; Hayasaka, D.; Matsutani, T.; Shirai, K.; Nagata, N.; Lim, C.K.; Suzuki, S.; Takasaki, T.; Suzuki, R.; et al. High clonality of virus-specific T lymphocytes defined by TCR usage in the brains of mice infected with West Nile virus. *J. Immunol.* **2011**, *187*, 3919–3930. [CrossRef]
12. Kitaura, K.; Shini, T.; Matsutani, T.; Suzuki, R. A new high-throughput sequencing method for determining diversity and similarity of T cell receptor (TCR) α and β repertoires and identifying potential new invariant TCR α chains. *BMC Immunol.* **2016**, *17*, 38. [CrossRef]
13. Kitaura, K.; Yamashita, H.; Ayabe, H.; Shini, T.; Matsutani, T.; Suzuki, R. Different Somatic Hypermutation Levels among Antibody Subclasses Disclosed by a New Next-Generation Sequencing-Based Antibody Repertoire Analysis. *Front. Immunol.* **2017**, *8*, 389. [CrossRef] [PubMed]
14. Kawano, M.; Nakayama, M.; Aoshima, Y.; Nakamura, K.; Ono, M.; Nishiya, T.; Nakamura, S.; Takeda, Y.; Dobashi, A.; Takahashi, A.; et al. NKG2D$^+$ IFN-γ^+ CD8$^+$ T cells are responsible for palladium allergy. *PLoS ONE* **2014**, *9*, e86810. [CrossRef] [PubMed]
15. Kobayashi, H.; Kumagai, K.; Eguchi, T.; Shigematsu, H.; Kitaura, K.; Kawano, M.; Horikawa, T.; Suzuki, S.; Matsutani, T.; Ogasawara, K.; et al. Characterization of T cell receptors of Th1 cells infiltrating inflamed skin of a novel murine model of palladium-induced metal allergy. *PLoS ONE* **2013**, *8*, e76385. [CrossRef]
16. Ward, J.P.; Franks, S.J.; Tindall, M.J.; King, J.R.; Curtis, A.; Evans, G.S. Mathematical modelling of contact dermatitis from nickel and chromium. *J. Math Biol.* **2019**, *79*, 595–630. [CrossRef] [PubMed]

17. Tanei, R.; Hasegawa, Y. Immunological Pathomechanisms of Spongiotic Dermatitis in Skin Lesions of Atopic Dermatitis. *Int. J. Mol. Sci.* **2022**, *23*, 6682. [CrossRef]
18. Yawalkar, N.; Hunger, R.E.; Buri, C.; Schmid, S.; Egli, F.; Brand, C.U.; Mueller, C.; Pichler, W.J.; Braathen, L.R. A comparative study of the expression of cytotoxic proteins in allergic contact dermatitis and psoriasis: Spongiotic skin lesions in allergic contact dermatitis are highly infiltrated by T cells expressing perforin and granzyme B. *Am. J. Pathol.* **2001**, *158*, 803–808. [CrossRef] [PubMed]
19. Uzzaman, A.; Cho, S.H. Chapter 28: Classification of hypersensitivity reactions. *Allergy Asthma Proc.* **2012**, *33* (Suppl. 1), 96–99. [CrossRef]
20. Saint-Mezard, P.; Berard, F.; Dubois, B.; Kaiserlian, D.; Nicolas, J.F. The role of CD4+ and CD8+ T cells in contact hypersensitivity and allergic contact dermatitis. *Eur. J. Derm.* **2004**, *14*, 131–138.
21. Sumiwi, Y.A.; Soesatyo, M.H.; Sosroseno, W. The role of CD4+ T cells in the induction of contact hypersensitivity to mercury in a murine model. *Cutan. Ocul. Toxicol.* **2010**, *29*, 30–33. [CrossRef] [PubMed]
22. Balato, A.; Unutmaz, D.; Gaspari, A.A. Natural killer T cells: An unconventional T-cell subset with diverse effector and regulatory functions. *J. Investig. Derm.* **2009**, *129*, 1628–1642. [CrossRef]
23. Godfrey, D.I.; Kronenberg, M. Going both ways: Immune regulation via CD1d-dependent NKT cells. *J. Clin. Investig.* **2004**, *114*, 1379–1388. [CrossRef] [PubMed]
24. Eguchi, T.; Kumagai, K.; Kobayashi, H.; Shigematsu, H.; Kitaura, K.; Suzuki, S.; Horikawa, T.; Hamada, Y.; Ogasawara, K.; Suzuki, R. Accumulation of invariant NKT cells into inflamed skin in a novel murine model of nickel allergy. *Cell Immunol.* **2013**, *284*, 163–171. [CrossRef]
25. Shigematsu, H.; Kumagai, K.; Kobayashi, H.; Eguchi, T.; Kitaura, K.; Suzuki, S.; Horikawa, T.; Matsutani, T.; Ogasawara, K.; Hamada, Y.; et al. Accumulation of metal-specific T cells in inflamed skin in a novel murine model of chromium-induced allergic contact dermatitis. *PLoS ONE* **2014**, *9*, e85983. [CrossRef]
26. Kumagai, K.; Matsubara, R.; Nakasone, Y.; Shigematsu, H.; Kitaura, K.; Suzuki, S.; Haneji, K.; Hamada, Y.; Suzuki, R. Possible involvement of invariant natural killer T cells and mucosal-associated invariant T cells in a murine model of titanium allergy. *J. Oral Maxillofac. Surg. Med. Pathol.* **2018**, *30*, 1–9. [CrossRef]
27. Kawai, S.; Takagi, Y.; Kaneko, S.; Kurosawa, T. Effect of three types of mixed anesthetic agents alternate to ketamine in mice. *Exp. Anim.* **2011**, *60*, 481–487. [CrossRef]
28. Fujii, Y.; Kitaura, K.; Nakamichi, K.; Takasaki, T.; Suzuki, R.; Kurane, I. Accumulation of T-cells with selected T-cell receptors in the brains of Japanese encephalitis virus-infected mice. *Jpn. J. Infect. Dis.* **2008**, *61*, 40–48.
29. Yoshida, R.; Yoshioka, T.; Yamane, S.; Matsutani, T.; Toyosaki-Maeda, T.; Tsuruta, Y.; Suzuki, R. A new method for quantitative analysis of the mouse T-cell receptor V region repertoires: Comparison of repertoires among strains. *Immunogenetics* **2000**, *52*, 35–45. [CrossRef]
30. Lefranc, M.P.; Giudicelli, V.; Ginestoux, C.; Jabado-Michaloud, J.; Folch, G.; Bellahcene, F.; Wu, Y.; Gemrot, E.; Brochet, X.; Lane, J.; et al. IMGT, the international ImMunoGeneTics information system. *Nucleic Acids Res.* **2009**, *37*, D1006–D1012. [CrossRef]
31. Kanda, Y. Investigation of the freely available easy-to-use software 'EZR' for medical statistics. *Bone Marrow Transpl.* **2013**, *48*, 452–458. [CrossRef] [PubMed]

International Journal of
Molecular Sciences

MDPI

Article

Characterization of Metal-Specific T-Cells in Inflamed Oral Mucosa in a Novel Murine Model of Chromium-Induced Allergic Contact Dermatitis

Takamasa Yoshizawa [1,2,†], Kenichi Kumagai [2,3,†], Ryota Matsubara [2,4], Keisuke Nasu [1,2], Kazutaka Kitaura [2,5], Motoaki Suzuki [2,6], Yoshiki Hamada [1,*] and Ryuji Suzuki [2,5]

1 Department of Oral and Maxillofacial Surgery, School of Dental Medicine, Tsurumi University, Yokohama 230-8501, Japan
2 Department of Rheumatology and Clinical Immunology, Clinical Research Center for Rheumatology and Allergy, Sagamihara National Hospital, National Hospital Organization, Sagamihara 252-0392, Japan
3 Department of Oral and Maxillofacial Surgery, Dentistry and Orthodontics, The University of Tokyo Hospital, Tokyo 113-8655, Japan
4 Department of Oral and Maxillofacial Surgery, Sendai Tokushukai Hospital, Sendai 981-3116, Japan
5 Repertoire Genesis Inc., Osaka 567-0085, Japan
6 Department of Anatomy and Physiology, Faculty of Medicine, Saga University, Saga 849-8501, Japan
* Correspondence: hamada-y@tsurumi-u.ac.jp; Tel./Fax: +81-45-580-8327
† These authors contributed equally to this work.

Abstract: The element chromium (Cr) is a component of several types of alloys found in the environment, or utilized in dentistry, that may cause intraoral metal contact allergy. However, the pathological mechanism of intraoral Cr allergy remains unclear because there is no established animal model of Cr allergy in the oral mucosa. In this study, we established a novel murine model of Cr-induced intraoral metal contact allergy and elucidated the immune response in terms of cytokine profiles and T-cell receptor repertoire. Two sensitizations with Cr plus lipopolysaccharide solution into the postauricular skin were followed by a single Cr challenge of the oral mucosa to generate the intraoral metal contact allergy model. Histological examination revealed that CD3+ T-cells had infiltrated the allergic oral mucosa one day after exposure to the allergen. The increase in T-cell markers and cytokines in allergic oral mucosa was also confirmed via quantitative PCR analysis. We detected Cr-specific T-cells bearing TRAV12D-1-TRAJ22 and natural killer (NK) T-cells in the oral mucosa and lymph nodes. Our model demonstrated that Cr-specific T-cells and potent NKT-cell activation may be involved in the immune responses of Cr-induced intraoral metal contact allergy.

Keywords: chromium allergy; contact dermatitis; allergic contact mucositis; metal allergy

Citation: Yoshizawa, T.; Kumagai, K.; Matsubara, R.; Nasu, K.; Kitaura, K.; Suzuki, M.; Hamada, Y.; Suzuki, R. Characterization of Metal-Specific T-Cells in Inflamed Oral Mucosa in a Novel Murine Model of Chromium-Induced Allergic Contact Dermatitis. *Int. J. Mol. Sci.* **2023**, *24*, 2807. https://doi.org/10.3390/ijms24032807

Academic Editor: Shun-Fa Yang

Received: 27 December 2022
Revised: 23 January 2023
Accepted: 29 January 2023
Published: 1 February 2023

1. Introduction

Metal allergy is categorized as a delayed-type hypersensitivity (DTH) reaction triggered by antigenic protein with haptens that exert antigenicity. It can be caused by metal ions released by jewelry, footwear, preservatives, and cosmetics [1]. In addition to nickel (Ni), cobalt (Co), palladium (Pd), zinc (Zn), and chromium (Cr) have been reported to cause allergic contact dermatitis [2–5]. Cr hypersensitivity is one of the most prevalent occupational metal skin diseases in cement workers [6]. Occupational allergic contact dermatitis was the most frequently occurring allergy in construction workers (45%), and the most frequent allergen was chromium (Cr) in cement [6,7]. In addition, the production of a chronic generalized eczematoid reaction has been reported as a causal intraoral Cr-allergic contact dermatitis [8]. Recently, Cr has been widely used in dental restorations such as implants, crown prostheses, and dentures. However, the pathological mechanism of intraoral Cr allergy remains unknown because there is no established animal model of Cr allergy in the oral mucosa. According to previous studies of murine models of DTH in the oral mucosa, various chemicals, such as oxazolone (4-ethoxymethylene-2-phenyloxazol-5-one) and 2,4-dinitro-1-fluorobenzene, induce allergic contact mucositis (ACM) with the local accumulation of antigen-presenting cells and T-cells [9].

Typically, metal allergy is associated with acquired immunity, which facilitates the migration of metal-specific T-cells to the site of allergic inflammation. T-cells recognize antigens on antigen-presenting cells through surface-expressed T-cell receptors (TCRs), heterodimers composed of an α- and β-chain (TRA and TRB) that determine the high specificity of T-cells [10]. In previous studies, the cells in the peripheral blood and skin of patients with metal allergy had limited TCR repertoires [11]. Several novel murine models of Ni, Pd, Cr, and titanium (Ti)-induced allergic contact dermatitis (ACD) have been generated using footpad skin, and these have aided in the characterization of antigen-specific immune responses in terms of TCR usage [12–15]. These models enabled us to identify the accumulation of metal-specific T-cells in inflamed skin and to demonstrate that the restricted usage of TCR genes in metal allergy is a result of the prolonged exposure of the host immune system to putative metal-associated antigens. The analysis of the TCR repertoire enables the identification of antigen-specific T-cells [16]. Recent advances in next-generation sequencing (NGS) have permitted the quantitative analyses of the TCR repertoire using large amounts of TCR sequencing data [17,18].

In the present study, we established a novel murine model of Cr-induced allergy in the oral mucosa to examine how the accumulation of T-cells at the site of allergic inflammation contributes to the development of Cr allergy in the oral mucosa and how TCR gene usage is regulated.

2. Results

2.1. Oral Mucosa Swelling in Cr-Induced Allergic Mice

At Day 1 post-challenge, maximal swelling in the buccal area of the oral mucosa occurred in all mice (Figure 1). At Day 7 post-challenge, the swelling of the oral mucosa was significantly greater in the ACM mice than in the control mice, but it was not significantly different in irritant contact mucositis (ICM) mice. From Days 1 to 12 post-challenge, oral mucosa swelling in the ACM mice was greater than in the control mice.

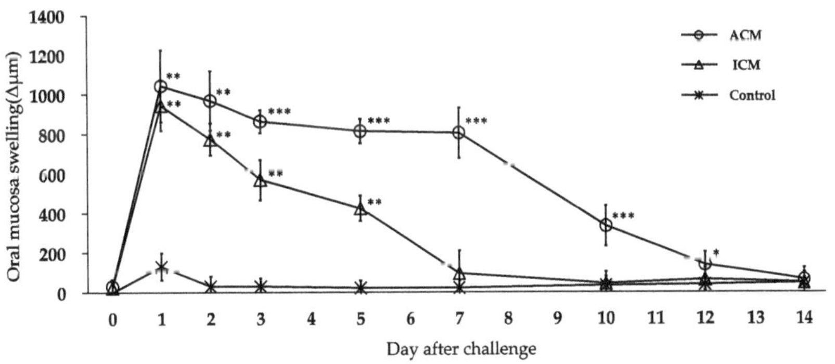

Figure 1. Swelling of the oral mucosa in Cr-induced allergic mice. The oral mucosa of all mice reached maximal swelling at Day 1 post-challenge. The oral mucosa swelled significantly more in the ACM mice from Days 1 to 12 post-challenge compared with the control mice. The bars and error bars represent the mean plus the standard deviation. The Kruskal–Wallis test was used to determine statistical significance, followed by Dunn's multiple comparison tests (* $p < 0.05$, ** $p < 0.01$, *** $p < 0.001$).

2.2. Histological and Immunohistochemical Analyses of F4/80 and CD3 in the Oral Mucosa of Cr-Induced Allergic Mice

To determine whether macrophages and T-cells infiltrated the inflamed oral mucosa, histological and immunohistochemical (IHC) analyses were performed on the oral mucosa of the control, ICM, and ACM mice at Days 1 and 7 post-challenge. Hematoxylin and eosin (HE) staining revealed dense infiltration of inflammatory cells in the basal epithelial layer and upper dermis, in addition to a swelling of the oral mucosa, in the ACM and ICM mice, but not in the control mice (Figure 2A–E). Inflammatory cells accumulated in the epithelium and upper dermis of the ACM mice (Figure 2D). The partial separation of epidermal keratinocytes produced spongiotic dermatitis (Figure 2D). In the ICM mice, the inflammatory response in the oral mucosa was reduced (Figure 2B,C). In contrast, inflammation of the oral mucosa persisted for Day 7 post-challenge in the ACM mice (Figure 2E). We performed the IHC staining of CD3 and F4/80 in the oral mucosa of the control, ICM, and ACM mice to determine whether T-cells and macrophages had infiltrated the inflamed oral mucosa of the ACM mice (Figure 2F–O). Significant CD3+ T-cell infiltration occurred into the basal epithelial layer and upper dermis of the ACM mice by Day 1 post-challenge compared with the control and ICM mice (Figure 2F,G,I). The ACM mice retained CD3+ T-cells in the basal epithelial layer and upper dermis Day 7 post-challenge (Figure 2J). There was little infiltration of CD3+ T-cells into the basal epithelial layer and upper dermis of the control and ICM mice (Figure 2F–H). At Day 1 post-challenge, F4/80+ macrophages were predominant in the layer and upper dermis of the ACM mice (Figure 2N), but not in the control and ICM mice (Figure 2K,L,M,O).

2.3. mRNA Expression of T-Cell Markers in the Oral Mucosa of Cr-Induced Allergic Mice

We performed quantitative polymerase chain reaction (qPCR) of CD3, CD4, and CD8 to verify the infiltration of T-cells into the inflamed oral mucosa and determine their relative mRNA expression. CD3 expression in the ACM mice was significantly higher than in the control and ICM mice at Days 1 and 7 post-challenge (Figure 3). At Day 1 post-challenge, the CD8/CD4 ratio was significantly higher in the ACM mice than in the control and ICM mice (Figure 3).

2.4. Relative mRNA Expression of T-Cell-Related Cytokines in the Oral Mucosa of Cr-Induced Allergic Mice

Subsequently, using qPCR analysis, the mRNA expression of inflammatory markers was analyzed to examine inflammation in the allergic oral mucosa. We compared the expression levels of a proinflammatory cytokine (IL-1β), a Th1-related gene (IFN-γ), a Th2-related gene (IL-4), and the serine protease(granzyme B) in the oral mucosa of the control, ICM, and ACM mice using qPCR (Figure 4). At Days 1 and 7 post-challenge, IL-1β and granzyme B expression levels were significantly higher in the ACM mice than in the control and ICM mice. IFN-γ, in contrast to IL-4, was significantly higher in the ACM mice relative to the ICM mice with similar expression maintained on Days 1 and 7.

2.5. TCR Repertoire Usage in the Oral Mucosa and Cervical Lymph Nodes of the ICM and ACM Mice

We performed an NGS-based TCR repertoire analysis to determine the diversity of T-cells that had infiltrated the oral mucosa and cervical lymph nodes of the ICM and ACM mice at Day 1 post-challenge. A 3D representation of the TRA repertoire revealed the dominance of particular combinations of TRAV and TRAJ genes, as well as the breadth of TCR usage diversity (Figure 5). The 3D images of the TRA repertoire revealed a low level of expression accompanied by a broad distribution of TRAV and TRAJ. The usage of TRAV11d-TRAJ18 was considerably higher in the oral mucosa and cervical lymph nodes of the ICM and ACM mice († in Figure 5).

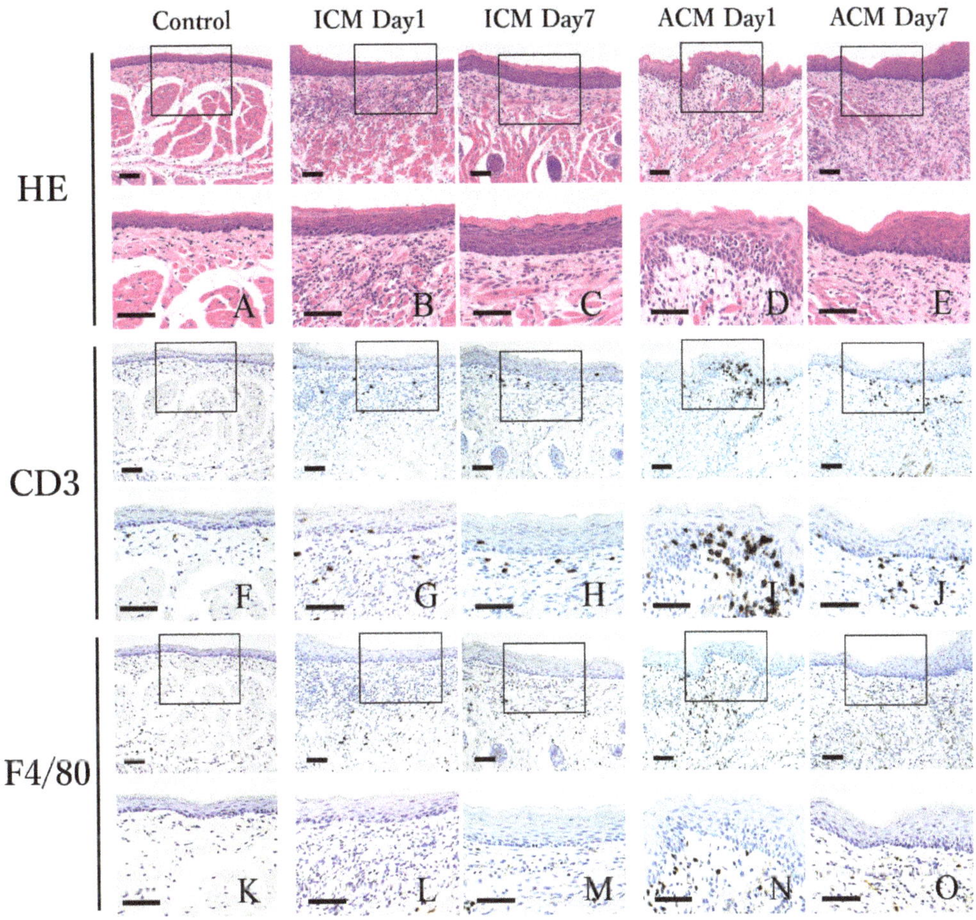

Figure 2. Histological and IHC analyses of CD3 and F4/80 in the oral mucosa of the control, ICM, and ACM mice. Histological and IHC analyses of CD3+ and F4/80+ T-cells in the oral mucosa at Days 1 and 7 post-challenge. At Day 1 post-challenge, the ACM mice exhibited abundant infiltration of mononuclear cells, swelling of the mucosal epithelium, and epidermal spongiosis (**D**), and IHC analyses revealed the presence of CD3+ T-cells in the epithelium of the ACM mice (**I**). Sections were stained with HE (**A**–**E**), CD3 (**F**–**J**), and F4/80 (**K**–**O**). Scale bar = 10 μm.

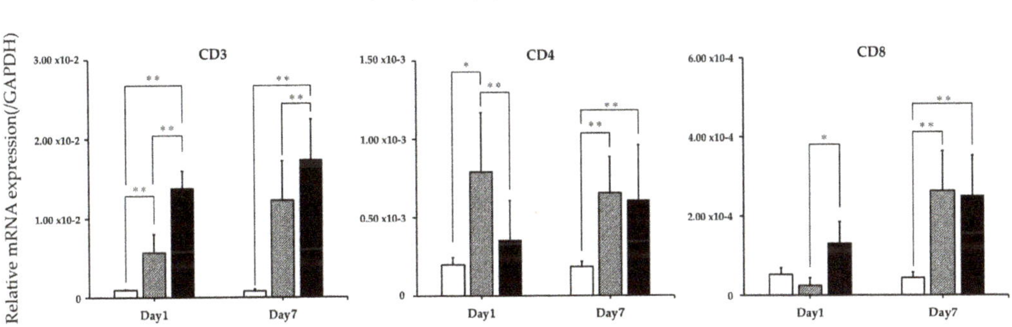

Figure 3. *Cont.*

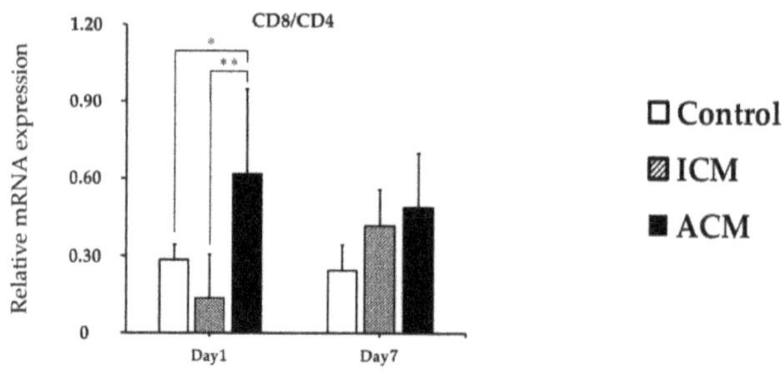

Figure 3. mRNA expression of T-cell phenotypes in the oral mucosa of Cr-induced allergic mice. At Days 1 and 7 post-challenge, the mRNA expression of CD3, CD4, CD8, and the CD8/CD4 ratio in the oral mucosa was evaluated (n = 8). The expression of the GAPDH gene served as an internal control. The mean and standard deviation are represented by bars and error bars. Kruskal–Wallis and Dunn's multiple comparison tests were used to determine statistical significance (* $p < 0.05$, ** $p < 0.01$).

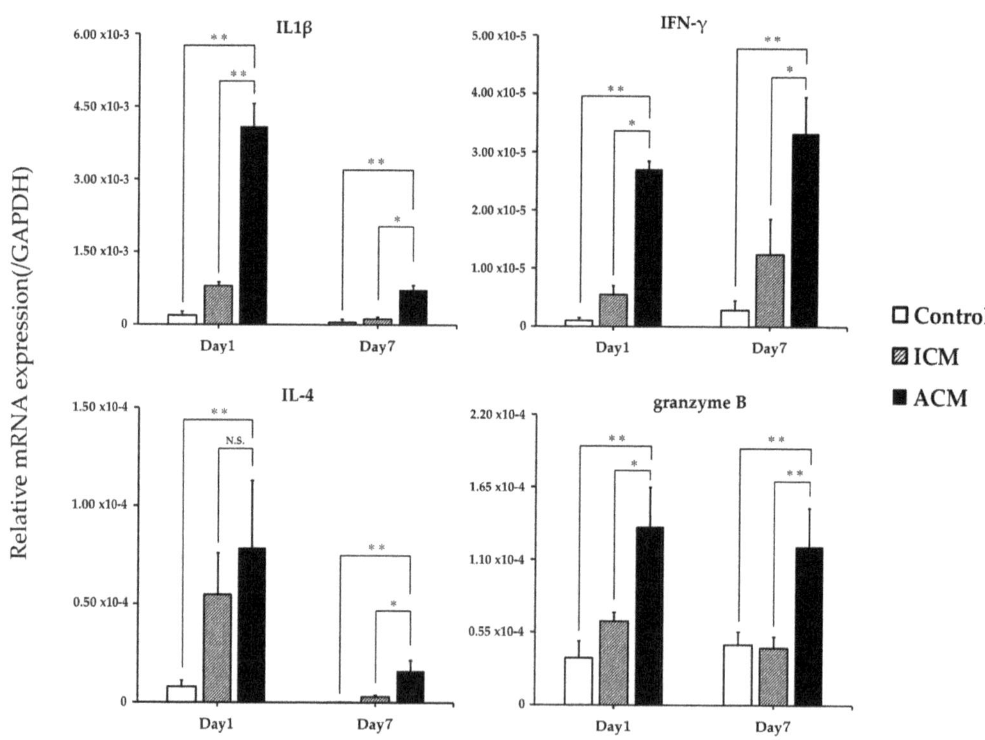

Figure 4. mRNA expression of T-cell-related cytokines in control, ICM, and ACM mice. At Days 1 and 7 post-challenge, the mRNA expression levels of IL-1β, IFN-γ, IL-4, and granzyme B in the oral mucosa (n = 8) were evaluated. The expression of the GAPDH gene served as an internal control. The mean and standard deviation are represented by bars and error bars, respectively. Kruskal–Wallis and Dunn's multiple comparison tests were used to determine the statistical significance of differences (* $p < 0.05$, ** $p < 0.01$, N.S. = not significant).

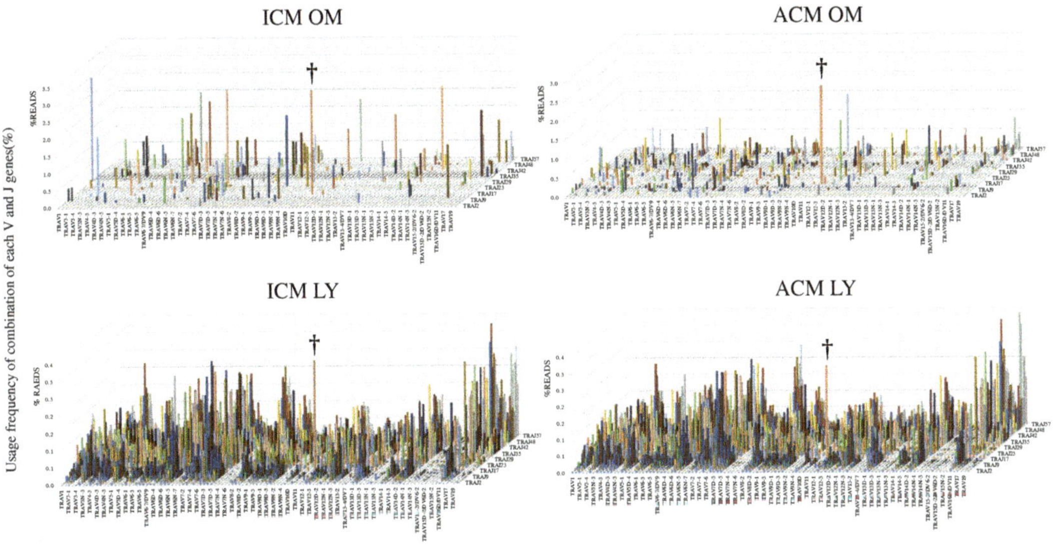

Figure 5. TCR repertoire in the oral mucosa and cervical lymph nodes of the ICM and ACM mice. NGS-based TCR repertoire analysis was performed on the oral mucosa and cervical lymph nodes of the ICM and ACM mice at Day 1 post-challenge (n = 6). Additionally, at Day 1 post-challenge, 3D images of the TCR repertoire depict the skewing of T-cells infiltrating the oral mucosa and cervical lymph nodes of the ICM and ACM mice. Combining TRAV on the X-axis and TRAJ on the Z-axis, with the frequency (percentage) of each clone on the Y-axis, the 3D images depict the TCR repertoire. †: TRAs bearing TRAV11d-TRAJ18. OM: oral mucosa. LY: lymph nodes of the cervical region.

Next, we analyzed the CDR3 amino acid sequences shared by the ICM and the ACM mice (Figure 6). Notably, the proportion of T-cells bearing TCR TRAV11d-TRAJ18 (CVVGDRGSALGRLHF) was highest in the oral mucosa and cervical lymph nodes (Figure 6, green shading) compared with other T-cells. In mice, invariant natural killer T (iNKT) cells express a TRA encoded by the gene segments TRAV11d-TRAJ18 [19]. T-cells bearing TRAV12D-1-TRAJ22 (Figure 6, yellow shading) with common CDR3 amino acid sequences (CALSEKSSGSWQLIF) were detected frequently in the oral mucosa and cervical lymph nodes of the ACM mice but in only a few cervical lymph nodes of the ICM mice. Furthermore, to identify changes in the T-cell diversity at Days 1, 3, and 7 post-challenge, we examined the TCR repertoire of T-cells in Cr allergy that had infiltrated the oral mucosa of the ACM mice at Days 3 and 7 post-challenge (Figure S1). The frequency ranking of TRA clonotypes for the top 30 read percentages indicated a high proportion of iNKT-cells in the oral mucosa of the ACM mice at Days 3 and 7 post-challenge (Figure S1, green shading). T-cells bearing TRAV1-TRAJ33 with common CDR3 amino acid sequences (CAVRDSNYQLIW) were detected in the oral mucosa of the ACM mice between Days 3 and 7 post-challenge (Figure S1, blue shading). Mucosal-associated invariant T (MAIT) cells in mice express a TRA encoded by the TRAV1-TRAJ33 gene segments [19]. Next, we examined the common CDR3 amino acid sequences in the TRB repertoire in the ACM mice (Figure S2). There was no shared TRB clone in the sequences from the inflamed oral mucosa of the ACM mice at Days 1, 3, and 7 post-challenge.

TRA Ranking Top1-30 of oral mucosa in the ICM Day1

Rank	TRAV	TRAJ	CDR3	%Reads
1	TRAV11D	TRAJ18	CVVGDRGSALGRLHF(iNKT)	2.57
2	TRAV7D-2	TRAJ30	CAASYDTNAYKVIF	2.48
3	TRAV14D-3/DV8	TRAJ43	CAASNNNNAPRF	2.24
4	TRAV1	TRAJ33	CAVRDSNYQLIW	2.12
5	TRAV6-7/DV9	TRAJ56	CALSTGGNNKLTF	1.90
6	TRAV9D-2	TRAJ27	CAVSARTNTGKLTF	1.87
7	TRAV17	TRAJ37	CALEAGNTRKLIF	1.79
8	TRAV6D-6	TRAJ49	CALTSNTGYQNFYF	1.70
9	TRAV12-2	TRAJ56	CALKTGGNNKLTF	1.69
10	TRAV4-2	TRAJ17	CAASNSAGNKLTF	1.59
11	TRAV13D-1	TRAJ43	CAMERNNNNAPRF	1.51
12	TRAV16D/DV11	TRAJ13	CAMREDDSGTYQRF	1.45
13	TRAV6-5	TRAJ52	CALSDDTGANTGKLTF	1.35
14	TRAV6D-7	TRAJ22	CALSSSSGSWQLIF	1.23
15	TRAV9D-3	TRAJ37	CAVSQLTGNTRKLIF	1.22
16	TRAV12-2	TRAJ43	CALSLNNNAPRF	1.12
17	TRAV17	TRAJ52	CALNTGANTGKLTF	1.11
18	TRAV5-1	TRAJ38	CSASRVGDNSKLIW	1.10
19	TRAV4-4/DV10	TRAJ34	CAAEAGDKVVF	1.09
20	TRAV13D-1	TRAJ31	CAMGNNNNRIFF	1.06
21	TRAV7D-5	TRAJ39	CAVSMRDAGAKLTF	1.01
22	TRAV9-3	TRAJ21	CAVKPNYNVLYF	0.95
23	TRAV6D-7	TRAJ31	CALSDKGYNNNRIFF	0.95
24	TRAV9-4	TRAJ39	CAVRINNAGAKLTF	0.93
25	TRAV10	TRAJ31	CAASKNNNNRIFF	0.92
26	TRAV7D-5	TRAJ33	CAVSNYQLIW	0.91
27	TRAV8D-1	TRAJ52	CAPCTGANTGKLTF	0.88
28	TRAV9D-1	TRAJ15	CAASPPTAYQGGRALIF	0.87
29	TRAV9D-4	TRAJ13	CALSPNSGTYQRF	0.82
30	TRAV7D-5	TRAJ17	CAVSTGNSAGNKLTF	0.81
-	TRAV12D-1	TRAJ22	CALSEKSSGSWQLIF	N.D.

TRA Ranking Top1-30 of cervical lymph nodes in the ICM Day1

Rank	TRAV	TRAJ	CDR3	%Reads
1	TRAV11D	TRAJ18	CVVGDRGSALGRLHF(iNKT)	0.26
2	TRAV4-4/DV10	TRAJ21	CAAEQPNYNVLYF	0.20
3	TRAV9-4	TRAJ27	CAVSKITNTGKLTF	0.16
4	TRAV7-5	TRAJ22	CAMEASGSWQLIF	0.08
5	TRAV7D-2	TRAJ27	CAASDTNTGKITF	0.08
6	TRAV6-5	TRAJ33	CALSGTDSNYQLIW	0.07
7	TRAV6D-7	TRAJ30	CALLGTNAYKIIF	0.07
8	TRAV9-3	TRAJ34	CAVGPSSSNTDKVVF	0.06
9	TRAV14-1	TRAJ40	CAARSVNTGNYKNVF	0.06
10	TRAV7-6	TRAJ52	CGGKANTGANTGKLTF	0.05
11	TRAV6-1	TRAJ33	CVLGHSNYQMIW	0.04
12	TRAV6N-5	TRAJ9	CARGSAMGYKLTF	0.04
13	TRAV7-3	TRAJ52	CAVSMITGANAGKLTF	0.04
14	TRAV6-7/DV9	TRAJ27	CALAYNTNTGKLTF	0.04
15	TRAV15D-1/DV6D-1	TRAJ52	CALWEPTGANTEKLTF	0.04
16	TRAV1	TRAJ33	CANMDSNYQLIW	0.04
17	TRAV6-5	TRAJ40	CALSPNTGNYKSVF	0.04
18	TRAV7D-2	TRAJ4	CAARLSGSFNKMTF	0.04
19	TRAV17	TRAJ50	CALETSSFSKLVF	0.03
20	TRAV1	TRAJ18	CAVRGGGSALGRLHF	0.03
21	TRAV7-3	TRAJ40	CAVTGNYKFVF	0.03
22	TRAV7-3	TRAJ40	CAVTGSYKYVF	0.03
23	TRAV4D-3	TRAJ57	CAADGDQGGSAKLIF	0.03
24	TRAV9-3	TRAJ44	CAVFTGRGGKLTL	0.03
25	TRAV8-2	TRAJ27	CATEPGTNTGKLTF	0.03
26	TRAV21/DV12	TRAJ57	CILRVVNQGGSAKLIF	0.03
27	TRAV16D/DV11	TRAJ31	CAMREYNNNRVFF	0.03
28	TRAV3D-3	TRAJ24	CAVRKTASLGKLQF	0.03
29	TRAV16D/DV11	TRAJ7	CAMRERDYSNNRLTL	0.03
30	TRAV19	TRAJ33	CAAGLDSNYQLIW	0.03
88132	TRAV12D-1	TRAJ22	CALSEKSSGSWQLIF	0.0088

TRA Ranking Top1-30 of oral mucosa in the ACM Day1

Rank	TRAV	TRAJ	CDR3	%Reads
1	TRAV11D	TRAJ18	CVVGDRGSALGRLHF(iNKT)	2.29
2	TRAV12D-1	TRAJ22	CALSEKSSGSWQLIF	1.49
3	TRAV10	TRAJ53	CAARNSGGSNYKLTF	1.23
4	TRAV6N-5	TRAJ45	CALVPNTGGADRLTF	0.99
5	TRAV13D-1	TRAJ32	CAMEDYGGSGNKLIF	0.92
6	TRAV14D-3/DV8	TRAJ22	CAASSSGSWQLIF	0.80
7	TRAV14-1	TRAJ40	CAASAGNYKYVF	0.69
8	TRAV13D-1	TRAJ45	CAMERTGGADRLTF	0.68
9	TRAV13D-1	TRAJ26	CAMEEDYAQGLTF	0.61
10	TRAV12-3	TRAJ44	CALTAGSGGKLTL	0.58
11	TRAV6-7/DV9	TRAJ53	CALGNSGGSNYKLTF	0.57
12	TRAV4-3	TRAJ57	CAADGDQGGSAKLIF	0.56
13	TRAV9-3	TRAJ44	CAVFTGSGGKLTL	0.54
14	TRAV4D-3	TRAJ57	CAADGDQGGSAKLIF	0.53
15	TRAV5-1	TRAJ7	CSASKGYSNNRLTL	0.52
16	TRAV4-4/DV10	TRAJ45	CAALLTGGADRLTF	0.52
17	TRAV7D-5	TRAJ40	CAPNTGNYKYVF	0.51
18	TRAV5-1	TRAJ12	CSASPGGYKVVF	0.50
19	TRAV12D-3	TRAJ30	CALSAPDAYKVIF	0.49
20	TRAV12-2	TRAJ43	CALNNNNAPRF	0.48
21	TRAV12-3	TRAJ38	CALSGRNVGDNSKLIW	0.48
22	TRAV9D-3	TRAJ45	CAVSAGGADRLTF	0.46
23	TRAV9-3	TRAJ26	CAVSMSYAQGLTF	0.46
24	TRAV14N-2	TRAJ49	CAADTGYQNFYF	0.45
25	TRAV7-1	TRAJ12	CAVSPGTGGYKVVF	0.45
26	TRAV14-1	TRAJ49	CAAYTGYQNFYF	0.45
27	TRAV13-1	TRAJ48	CAMVPYGNEKITF	0.43
28	TRAV12N-2	TRAJ26	CAPNYAQGLTF	0.43
29	TRAV6-7/DV9	TRAJ49	CALGENTGYQNFYF	0.42
30	TRAV14D-3/DV8	TRAJ58	CAASAGTGSKLSF	0.42

TRA Ranking Top1-30 of cervical lymph nodes in the ACM Day1

Rank	TRAV	TRAJ	CDR3	%Reads
1	TRAV11D	TRAJ18	CVVGDRGSALGRLHF(iNKT)	0.22
2	TRAV7-5	TRAJ2	CAMSMWRTGGLSGKLTF	0.06
3	TRAV12-2	TRAJ57	CALLNQGGSAKLIF	0.06
4	TRAV16D/DV11	TRAJ33	CAMREGRVIW	0.06
5	TRAV6-6	TRAJ40	CALGDPGNYKYVF	0.05
6	TRAV16D/DV11	TRAJ31	CAMREYNNNRVFF	0.04
7	TRAV1	TRAJ33	CANMDSNYQLIW	0.04
8	TRAV17	TRAJ50	CALETSSFSKLVF	0.04
9	TRAV6D-7	TRAJ33	CALDSNYQLIW	0.04
10	TRAV14D-3/DV8	TRAJ58	CAASAGTGSKLTF	0.04
11	TRAV9-3	TRAJ26	CAVRMSYAQGLTF	0.03
12	TRAV7D-2	TRAJ35	CAARRFVSALTF	0.03
13	TRAV6-7/DV9	TRAJ27	CALAYNTNTGKLTF	0.03
14	TRAV6D-3	TRAJ31	CAMRDRHNNNRIFF	0.03
15	TRAV14D-3/DV8	TRAJ22	CAASSSDSWQLIF	0.03
16	TRAV12-3	TRAJ50	CALRIASSSFSKLVF	0.03
17	TRAV8-2	TRAJ27	CATEPGTNTGKLTF	0.03
18	TRAV4-3	TRAJ57	CAADGEQGGSAKLIF	0.03
19	TRAV13D-1	TRAJ32	CAMEDYGGRGNKLIF	0.03
20	TRAV7D-2	TRAJ4	CAARLSGSFNKMTF	0.03
21	TRAV13D-2	TRAJ15	CAIVYQGGRAMIF	0.03
22	TRAV12D-1	TRAJ22	CALSEKSSGSWQLIF	0.03
23	TRAV7-3	TRAJ52	CAGSMITGANTGKLTF	0.03
24	TRAV16D/DV11	TRAJ42	CAMREKTGGSNEKLTF	0.03
25	TRAV14D-3/DV8	TRAJ15	GAASSGGGRALIF	0.02
26	TRAV9-3	TRAJ26	CAVSMSDAQGLTF	0.02
27	TRAV12-2	TRAJ57	CALSPGGGGSAKLIF	0.02
28	TRAV21/DV12	TRAJ57	CILRVVNQGGSAKLIF	0.02
29	TRAV9-1	TRAJ32	CAGSAWYGGSGNKLIF	0.02
30	TRAV14-1	TRAJ40	CAASTGNYKYVF	0.02

Figure 6. Ranking of the top 30 most frequently read TRA clonotypes (as a percentage) in the oral mucosa and cervical lymph nodes of the ICM and ACM mice at Day 1 post-challenge. Distributions of frequency (%) reads of amino acid sequences of CDR3 regions in the oral mucosa and cervical lymph nodes of the ACM mice revealed the presence of Cr-specific T-cells bearing TRAV12D-1-TRAJ22 (yellow shading). In both the oral mucosa and the cervical lymph nodes of the ICM and ACM mice, iNKT-cells (green shading) are highly expressed. N.D. = not detected.

3. Discussion

In this study, we successfully established a mouse model of intraoral Cr allergy and induced a delayed allergic response in the oral mucosa, building on previous studies documenting a Cr-induced allergic mouse model using the footpad skin or the auricle [15,20]. Recently, the usage of Cr in dental treatments, such as implants, and other intraoral treatments has increased. In a previous study, adverse effects were observed when local chromium concentrations were high owing to the release of particulate and soluble chromium released from implants [21]. However, the pathological mechanism of intraoral Cr allergy remains unclear because there is no established animal model of Cr allergy in the oral mucosa.

To our knowledge, this is the first study to clarify Cr-specific immune responses in the oral mucosa using a murine model of intraoral Cr allergy. In the ICM group, which was exposed to the external stimuli of Cr injections without sensitization, histopathological examination revealed a significant amount of inflammatory cell infiltration under the epithelium, as well as inflammatory reactions in the dermis and muscle layer at Day 1 post-challenge. In the ICM group at Day 7 post-challenge, there was little irritated inflammation at the site. The ACM group showed obvious spongiform edema and T-cell infiltration at Day 1 post-challenge in the injected epithelium. Intraepithelial spongiform edema is a defining feature of DTH [22]. By Day 7 post-challenge, the spongiform edema had decreased and T-cells had infiltrated the subepithelial layer. In the ACM group, F4/80+ macrophages infiltrated the basal epithelial and subepithelial layers at Day 1 post-challenge but then decreased until Day 7 post-challenge. Our Cr-induced allergic mouse model in the oral mucosa has characteristics resembling DTH; therefore, it may be appropriate for studying delayed allergic reactions in the oral mucosa. As regards the duration of acute ACD, previous studies have indicated that it develops between 24 and 48 h after initiation [23]. Initially, skin lesions are asymmetric and limited to the area of contact; later, they often spread or disseminate. In the case of severe reactions, swelling and blistering are observed. The major clinical differences between non-sensitized irritant contact dermatitis (ICD) and sensitized ACD are the more rapid onset of ICD and the tendency of ACD to spread. Characteristic widespread reactions are usually symmetric, although the primary reaction is not [23]. Metal allergy can be induced by either CD4+ or CD8+ T-cells, depending on the antigen processing pathway [9,24]. In this mouse model of intraoral Cr allergy, CD8+ T-cells accumulated in large numbers in ACM at Day 1 post-challenge, but this was not observed in control or ICM mice (Figure 3). These results suggest that sensitization had occurred and that CD8+ T-cells specific to Cr were induced in intraoral Cr-induced allergic mice. Another mouse model of metal allergy found that CD8+ T-cells accumulated at inflammatory sites, suggesting that these T-cells promote inflammation during the DTH induction stage [14].

We also compared the IL-1β, IFN-γ, IL-4, and granzyme B expression levels in the oral mucosa of the control, ICM, and ACM mice (Figure 4). At Day 1 post-challenge, the levels of IFN-γ expression in the ACM mice were significantly higher than those seen in the ICM mice. However, there were no differences in the levels of IL-4 expression between the ACM and ICM mice at Day 1 post-challenge. This indicates that the Cr-allergic immune response in the oral mucosa may be Th-1-biased at Day 1 post-challenge. CD8+ T-cells were reported as infiltrating the inflammatory site during metal allergy induction and producing IFN-γ [14]. Furthermore, IL-1β induces Th-1 and Th-17 differentiation and stimulates cytokine production in T-cells and CD8+ T-cells residing in the epidermal tissue, which may increase IL-1β production via a positive feedback loop [25]. Our results were consistent with those of this study. According to a separate study, granzyme B expression was elevated in the skin of patients with metal allergy [26]. CD8+ T-cells are cytotoxic and kill target T-cells by releasing cytotoxic granules. Our findings indicate that IFN-γ is the principal effector cytokine of Cr allergy in the oral mucosa. The apoptosis of keratinocytes induced by macrophages and CD8+ T-cells may also play a role in the pathogenesis of Cr allergy in the oral mucosa. We previously developed a mouse model of Cr-induced ACD in the footpad of mice and found that allergic-specific T-cells used a specific TCR repertoire in Cr-induced ACD [15]. The infiltrating T-cells included iNKT-cells and Cr-specific T-cells with VA11-1/VB14-1 usage. In this study, we observed an accumulation of iNKT-cells in the inflamed mucosa of the ICM and ACM mice. These iNKT-cells help to amplify early immune responses [19,27]. The accumulation of iNKT-cells in the inflamed oral mucosa in Ni allergy suggests that iNKT-cells are involved in metal allergy [28]. In this study, we also observed the accumulation of Cr-specific T-cells bearing TRAV12D-1-TRAJ22 and iNKT-cells in the inflamed oral mucosa of Cr-allergic mice (Figures 5 and 6). This result contributes significantly to our understanding of antigen recognition by Cr-specific T-cells. Th1-type cytokines and cytotoxic molecules from iNKT-cells and Cr-specific T-cells are likely to be positively correlated with the pathogenesis of metal allergy in the footpads of mice. TCR repertoire and CDR3 sequencing analyses revealed a shared TCR repertoire expressing TRAV12D-1-TRAJ22 with the CDR3 sequence (CALSEKSSGSWQLIF) in the oral mucosa and lymph nodes at Day 1 post-challenge (Figure 6). In contrast to the highly restricted TRAV repertoire, oral mucosa-infiltrating T-cells displayed a relatively broad TRBV repertoire. As the TCR repertoire involved in Cr allergy differs between the oral mucosa, the footpad, and the ears, it is likely that organ-specific immune responses may occur during metal allergy. Furthermore, iNKT-cells were abundant in the ICM and ACM mice, while MAIT-cells were detected in the oral mucosa of the ACM mice at Days 3 and 7 post-challenge, whereas TRAV12D-1-TRAJ22-bearing T-cells were not (Figure S1). iNKT and MAIT-cells were found in high numbers in our previous research on cross-reactive metal allergy, suggesting that they may be involved in the development of cross-reactive metal allergy [29]. Consequently, the data on the TCR repertoire contribute to our understanding of the structural identity of antigenic determinants recognized by Cr-induced intraoral metal contact allergy.

In conclusion, we demonstrated that CD8+ T-cells infiltrated the oral mucosa during an allergic inflammatory response in a mouse model of metal allergy. We also identified clones of T-cells specific to Cr-induced allergy. Our analysis of the TCR repertoire demonstrated that restricted TRAV12D-1-TRAJ22 usage with the CDR3 amino acid sequence (CALSEKSSGSWQLIF) and broad TRBV might specifically recognize Ag in Cr allergy of the oral mucosa. In that case, the direct cloning of TCR genes from local sites of inflammation using this model would be a powerful tool for advancing our understanding of T-cell-mediated immune disease in metal allergy, and it would also provide new insights into Ag recognition by Cr-specific TCR in the oral mucosa.

4. Materials and Methods

4.1. Animals

BALB/cAJcl mice (4-week-old females) were purchased from CLEA Japan (Tokyo, Japan). All of the mice were in good health throughout the duration of the study and were given 1 week to acclimatize to their surroundings before the study began. At the beginning of the experiment, the mice were 5 weeks old. All of the mice were kept in plastic cages (with lids made of stainless-steel wire) and given ad libitum access to food and water. The mice were kept in our conventional animal facility with a temperature of 19–23 °C, a humidity of 30–70%, and a 12 h day/night cycle. During experiments, their appearance, their behavior, and the amount of food and water they consumed were monitored daily to ensure the health and comfort of all of the mice. All surgeries were performed using three different types of mixed anesthetics, and every effort was made to minimize animal suffering. All of the mice were euthanized by cervical dislocation under three different types of mixed anesthetic agents to prevent pain caused by tissue harvesting.

4.2. Reagents

FUJIFILM Wako Pure Chemical Co., Ltd. was the supplier of high-purity $CrCl_2$ (>95%) (Osaka, Japan). Lipopolysaccharide (LPS) from *Escherichia coli* (O55:B5) prepared by phenol–water extraction was purchased from Sigma-Aldrich (St Louis, MO, USA). $CrCl_2$ and LPS were dissolved in sterile saline (Otsuka Normal Saline, Otsuka Pharmaceutical Factory, Inc., Tokushima, Japan).

4.3. Anesthetic Agents

The anesthetic was created by combining three medications. We purchased medetomidine hydrochloride from Nippon Zenyaku Kogyo Co., Ltd. (Fukushima, Japan), midazolam from Sandoz (Tokyo, Japan), and butorphanol tartrate from Meiji Seika Pharma Co., Ltd. (Tokyo, Japan). These drugs were kept at room temperature (RT). We combined doses of 0.3 mg/kg medetomidine hydrochloride, 4 mg/kg midazolam, and 5 mg/kg butorphanol tartrate. The concentration ratio of the three types of mixed anesthetic agents was determined according to data from a previous study [30]. Typically, 25 mL of anesthetic agent was prepared by combining 0.75 mL of medetomidine hydrochloride, 2 mL of midazolam, 2.50 mL of butorphanol tartrate, and 19.75 mL of sterile saline. All of the agents were diluted in sterile saline and stored in the dark at 4 °C. The mice were administered a volume of 10 μL/g of body weight of the anesthetic mixture. The mixture of the three types of anesthetic agents was injected intraperitoneally into every mouse.

4.4. Experimental Protocol for the Mouse Model of Cr-Induced Intraoral Metal Contact Allergy

We have developed an experimental protocol for the induction of metal allergy in the oral mucosa [28] based on a previous protocol for the induction of metal allergy in footpad skin. Mice were separated into three groups: ACM (n = 15), ICM (n = 15), and control (n = 15), with each group consisting of randomly selected mice. All experiments were performed in another room upon transfer from the animal holding area.

Sensitization: The mice received two intradermal injections of 125 μL of 10 mM $CrCl_2$ and 10 g/mL LPS in sterile saline with a 7-day interval between injections. The ACM group was sensitized on the postauricular skin in the left and right ears. Seven days after the second sensitization, the mice were challenged for the first time.

Challenge for elicitation: 25 μL of 10 mM $CrCl_2$ without LPS in sterile saline was used to elicit an immune response. The ACM mice were challenged by submucosal injection in the left and right buccal regions of the oral mucosa. The ICM mice were not sensitized, but they were challenged with $CrCl_2$. The control mice were sensitized and then challenged with sterile saline.

4.5. Measurement of Oral Mucosa Swelling

The swelling of the buccal region was measured before the challenge and at 1, 2, 3, 5, 7, 10, 12, and 14 days after the initial challenge using a Peacock Dial Thickness Gauge (Ozaki MFG Co., Ltd., Tokyo, Japan). The thickness of the buccal area oral mucosa was measured before and after the challenge, and the difference was recorded. The same experimenter performed all procedures on the mice while they were under anesthesia.

4.6. Histological and IHC Analysis

For histological and immunohistochemical analyses, buccal oral mucosa samples were obtained from the control, ICM, and ACM mice. Furthermore, the tissue samples were immersed in 4% paraformaldehyde–lysine–periodate for 48 h at 4 °C. After a 10-min wash in PBS, fixed tissues were soaked in 5% sucrose/PBS for 1 h at 4 °C, 15% sucrose/PBS for 3 h at 4 °C, and finally, 30% sucrose/PBS overnight at 4 °C. The buccal mucosa tissues were snap-frozen by immersion in a mixture of acetone and dry ice in Tissue Mount (Chiba Medical, Saitama, Japan). Frozen sections were cut into 6 μm thick cryosections and air-dried on poly-L-lysine-coated glass slides. The HE stain was applied to cryosections for histological analysis. The cryosections were stained with anti-mouse F4/80 (1:1000; Cl-A3-1, Abcam, Cambridge, UK) and anti-mouse CD3 (1:500; SP7, Abcam) monoclonal antibodies for IHC analysis (mAbs). The F4/80 monoclonal antibody was used to detect mouse macrophage populations in many buccal oral mucosal tissues. The sections were incubated at RT for 30 min in PBS containing 5% normal goat/rabbit serum, 0.025% Triton X-100

(FUJIFILM Wako Pure Chemical, Osaka, Japan), and 5% bovine serum albumin (Sigma-Aldrich). Sections were incubated with primary mAbs for 1 h at RT. After three 5-min washes with PBS, intrinsic peroxidase was inhibited with 3% H_2O_2 in methanol. After the tissue sections were soaked in distilled water, they were washed twice and incubated for 1 h at RT with a secondary antibody (biotinylated goat anti-hamster immunoglobulin G or biotinylated rabbit anti-rat immunoglobulin G). The sections were treated with Vectastain ABC Reagent (Vector Laboratories, Burlingame, CA, USA) for 30 min at RT, followed by 3,3-diaminobenzidine staining (0.06% diaminobenzidine and 0.03% H_2O_2 in 0.1 M Tris–HCl, pH 7.6; FUJIFILM Wako Pure Chemical). Hematoxylin was used to counterstain tissue sections to visualize cell nuclei.

4.7. RNA Extraction

The buccal region of the oral mucosa in each mouse was freshly obtained and immersed immediately in RNAlater RNA Stabilization Reagent (Invitrogen, Carlsbad, CA, USA). Total RNA was extracted from the buccal region of the oral mucosa using the RNeasy Lipid Tissue Mini Kit (Qiagen, Hilden, Germany) as directed by the manufacturer. Complementary DNA (cDNA) was synthesized from DNA-free RNA.

4.8. Quantitative Polymerase Chain Reaction

The expression levels of immune response-related genes, including T-cell-related CD antigens, cytokines, cytotoxic granules, and regulatory T-cell transcription factors, were measured using qPCR on a Bio-Rad CFX96 instrument (Bio-Rad, Hercules, CA, USA). Previously, specific primers for GAPDH, CD3, CD4, CD8, IFN-γ, Granzyme B, IL-4, and IL-1β were described [15,16,31]. The Prime Script RT reagent Kit (Takara Bio, Shiga, Japan) was used to convert newly isolated total RNA from the oral mucosa and submandibular lymph nodes into complementary DNA (cDNA). In a final volume of 10 μL, the PCR contained 5 μL SsoFast EvaGreen Supermix (Bio-Rad), 3.5 μL RNase/DNase-free water, 0.5 μL of 5 μM primer mix, and 1 μL cDNA. The following cycling conditions were used: 30 s at 95 °C, followed by 50 cycles of 1 s at 95 °C and 5 s at 60 °C. At the conclusion of each protocol, a melting curve analysis from 70 °C to 90 °C was performed to confirm the homogeneity of the PCR products. All tests were performed three times, and the mean values were used to determine gene expression levels. Five 10-fold serial dilutions of each standard transcript were utilized to determine the absolute amount, specificity, and amplification efficiency of each primer set. Standard transcripts were generated through the in vitro transcription of the corresponding PCR product in a plasmid. DNA sequencing confirmed the nucleotide sequences using the CEQ8000 Genetic Analysis System (Beckman Coulter, Fullerton, CA, USA). Next, using an Agilent DNA 7500 Kit on an Agilent 2100 Bioanalyzer, their quality and concentration were determined (Agilent, Santa Clara, CA, USA). The expression of the GAPDH gene served as an internal control.

4.9. Mouse TCR Repertoire Analysis

Total RNA was extracted from the oral mucosa and cervical lymph nodes of the ICM and ACM mice at Day 1 post-challenge and from the oral mucosa of the ACM mice at Days 3 and 7 post-challenge. NGS was used to perform TCR repertoire analysis developed by Repertoire Genesis Inc. (Osaka, Japan [18]). As detailed in a previous report, an unbiased adaptor-ligation PCR was performed [32]. Superscript III reverse transcriptase was utilized to convert total RNA to cDNA (Invitrogen). Subsequently, double-stranded (ds) cDNA was synthesized, and an adaptor was ligated to the 5′ end of the ds-cDNA before it was cut with the SphI restriction enzyme. For TCRα, PCR was performed using a P20EA adaptor primer and a TCR α-chain constant region-specific primer (mCA1). The second PCR was conducted using the same PCR conditions and primers, mCA2 and P20EA. As regards TCRβ, the first and second primers used for PCR were mCB1 and mCB2, respectively. After Tag PCR amplification, index (barcode) sequences were amplified using a Nextera XT Index Kit v2 setA (Illumina, San Diego, CA, USA). Sequencing was performed using the paired-end Illumina MiSeq platform (2 × 300 base pairs [bp]). The repertoire analysis software of Repertoire Genesis, Inc. was used for automatic data processing, assignment, and aggregation. TCR (TRA and TRB) sequences were mapped to a reference sequence dataset from the international ImMunoGeneTics information system (IMGT) database (http://www.imgt.org, accessed on 1 November 2022 [33]). Nucleotide sequences of CDR3 regions ranged from a conserved cysteine at position 104 (Cys104) to a conserved phenylalanine at position 118 (Phe118), and the following glycine (Gly119) was translated into an amino acid sequence. A unique sequence read was defined as a sequence read with no identity in TRAV, TRAJ, and the deduced amino acid sequence of CDR3. The copy number of identical unique sequence reads in each sample was automatically counted and ranked by copy number using software for repertoire analysis. The percentage occurrence frequencies were calculated for sequence reads containing TRAV and TRAJ and for genes.

4.10. Statistical Analysis

Statistically significant differences between the mean values of each experimental group were analyzed using the Kruskal–Wallis test followed by Dunn's multiple comparison tests. All analyses were performed using IBM SPSS Statistics version 24 (IBM, Armonk, NY, USA). A p-value of <0.05 was considered significant.

Supplementary Materials: The following supporting information can be downloaded at: https://www.mdpi.com/article/10.3390/ijms24032807/s1.

Author Contributions: T.Y., K.K. (Kenichi Kumagai), Y.H. and R.S. conceived and designed the experiments; T.Y., K.K. (Kenichi Kumagai), R.M., K.N. and R.S. performed the experiments; T.Y., K.N., K.K. (Kazutaka Kitaura) and M.S. analyzed the data; T.Y., K.K. (Kazutaka Kitaura) and K.N. contributed reagents/materials/analysis tools; T.Y., K.K. (Kenichi Kumagai), M.S. and R.S. wrote the paper. All authors have read and agreed to the published version of the manuscript.

Funding: This work was supported by the Japan Society for the Promotion of Science KAKENHI Grant-in-Aid for Scientific Research C Grant No. 19K10371.

Institutional Review Board Statement: All animal experiments in this study were carried out according to the relevant ethical requirements with approval from the committees for animal experiments at Tsurumi University (approval number: 21A025).

Informed Consent Statement: Not applicable.

Data Availability Statement: The data presented in this study are available on request from the corresponding author.

Conflicts of Interest: The authors declare no conflict of interest.

Abbreviations

Cr	chromium
ACM	allergic contact mucositis
ICM	irritant contact mucositis
TCR	T-cell receptor
TRAV	TCRα-chain variable region
TRAJ	TCRα-chain joining region
TRBV	TCRβ-chain variable region
TRBJ	TCRβ-chain joining region
CDR3	complementarity-determining region 3
CD	cluster of differentiation

References

1. Alinaghi, F.; Bennike, N.H.; Egeberg, A.; Thyssen, J.P.; Johansen, J.D. Prevalence of contact allergy in the general population: A systematic review and meta-analysis. *Contact Dermat.* **2019**, *80*, 77–85. [CrossRef] [PubMed]
2. Zhang, X.; Wei, L.C.; Wu, B.; Yu, L.Y.; Wang, X.P.; Liu, Y. A comparative analysis of metal allergens associated with dental alloy prostheses and the expression of HLA-DR in gingival tissue. *Mol. Med. Rep.* **2016**, *13*, 91–98. [CrossRef] [PubMed]

3. Olms, C.; Schor, J.; Yahiaoui-Doktor, M. Potential Co-Factors of an Intraoral Contact Allergy-A Cross-Sectional Study. *Dent. J.* **2020**, *8*, 83. [CrossRef] [PubMed]
4. Kitagawa, M.; Murakami, S.; Akashi, Y.; Oka, H.; Shintani, T.; Ogawa, I.; Inoue, T.; Kurihara, H. Current status of dental metal allergy in Japan. *J. Prosthodont. Res.* **2019**, *63*, 309–312. [CrossRef]
5. Alnazzawi, A.A. Oral diseases associated with fixed prosthodontic restorations. *Saudi Med. J.* **2017**, *38*, 322–324. [CrossRef]
6. Zemelka-Wiacek, M. Metal Allergy: State-of-the-Art Mechanisms, Biomarkers, Hypersensitivity to Implants. *J. Clin. Med.* **2022**, *11*, 6971. [CrossRef]
7. Özkaya, E.; Elinç Aslan, M.S. Occupational allergic contact dermatitis: A 24-year, retrospective cohort study from Turkey. *Contact Dermat.* **2021**, *85*, 503–513. [CrossRef]
8. Hubler, W.R., Jr.; Hubler, W.R., Sr. Dermatitis from a chromium dental plate. *Contact Dermat.* **1983**, *9*, 377–383. [CrossRef]
9. van Wilsem, E.J.; Breve, J.; Savelkoul, H.; Claessen, A.; Scheper, R.J.; Kraal, G. Oral tolerance is determined at the level of draining lymph nodes. *Immunobiology* **1995**, *194*, 403–414. [CrossRef]
10. Davis, M.M.; Bjorkman, P.J. T-cell antigen receptor genes and T-cell recognition. *Nature* **1988**, *334*, 395–402. [CrossRef]
11. Kumagai, K.; Horikawa, T.; Shigematsu, H.; Matsubara, R.; Kitaura, K.; Eguchi, T.; Kobayashi, H.; Nakasone, Y.; Sato, K.; Yamada, H.; et al. Possible Immune Regulation of Natural Killer T Cells in a Murine Model of Metal Ion-Induced Allergic Contact Dermatitis. *Int. J. Mol. Sci.* **2016**, *17*, 87. [CrossRef] [PubMed]
12. Eguchi, T.; Kumagai, K.; Kobayashi, H.; Shigematsu, H.; Kitaura, K.; Suzuki, S.; Horikawa, T.; Hamada, Y.; Ogasawara, K.; Suzuki, R. Accumulation of invariant NKT cells into inflamed skin in a novel murine model of nickel allergy. *Cell. Immunol.* **2013**, *284*, 163–171. [CrossRef] [PubMed]
13. Kobayashi, H.; Kumagai, K.; Eguchi, T.; Shigematsu, H.; Kitaura, K.; Kawano, M.; Horikawa, T.; Suzuki, S.; Matsutani, T.; Ogasawara, K.; et al. Characterization of T cell receptors of Th1 cells infiltrating inflamed skin of a novel murine model of palladium-induced metal allergy. *PLoS ONE* **2013**, *8*, e76385. [CrossRef] [PubMed]
14. Kumagai, K.; Matsubara, R.; Nakasone, Y.; Shigematsu, H.; Kitaura, K.; Suzuki, S.; Haneji, K.; Hamada, Y.; Suzuki, R. Possible involvement of invariant natural killer T cells and mucosal-associated invariant T cells in a murine model of titanium allergy. *J. Oral Maxillofac. Surg. Med. Pathol.* **2018**, *30*, 1–9. [CrossRef]
15. Shigematsu, H.; Kumagai, K.; Kobayashi, H.; Eguchi, T.; Kitaura, K.; Suzuki, S.; Horikawa, T.; Matsutani, T.; Ogasawara, K.; Hamada, Y.; et al. Accumulation of metal-specific T cells in inflamed skin in a novel murine model of chromium-induced allergic contact dermatitis. *PLoS ONE* **2014**, *9*, e85983. [CrossRef]
16. Kitaura, K.; Fujii, Y.; Hayasaka, D.; Matsutani, T.; Shirai, K.; Nagata, N.; Lim, C.K.; Suzuki, S.; Takasaki, T.; Suzuki, R.; et al. High clonality of virus-specific T lymphocytes defined by TCR usage in the brains of mice infected with West Nile virus. *J. Immunol.* **2011**, *187*, 3919–3930. [CrossRef]
17. Kitaura, K.; Shini, T.; Matsutani, T.; Suzuki, R. A new high-throughput sequencing method for determining diversity and similarity of T cell receptor (TCR) alpha and beta repertoires and identifying potential new invariant TCR alpha chains. *BMC Immunol.* **2016**, *17*, 38. [CrossRef]
18. Kitaura, K.; Yamashita, H.; Ayabe, H.; Shini, T.; Matsutani, T.; Suzuki, R. Different Somatic Hypermutation Levels among Antibody Subclasses Disclosed by a New Next-Generation Sequencing-Based Antibody Repertoire Analysis. *Front. Immunol.* **2017**, *8*, 389. [CrossRef]
19. de Lima Moreira, M.; Souter, M.N.T.; Chen, Z.; Loh, L.; McCluskey, J.; Pellicci, D.G.; Eckle, S.B.G. Hypersensitivities following allergen antigen recognition by unconventional T cells. *Allergy* **2020**, *75*, 2477–2490. [CrossRef]
20. Sitalaksmi, R.M.; Ito, K.; Ogasawara, K.; Suto, Y.; Itabashi, M.; Ueda, K.; Hirasawa, N.; Narushima, T.; Hendrijantini, N.; Kresnoadi, U.; et al. COX-2 induces T cell accumulation and IFN-gamma production during the development of chromium allergy. *Autoimmunity* **2019**, *52*, 228–234. [CrossRef]
21. Jumina, J.; Harizal, H. Dermatologic Toxicities and Biological Activities of Chromium. In *Trace Metals in the Environment-New Approaches and Recent Advances*; IntechOpen: London, UK, 2019; pp. 1–22. [CrossRef]
22. Tanei, R.; Hasegawa, Y. Immunological Pathomechanisms of Spongiotic Dermatitis in Skin Lesions of Atopic Dermatitis. *Int. J. Mol. Sci.* **2022**, *23*, 6682. [CrossRef]
23. Novak-Bilic, G.; Vucic, M.; Japundzic, I.; Mestrovic-Stefekov, J.; Stanic-Duktaj, S.; Lugovic-Mihic, L. Irritant and Allergic Contact Dermatitis-Skin Lesion Characteristics. *Acta Clin. Croat.* **2018**, *57*, 713–720. [CrossRef]
24. Scadding, G.; Durham, S. Mechanisms of sublingual immunotherapy. *J. Asthma* **2009**, *46*, 322–334. [CrossRef]
25. Yeung, K.; Mraz, V.; Geisler, C.; Skov, L.; Bonefeld, C.M. The role of interleukin-1beta in the immune response to contact allergens. *Contact Dermat.* **2021**, *85*, 387–397. [CrossRef]
26. Yawalkar, N.; Hunger, R.E.; Buri, C.; Schmid, S.; Egli, F.; Brand, C.U.; Mueller, C.; Pichler, W.J.; Braathen, L.R. A Comparative Study of the Expression of Cytotoxic Proteins in Allergic Contact Dermatitis and Psoriasis. *Am. J. Pathol.* **2001**, *158*, 803–808. [CrossRef]
27. Kanda, M.; Yamanaka, H.; Kojo, S.; Usui, Y.; Honda, H.; Sotomaru, Y.; Harada, M.; Taniguchi, M.; Suzuki, N.; Atsumi, T.; et al. Transcriptional regulator Bhlhe40 works as a cofactor of T-bet in the regulation of IFN-gamma production in iNKT cells. *Proc. Natl. Acad. Sci. USA* **2016**, *113*, E3394–E3402. [CrossRef]

28. Nakasone, Y.; Kumagai, K.; Matsubara, R.; Shigematsu, H.; Kitaura, K.; Suzuki, S.; Satoh, M.; Hamada, Y.; Suzuki, R. Characterization of T cell receptors in a novel murine model of nickel-induced intraoral metal contact allergy. *PLoS ONE* **2018**, *13*, e0209248. [CrossRef]
29. Shigematsu, H.; Kumagai, K.; Suzuki, M.; Eguchi, T.; Matsubara, R.; Nakasone, Y.; Nasu, K.; Yoshizawa, T.; Ichikawa, H.; Mori, T.; et al. Cross-Reactivity of Palladium in a Murine Model of Metal-Induced Allergic Contact Dermatitis. *Int. J. Mol. Sci.* **2020**, *21*, 4061. [CrossRef]
30. Kawai, S.; Takagi, Y.; Kaneko, S.; Kurosawa, T. Effect of three types of mixed anesthetic agents alternate to ketamine in mice. *Exp. Anim.* **2011**, *60*, 481–487. [CrossRef]
31. Fujii, Y.; Kitaura, K.; Nakamichi, K.; Takasaki, T.; Suzuki, R.; Kurane, I. Accumulation of T-cells with selected T-cell receptors in the brains of Japanese encephalitis virus-infected mice. *Jpn. J. Infect. Dis.* **2008**, *61*, 40–48.
32. Yoshida, R.; Yoshioka, T.; Yamane, S.; Matsutani, T.; Toyosaki-Maeda, T.; Tsuruta, Y.; Suzuki, R. A new method for quantitative analysis of the mouse T-cell receptor V region repertoires: Comparison of repertoires among strains. *Immunogenetics* **2000**, *52*, 35–45. [CrossRef] [PubMed]
33. Lefranc, M.P.; Giudicelli, V.; Ginestoux, C.; Jabado-Michaloud, J.; Folch, G.; Bellahcene, F.; Wu, Y.; Gemrot, E.; Brochet, X.; Lane, J.; et al. IMGT, the international ImMunoGeneTics information system. *Nucleic Acids Res.* **2009**, *37*, D1006–D1012. [CrossRef] [PubMed]

International Journal of
Molecular Sciences

MDPI

Article

Human β-Defensin 3 Inhibits *Porphyromonas Gingivalis* Lipopolysaccharide-Induced Oxidative and Inflammatory Responses of Microglia by Suppression of Cathepsins B and L

Erika Inoue [1,†], Shiyo Minatozaki [1,†], Yui Katsuta [1], Saori Nonaka [2] and Hiroshi Nakanishi [2,*]

1 Faculty of Pharmacy, Yasuda Women's University, Hiroshima 731-0153, Japan
2 Department of Pharmacology, Faculty of Pharmacy, Yasuda Women's University, Hiroshima 731-0153, Japan
* Correspondence: nakanishi-h@yasuda-u.ac.jp
† These authors contributed equally to this work.

Abstract: Recently, the effects of antibacterial peptides are suggested to have therapeutic potential in Alzheimer's disease. Furthermore, systemic treatment of *Porphyromonas gingivalis* (*Pg*) lipopolysaccharide (LPS) induced Alzheimer's disease-like neuropathological changes in middle-aged mice. Then, we examined whether human β-defensins (hBDs), antimicrobial peptides produced by the oral mucosa and salivary glands, can suppress *Pg* LPS-induced oxidative and inflammatory responses by microglia. hBD3 (1 μM) significantly suppressed *Pg* LPS-induced production of nitric oxide and interleukin-6 (IL-6) by MG6 cells, a mouse microglial cell line. hBD3 (1 μM) also significantly inhibited *Pg* LPS-induced expression of IL-6 by HMC3 cells, a human microglial cell line. In contrast, neither hBD1, hBD2 nor hBD4 failed to inhibit their productions. Furthermore, hBD3 suppressed *Pg* LPS-induced p65 nuclear translocation through the IκBα degradation. *Pg* LPS-induced expression of IL-6 was significantly suppressed by E64d, a cysteine protease inhibitor, and CA-074Me, a known specific inhibitor for cathepsin B, but not by pepstatin A, an aspartic protease inhibitor. Interestingly, hBD3 significantly inhibited enzymatic activities of recombinant human cathepsins B and L, lysosomal cysteine proteases, and their intracellular activities in MG6 cells. Therefore, hBD3 suppressed oxidative and inflammatory responses of microglia through the inhibition of cathepsins B and L, which enzymatic activities are necessary for the NF-κB activation.

Keywords: antibacterial peptide; cathepsin B; cathepsin L; human β-defensins; interleukin-6; lipopolysaccharide; microglia; NF-κB p65; nitric oxide; *Porphyromonas gingivalis*

Citation: Inoue, E.; Minatozaki, S.; Katsuta, Y.; Nonaka, S.; Nakanishi, H. Human β-Defensin 3 Inhibits *Porphyromonas Gingivalis* Lipopolysaccharide-Induced Oxidative and Inflammatory Responses of Microglia by Suppression of Cathepsins B and L. *Int. J. Mol. Sci.* **2022**, *23*, 15099. https://doi.org/10.3390/ijms232315099

Academic Editor: Kenichi Kumagai

Received: 25 October 2022
Accepted: 29 November 2022
Published: 1 December 2022

1. Introduction

Neuroinflammation is a reaction of the host to infectious or sterile tissue damage, and has the physiological purpose of restoring tissue homeostasis. On the other hand, uncontrolled or unresolved chronic neuroinflammation can lead to neuronal damage. It is well known that chronic neuroinflammation is associated with brain pathologies including Alzheimer's disease (AD). However, there is limited knowledge regarding the mechanism by which acute neuroinflammation turns into a chronic one. Recently, much attention has been paid to the role of microbial infection in the pathogenesis of sporadic AD [1]. Periodontitis is a common oral chronic multi-bacterial infection which results in a persistent bacterial and inflammatory load in the body. An increasing number of clinical studies have demonstrated the impact of periodontitis on AD, and recent experimental studies clarified the route of transduction of inflammatory signals from periodontitis to the brain. *Porphyromonas gingivalis* (*Pg*), a major pathogen of chronic periodontitis, has virulence factors including lipopolysaccharide (LPS), cysteine proteases called gingipains and outer membrane vesicles [1–4]. Recently, LPS and gingipains, two major virulence factors of *Pg*, have been detected in the brain tissue of patients with AD on autopsy [5,6]. Cohort studies showed that periodontitis is associated with an increase in cognitive decline in AD patients [7,8]. Therefore, much attention has been paid to the causal relationship between *Pg* virulence factors and cognitive dysfunction in AD patients [9–11]. We have first reported that chronic systemic exposure of *Pg* LPS induced AD-like pathologies, including microglia-mediated neuroinflammation and cognitive decline in middle-aged wild-type, but not cathepsin B (CatB)-deficient mice [12]. Furthermore, neuroinflammation is responsible for cognitive impairment following chronic systemic exposure of *Pg* LPS in young adult mice [13] and periodontitis induced by *Pg* LPS in young adult rats [14].

Besides the ubiquitin-proteasome system, there is accumulating evidence demonstrating that the endosomal-lysosomal system is responsible for delayed degradation of IκBα, an endogenous inhibitor of nuclear factor-κB (NF-κB), at the late phase of oxidative and inflammatory responses by macrophages and microglia [15–20]. CatB was suggested to promote the autolysosomal degradation of IκBα and subsequent NF-κB nuclear translocation following hypoxic-ischemic brain injury of neonatal mice [18]. These observations suggest that *Pg* LPS has a significant influence over AD pathologies through oxidative and inflammatory responses mediated by activated microglia in a lysosome-dependent manner.

The emerging role of microbes and innate immune pathways in AD pathology suggests that antimicrobial peptides may be considered for early therapeutic interventions in future clinical trials [21,22]. Human β-defensins (hBDs) are cationic and small antimicrobial peptides produced by the oral mucosa and salivary glands. hBD1 and hBD2 are also expressed by microglia and astrocytes of both mouse and human brains [23–25]. With the exception of constitutively expressed hBD1, inducible hBD2 and hBD3 are generally up regulated by an inflammatory environment [25]. hBD1 is constitutively expressed in the choroid plexus, which acts as a primary immuno surveillance and defense mechanism against invading pathogens and systemic diseases [26]. Furthermore, the expression of hBD1 is increased within granulovacuolar degeneration structures localized in the cytoplasm of hippocampal pyramidal neurons and astrocytes of AD brain compared with the control brain [27]. A higher level of hBD1 was also seen in the choroid plexus of AD brain in comparison with an age-matched control brain. Moreover, AD patients show higher copy numbers of polymorphism of the DEFB4 gene that encodes for hBD2 and influences the production of hBD2, thus explaining the increased levels of hBD2 reported in the serum and cerebrospinal fluid of AD patients [28]. Human cultured astrocytes have been shown to express hBD2 following treatment with LPS and cytokines [23]. hBD3 is a potent antimicrobial peptide and exhibits immunosuppressive and anti-inflammatory effects on LPS-stimulated macrophages with relatively low concentrations [29,30]. These findings suggest that increased expression of hBDs may ameliorate the pathological progression of AD through regulation of activated microglia.

It remains, however, unclear whether hBDs are able to suppress oxidative and in flammatory responses induced by *Pg* LPS-stimulated microglia. In this study, we have thus sought to elucidate possible inhibitory effects of hBDs on oxidative and inflammatory responses induced by *Pg* LPS-stimulated microglia.

2. Results

2.1. Effects of hBDs on Cell Viability of MG6

We first examined effects of hBDs on the cell viability of MG6 cells using the cell counting kit; 8 at 26 h after treatment with hBDs with the concentration ranging from 100 nM to 10 μM. hBD1, 2 and 4 with the concentration up to 10 μM had no significant toxic effect on MG6 cells. On the other hand, hBD3 did not exhibit a significant toxic effect on MG6 cells up to 1 μM, but 10 μM hBD3 significantly reduced the cell viability (Figure 1A).

2.2. Effects of hBDs on *Pg* LPS-Induced Nitric Oxide (NO) Production and Inducible NO Synthase (iNOS) Expression by Microglia

To address whether hBDs could reduce *Pg* LPS-induced NO production by microglia, we next utilized a Griess assay to determine levels of nitrite, the major NO metabolite, in MG6 cells. hBD3 with the concentration of 100 nM and 1 μM significantly reduced nitrite production in MG6 cells after treatment with *Pg* LPS (10 μg/mL) for 24 h (Figure 1B). In contrast, neither hBD1, hBD2 nor hBD4 with the concentration up to 1 μM showed significant effect on the mean level of nitrite produced by MG6 cells after treatment with *Pg* LPS. The mean level of *Pg* LPS-induced iNOS mRNA was also significantly reduced by hBD3 (1 μM), but not by either hBD1, hBD3 or hBD4 (Figure 1C).

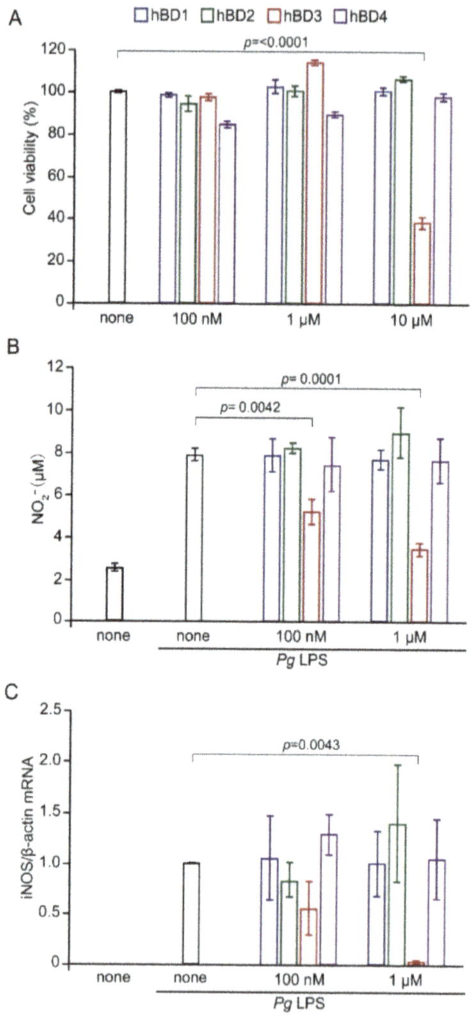

Figure 1. Effect of hBDs on cell viability, NO production and iNOS expression in MG6 cells after treatment of *Pg* LPS. (**A**) Cell viability of MG6 cells was evaluated using CCK-8 assay at 26 h after treatment with hBDs. MG6 cells were incubated with hBD1, 2, 3 and 4 with the concentrations ranging from 100 nM to 10 μM. (**B**) the mean level of nitrite, the major NO metabolite, was measured by the Griess assay. MG6 cells were pre-treated with hBDs at the indicated concentrations (100 nM and 1 μM) for 2 h and then treated with *Pg* LPS (10 μg/mL) for 24 h. **C**. The mean mRNA expression level of iNOS was determined by qRT-PCR analysis and β-actin mRNA served as the internal control for the normalization. MG6 cells were pre-treated with hBDs at the indicated concentrations (100 nM and 1 μM) for 2 h and then treated with *Pg* LPS (10 μg/mL) for 6 h. (**A–C**) The data are presented as the mean ± SE of three independent experiments, and *p* values were calculated using a one-way ANOVA with a post-hoc Tukey's test. A value of $p < 0.05$ was considered to indicate statistical significance.

2.3. Effects of hBDs on *Pg* LPS-Induced Interleukin-6 (IL-6) mRNA and Protein by Microglia

To determine cytokines and chemokines induced by stimulation with *Pg* LPS, we examined the cytokine expression in MG6 cells by using a cytokine antibody array. Untreated MG6 cells (none, distilled water only) were found constitutively to express relatively high levels of insulin growth factor-1 (IGF-1) and macrophage inflammatory protein-1α (MIP-1α), while all of the other cytokines and chemokines were undetectable. Both IL-6 and granulocyte-colony stimulating factor (G-CSF) were most intensively expressed in MG6 cells following stimulation with *Pg* LPS (10 μg/mL) for 24 h (Figure 2A,B). It was also noted that secretion levels of monocyte chemoattractant protein-1 (MCP-1), IL-1α, MIP-1α and vascular endothelial growth factor-A (VEGF-A) were also markedly increased following treatment with *Pg* LPS.

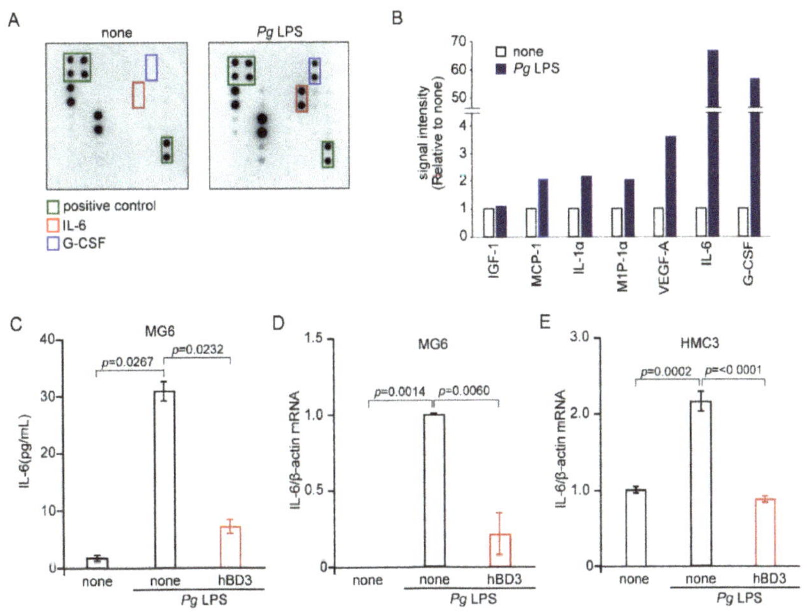

Figure 2. Effects of hBDs on the cytokine expression by microglia after exposure to *Pg* LPS. (**A**) Cytokines secreted from MG6 cells after exposure to *Pg* LPS were detected by cytokine antibody array. Each cytokine is represented by duplicate spots in the locations shown. (**B**) Densitometric analyses of cytokine spots shown in A. Each column represents fold change of optical intensity for average of two spots detected. (**C**) The mean protein level of IL-6 secreted in the culture medium of MG6 cells was determined by ELISA. MG6 cells were pre-treated with hBD3 (1 μM) for 2 h and then treated with *Pg* LPS (10 μg/mL) for 6 h. (**D**) The mean mRNA expression level of IL-6 in MG6 cells was determined by qRT-PCR analysis and β-actin mRNA served as the internal control for the normalization. MG6 cells were pre-treated with hBD3 (1 μM) for 2 h and then treated with *Pg* LPS (10 μg/mL) for 6 h. (**E**) The mean mRNA expression level of IL-6 in HMC3 cells was determined by qRT-PCR analysis and β-actin mRNA served as the internal control for the normalization. HMC3 cells were pre-treated with hBD3 (1 μM) for 2 h and then treated with *Pg* LPS (30 μg/mL) for 3 h. (**C–E**) The data are presented as the mean ± SE of three independent experiments, and *p* values were calculated using a one-way ANOVA with a post-hoc Tukey's test. A value of $p < 0.05$ was considered to indicate statistical significance.

IL-6 is a pleiotropic cytokine that regulates multiple biological processes, including the development of the nervous and hematopoietic systems, acute-phase responses, inflammation, and immune responses [31]. On the other hand, G-CSF is essential for the protective inflammatory response and for maintaining the balance between anti- and pro inflammatory reactions in an inflammatory condition [32]. We then focused on the *Pg* LPS-induced expression of IL-6. Effects of hBD3 on the *Pg* LPS-induced IL-6 expression in MG6 cells were examined at both the protein and transcriptional levels using enzyme-linked immunosorbent assay (ELISA) and quantitative real-time polymerase chain reaction (qRT-PCR), respectively. We examined thee effects of only hBD3 on the expression of IL-6, because the production of NO and IL-6 in microglia uses common signaling pathways. hBD3 (1 μM) significantly inhibited *Pg* LPS-induced expression of IL-6 in MG6 cells at both protein (Figure 2C) and mRNA (Figure 2D) levels. To further generalize the above observations, we used human embryonic microglia clone 3 (HMC3). The responsiveness of HMC3 cells to stimulation with *Pg* LPS (10 μg/mL) are relatively weak when compared with that of MG6 cells. *Pg* LPS (10 μg/mL) could induce expression of IL-6 mRNA in HMC3 cells, but there was a significant variation in its effectiveness. We then used *Pg* LPS with the concentration of 30 μg/mL. *Pg* LPS (30 μg/mL for 3 h) induced a significant increase in the expression of IL-6 mRNA in HMC3 cells (Figure 2E). hBD3 (1 μM) significantly inhibited *Pg* LPS-induced mRNA expression of IL-6 in HMC3 cells (Figure 2E).

2.4. Effects of hBD3 on Pg LPS-Induced Nuclear Translocation of p65 and Proteolytic Degradation of IκBα

The promoter regions of both iNOS and IL-6 genes have a putative NF-κB binding site [33,34]. Therefore, abrogation of NF-κB activation is one of the plausible mechanisms underlying the inhibitory effect of hBD3 on the *Pg* LPS-induced NO and IL-6 production in MG6 cells. We thus examined effects of hBD3 on the *Pg* LPS-induced p65 nuclear translocation and degradation of IκBα, an endogenous inhibitor of NF-κB. hBD3 (1 μM) significantly inhibited the *Pg* LPS-induced p65 nuclear translocation in MG6 cells (Figure 3A,B). Furthermore, the mean level of IκBα was significantly decreased in MG6 cells after treatment with *Pg* LPS (10 μg/mL) at 6–12 h (Figure 3C). hBD3 (1 μM) significantly inhibited *Pg* LPS-induced delayed reduction of IκBα (Figure 3D,E).

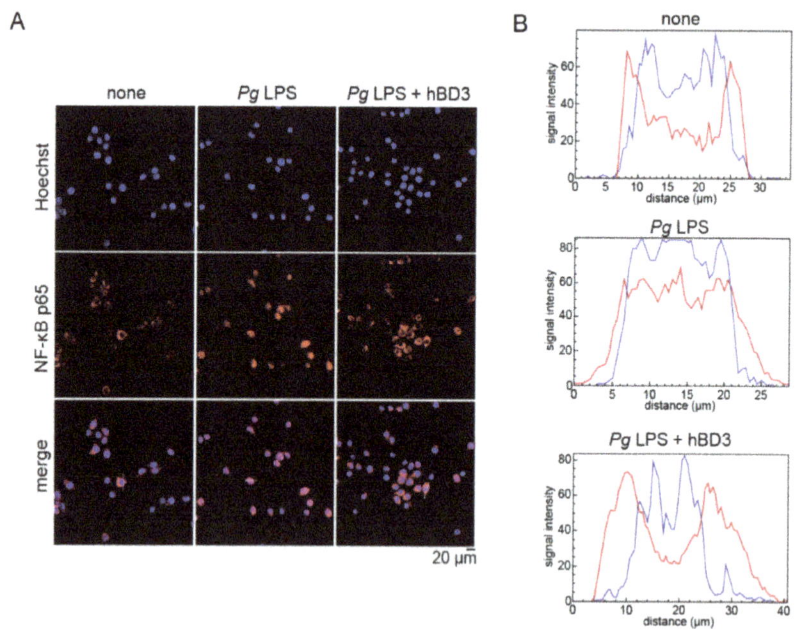

Figure 3. *Cont.*

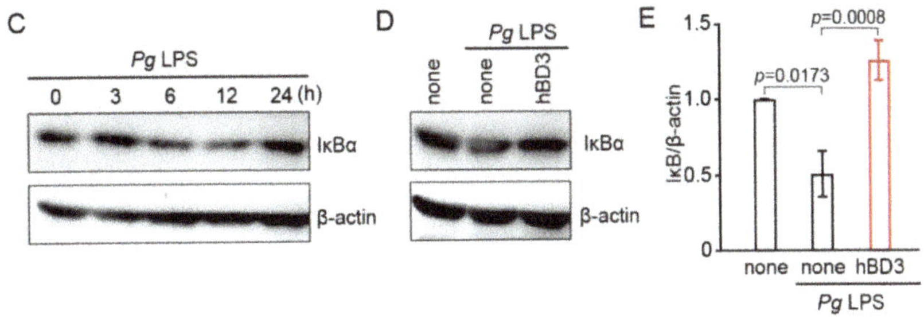

Figure 3. Effects of hBD3 on the *Pg* LPS-induced nuclear translocation of NF-κB p65 and degradation of IκBα. (**A**) Immunofluorescence CLMS images indicating the nuclear translocation of p65 (red) in MG6 cells with Hoechst-stained nuclei (blue) at 3 h after stimulation with *Pg* LPS (10 μg/mL) in the presence or absence of hBD3 (1 μM). (**B**) The typical cells were analyzed by line plot profile to show the cytosol and nuclear location of p65. (**C**) The protein expression of IκBα in MG6 cells after stimulation with *Pg* LPS (10 μg/mL) at the indicated time. (**D**) The protein expression of IκBα in MG6 cells after stimulation with *Pg* LPS (10 μg/mL) for 6 h in the presence or absence of hBD3 (1 μM). (**E**) The mean intensity of IκBα, which were detected by the immunoblots shown in (**D**), were measured and normalized against the signal of β-actin. They are shown here relative to the values in untreated cells. The data are presented as the mean ± SE of five independent experiments from MG6 cells after stimulation with *Pg* LPS for 6–24 h in the presence or absence of hBD3, and *p* values were calculated using a one-way ANOVA with a post-hoc Tukey's test. A value of $p < 0.05$ was considered to indicate statistical significance.

2.5. Inhibitory Effects of hBD3 on Activities of Human CatB and Cathepsin L (CatL)

We previously reported that CatB played a critical role in AD-like pathologies, including microglia-mediated neuroinflammation following a chronic systemic exposure of *Pg* LPS [12]. Furthermore, growing evidence suggests that inflammatory and oxidative responses of macrophages/microglia at the late phase depends on the lysosomal proteolytic machinery [15–20]. We thus examined the effects of lysosomal protease inhibitors on the *Pg* LPS-induced expression of IL-6 mRNA. *Pg* LPS-induced expression of IL-6 mRNA in MG6 cells was significantly inhibited by E64d (50 μM), a broad cysteine protease inhibitor, and CA-074Me (30 μM), a known specific inhibitor for CatB, but not by pepstatin A (30 μM), an aspartic protease inhibitor (Figure 4A).

Therefore, possible inhibitory effects of hBD3 on the enzymatic activities of CatB and CatL, typical cysteine cathepsins, were examined by use of cell-permeable, fluorescently labeled substrates, z-Arg-Arg-cresyl violet and z-Phe-Arg-cresyl violet, respectively. The fluorescent cresyl violet group of which is designed to be dequenched upon the cleavage of dipeptides by CatB or CatL. The enzymatic activities of CatB and CatL, visualized as punctuate bright signals in MG6 cells, were increased after stimulation with *Pg* LPS (10 μg/mL). hBD3 (1 μM) markedly reduced the increased fluorescent signals of both CatB and CatL after treatment with *Pg* LPS (Figure 4B,C), suggesting that hBD3 penetrated into MG6 cells to inhibit their enzymatic activities in the lysosomes.

To quantify the inhibitory effects of hBD3 on activities of CatB and CatL, we examined effects of hBDs on the enzymatic activities of recombinant human CatB and CatL using the cleaved synthetic AFC based peptide substrates, Ac-Arg-Arg-AFC and Ac-Phe-Arg-AFC, respectively. hBD3 (1 μM) significantly inhibited enzymatic activities of recombinant human CatB and CatL (Figure 4D,E). In contrast, either hBD1, 2 or 4 showed no significant effect on their enzymatic activities, suggesting that their lack of inhibitory effect on the CatB and CatL activities is responsible for their inability of inhibiting *Pg* LPS-induced oxidative and inflammatory responses.

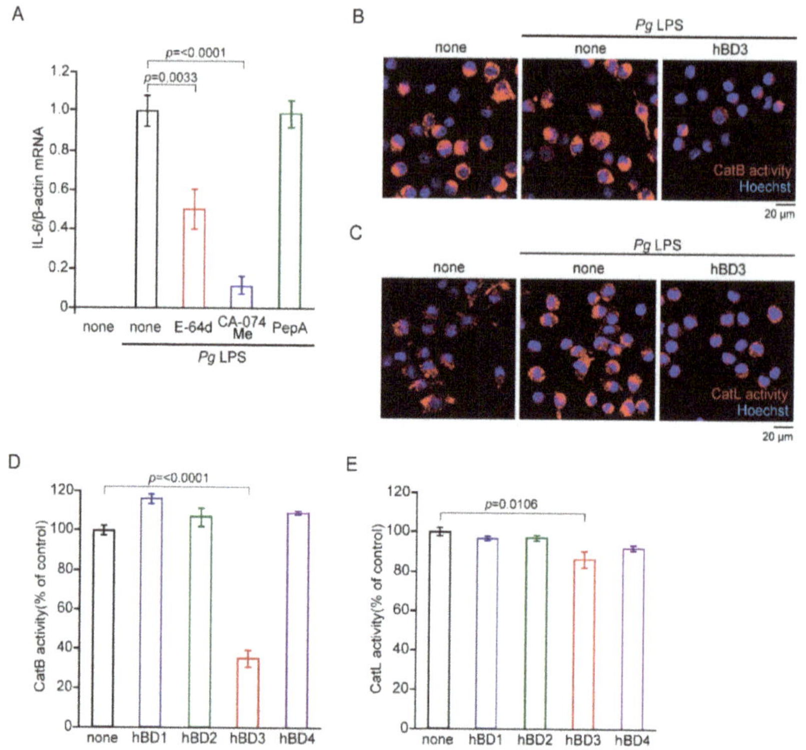

Figure 4. Effects of hBD3 on the enzymatic activity of human CatB and CatL. (**A**) The mean level of *Pg* LPS-induced IL-6 mRNA was significantly reduced by E64d (50 μM) and CA-074Me (30 μM), but not by pepstatin A (30 μM). (**B**,**C**) CLSM images of z-Arg-Arg-cresyl violet (**B**) and z-Phe-Arg-cresyl violet (**C**) in non-treated and *Pg* LPS-treated MG6 cells in the absence and presence of hBD3 (1 μM). (**D**,**E**) The mean enzymatic activity levels of human CatB (**D**) and human CatL (**E**) were measured by measured by AFC based peptide substrates to release AFC. (**A**,**D**,**E**) The data are presented as the mean ± SE of three independent experiments, and *p* values were calculated using a one-way ANOVA with a post-hoc Tukey's test. A value of $p < 0.05$ was considered to indicate statistical significance.

3. Discussion

In this study, we sought to determine whether hBDs could reduce *Pg* LPS-induced oxidative and inflammatory responses in microglia. Our observations indicate that hBD3 with a relatively low concentration (i.e., 1 μM) significantly suppressed *Pg* LPS-induced production of NO and iNOS by MG6 cells. Furthermore, hBD3 also significantly suppressed *Pg* LPS-induced expression of IL-6 by MG6 cells in both mRNA and protein levels. The inhibitory effect of hBD3 on *Pg* LPS-induced mRNA expression of IL-6 was also observed in HMC3 cells. In contrast, either hBD1, hBD2 or hBD4 failed to inhibit the production of NO and IL-6 by MG6 cells.

Chronic systemic exposure of *Pg* LPS induces AD-like pathology, including microglia-mediated neuroinflammation and cognitive decline in both mice and rats [12–14]. There is increasing evidence of lysosome-related degradation of IκBα, an endogenous inhibitor of NF-κB, at the late stage of inflammation [15–20]. Several studies have revealed that hBD3 inhibits *E. coli* LPS-induced inflammatory responses of the macrophage cell line RAW 264.7 cells by down regulation of NF-κB and inflammatory gene transcription [35], or directly suppressing the degradation of phosphorylated IκBα [36]. In this study, hBD3 significantly suppressed the *Pg* LPS-induced p65 nuclear translocation. It has been reported that standard

Pg LPS, which contains other bacterial components including lipoproteins, could activate the NF-κB/STAT3 signaling pathways in BV-2 microglial cells through Toll-like receptor 2/4 (TLR2/4) [37,38]. Therefore, it is considered that hBD3 inhibits *Pg* LPS-induced p65 nuclear translocation through the IκBα degradation and subsequent NF-κB activation at the late stage of neuroinflammation, which may drive delayed and/or chronic state.

hBD3 is highly membrane-permeable, because it has a cationic amino acid sequence in the C-terminus and a hydrophobic amino acid sequence in the N-terminus (Supplementary Figure S1) [36,39]. Therefore, the potent cell-penetrating activity of hBD3 is partly responsible for the antioxidant and anti-inflammatory effects in microglia. In this study, *Pg* LPS-induced expression of IL-6 by MG6 cells was significantly inhibited by E64d, a cysteine cathepsin inhibitor, and CA-074Me, a known specific inhibitor for CatB, but not by pepstatin A, a specific aspartic protease inhibitor. Moreover, it has been reported that hBD2 and hBD3 could be substrates for cysteine cathepsins such as CatB and CatL [40]. These observations further prompted us to examine the effects of hBD3 on the enzymatic activities of cysteine cathepsins. As expected, hBD3 (1 μM) significantly inhibited the enzymatic activities of recombinant human CatB and CatL, probably through the suicide substrate-based inhibition mechanism. hBD3 (1 μM) also inhibited enzymatic activities of both CatB and CatL in MG6 cells by use of cell-permeable fluorescently labeled substrates of CatB and CatL (Figure 4B,C). The punctate fluorescent signals, which indicate enzymatic activity in the endosomes/lysosomes, were markedly reduced by treatment with hBD3, suggesting that hBD3 can penetrate into the endosomes/lysosomes. In contrast, either hBD1, 2 or 4 showed no significant effects on their activities. On the basis of these observations, their lack of an inhibitory effect on the CatB and CatL activities, rather than their weak membrane permeability, is responsible for the inability of inhibiting *Pg* LPS-induced oxidative and inflammatory responses. Therefore, it may be concluded that hBD3 suppresses the *Pg LPS*-induced oxidative and inflammatory responses in microglia through inhibiting CatB and CatL, which enzymatic activities are necessary for degradation of IκBα at the late stage of inflammation. The present results are consistent with our previous findings. Both the genetic deficiency of CatB and its pharmacological inhibition by CA-074Me ameliorated microglia-mediated neuroinflammation and cognitive impairment of *Pg* LPS-treated middle-aged mice [12]. Furthermore, CA-074Me significantly inhibited the *Pg* infection-induced Aβ peptide production by macrophages [41]. It is worthwhile to mention that CA-074Me with relatively high concentrations inhibits not only CatB, but also other cysteine cathepsins such as CatL in mammalian cells [42,43].

In the brain, astrocytes, epithelium of the choroid plexus and meningothelial cells express and likely secrete immuno-modulatory antimicrobial peptides, including hBD1 and hBD2, that could influence inflammation within the brain [23,26]. It has been reported that salivary lactoferrin is transferred into the brain by absorption from the sublingual mucosa, in which the favorable effects of salivary lactoferrin on brain will be expected via the sublingual mucosa [44]. hBDs including hBD3 are produced by the oral mucosa and salivary glands. Therefore, it is interesting to speculate that hBD3 can be also transferred into the brain through the sublingual route. The relationship between salivary glands and brain function might provide new insight into potential therapeutic approaches for brain neurological disorders [45]. Further study will be necessary to elucidate the absorption and transportation of salivary hBD3 into the brain.

4. Materials and Methods

4.1. Reagents

hBD1-4, E64d, CA-074Me and pepstatin A were purchase from Peptide Institute Inc. (Osaka, Japan). Standard *Pg* LPS was purchased from Invivogen (San Diego, CA, USA).

4.2. Cell Culture

The c-myc-immortalized mouse microglial cell line, MG6 (RIKEN Cell Bank), was maintained at 37 °C in DMEM containing 10% heat-inactivated fetal bovine serum (FBS,

ICN Biomedicals) supplemented with 3.5 mg/mL glucose, 100 μM β-mercaptoethanol, 10 μg/mL insulin, 100 μg/mL streptomycin, and 100 U/mL penicillin (ThermoFischer Scientific, Walthan, MA, USA). HMC3 (CRL-3304) cells obtained from American Type Culture Collection (ATCC, Manassas, VA, USA) were maintained at 37 °C in DMEM containing 10% heat-inactivated FBS supplemented with 2 mg/mL glucose, 2 mM glutamine, 100 units/mL penicillin, and 100 μg/mL streptomycin.

4.3. Assay for Cell Survival

Cell viability was measured by WST-8 conversion to water-soluble formazan by mitochondrial dehydrogenase (Cell Counting Kit-8, Dojindo, Japan) following the protocol provided by the manufacturer. Briefly, WST-8 was added to MG6 cells in 96-well plates (3×10^4 cells/well) and incubated at 37 °C for 1 h. The optical density was read at a wavelength of 450 nm with a microplate reader.

4.4. Assay of NO Production

MG6 cells cultured in a 96-well plate (3×10^4 cells/well) were pre-treated with hBD1, 2, 3, and 4 at 100 nM and 1 μM for 2 h and then exposed to *Pg* LPS (10 μg/mL) for 24 h. The supernatant was transferred to a new 96-well plate with Griess reagent (Griess Reagent Kit; Dojindo, Kumamoto, Japan) and incubated at room temperature for 15 min. The amount of nitrite, the major NO metabolite, in the cytosol was measured spectrophotometrically using a microplate reader at a wavelength of 540 nm.

4.5. qRT-PCR

Total RNA was extracted using TRIzol regent (ThermoFischer Scientific) according to the manufacturer's instructions. 1–2 μg of total RNA was used for cDNA synthesis using the ReverTra Ace (TOYOBO, Osaka, Japan). After a denaturation step at 95 °C for 1 min, 45 cycles of denaturation at 95 °C for 15 s, annealing at 61 °C (iNOS (mouse)), 57 °C (IL-6 (mouse)), 55 °C (β-actin (mouse)), or 52°C (IL-6 (human) and β-actin (human)) for 15 s and extension at 72 °C for 29 s (iNOS (mouse)), 20 s (IL-6 (mouse)), or 1 min (β-actin (mouse), IL-6 (human), β-actin (human)) were carried out to amplify each cDNA. The cDNA was amplified in triplicate using a Thunderbird next SYBR qPCR mix (TOYOBO) with Real-Time PCR System TP800 (TaKaRa-bio, Shiga, Japan) or TP990 (TaKaRa-bio). The sequences of primer pairs were described as follows: iNOS (mouse): 5′-CTCGGAACTGTAGCACAGCAC-3′ and 5′-AAGACCAGAGGCAGCACATCAA-3′; IL-6 (mouse): 5′-TCTTGGGACTGATGCTGGTG-3′ and 5′-GCA CAA CTC TTT TCT CATTTCC-3′; β-actin (mouse): 5′-GGCATTGTGATGGACTCCG-3′ and 5′-GCT GGA AGGTGGACAGTGA-3′; IL-6 (human): 5′-GAACTCCTTCTCCACAGCG-3′ and 5′-TTTTCTGCCAGTGAATCTTT-3′; β-actin (human): 5′-CAT CTCTTGCTCGAAGT CCA-3′ and 5′-ATCATGTTTGAGACCTT CAACA-3′. For data normalization, an endogenous control (β-actin) was assessed to control for the cDNA input, and the relative units were calculated by a calibration curve method. All qRT-PCR experiments were repeated three times, and the results are presented as the means of the ratios ± SE.

4.6. Cytokine Antibody Array

MG6 cells were exposed to *Pg* LPS (10 μg/mL) at 37 °C for 24 h. Culture supernatant were then analyzed with cytokine antibody array by using the C-Series Mouse Neuro Discovery Array C1 (RayBiotech, Norcross, GA, USA) according to the manufacturer's instructions. Signal intensities were detected with LAS-4000 (Fujifilm, Tokyo, Japan) and analyzed with Image Lab 6.0.1 software program (Bio-Rad, Hercules, CA, USA). Biotin-conjugated immunoglobulin G served as a positive control at six spots, where it was used to identify membrane orientation and to normalize the results from different membranes that were being compared. Following, 23 molecules containing cytokines and chemokines were detected and semi quantified on membrane arrays containing different cytokine antibodies: Fas ligand, fractalkine, G-CSF, interferon-γ, IGF-1, IL-10, IL-1α, IL-1β, IL-4, IL-6, keratinocyte chemoattractant, LPS-induced CXC chemokine, MCP-1, MIP-1α, macrophage-colony stimulating factor, matrix metaloprotease-2, matrix metaloprotease-3, receptor for advanced glycation end products, thymus and activation-regulated chemokine, stromal cell-derived factor-1α, transforming growth factor-β, tumor necrosis factor-α, VEGF-A.

4.7. ELISA

The amounts of secreted IL-6 from MG6 cells were measured by ELISA (RayBiotech Inc., Norcross, GA, USA) following the protocol provided by the manufacturer. The absorbance at 450 nm was determined using a microplate reader.

4.8. Immunoblotting Analyses

MG6 cells were seeded on a petri dish at a density of 7.5–10 $\times$ 10^5 cells/dish for 1–2 days. After treatment with *Pg* LPS for 3, 6, 12, 24 h, the cells were then lysed with RIPA buffer, consisting of 10 mM Tris-HCl, pH 7.5, 1% (*v*/*v*) NP-40, 0.1% (*w*/*v*) sodium deoxycholate, 0.1% (*w*/*v*) SDS, 150 mM NaCl, 1 mM EDTA, protease inhibitor cocktail (Nacalai Tesque, Kyoto, Japan), and then cell lysates were subjected to 10% (*w*/*v*) sodium dodecyl sulfate-polyacrylamide gel electrophoresis. The proteins on the SDS–polyacrylamide gels were then transferred to nitrocellulose membranes. After blocking with Blocking one (Nacalai Tesque), the membranes were incubated with a rabbit anti-IκBα (Abcam, Cambridge, UK) and rabbit-anti β actin (GENETEX, Irvine, CA, USA) at 4 °C overnight. After being washed, the membranes were incubated with horseradish peroxidase (HRP)-labeled anti-rabbit IgG antibodies (GE Healthcare, Tokyo, Japan) for 1 h at room temperature. Subsequently, the membrane-bound HRP-labeled antibodies were detected using an Amasham ECL western blotting detection reagent and analysis system (GE Healthcare) with an imaging analyzer (LAS-4000, Fujifilm, Tokyo, Japan). Signal intensities were determined using the Image Lab 6.0.1 software program (Bio-Rad, Hercules, CA, USA).

4.9. Enzymatic Activity Assay of CatB and CatL

The enzymatic activities of CatB or CatL in the absence and presence of hBDs were measured by the cleaved synthetic 7-amino-4-trifluoromethylcoumarin (AFC) based peptide substrates, Ac-Arg-Arg-AFC and Ac-Phe-Arg-AFC, respectively (BioVision, Milpitas, CA, USA). The fluorescence intensities of AFC were measured by a microplate reader at an excitation wavelength of 400 nm and an emission wavelength of 505 nm.

4.10. Fluorescence Imaging of Enzymatic Activities of CatB and CatL

MG6 microglia were stained with the cell-permeable fluorescently labeled CatB substrate z-Arg-Arg-cresyl violet or CatL substrate z-Phe-Arg-cresyl violet according to the manufacturer's instructions (cv-CatB detection kit and cv CatL detection kit, Enzo Life Sciences, Inc., Farmingdale, NY, USA). Chamber slides containing the stained live cells were then mounted in PBS. Fluorescent images were taken using a confocal laser-scanning microscope (CLSM, FV1000, Olympus, Tokyo, Japan).

4.11. Immunostaining

MG6 cells were treated with *Pg* LPS (10 μg/mL) for 3 h in the absence and presence of hBD3 (1 μM) and were fixed with 4% paraformaldeyte. They were then incubated with rabbit anti-NF-κB p65 IgG (Abcam). After washing with PBS, cells were incubated with donkey anti-rabbit Alexa 555 (ThermoFisher Scientific, Waltham, MA, USA), then incubated with Hoechst and mounted in Vectashield anti fading medium (Vector Laboratories, Newark, CA, USA). Fluorescent images were taken using a confocal laser-scanning microscope (FV1000). The line plot profile was analyzed using Image J.

4.12. Statistical Analysis

The data are represented as the mean $\pm$ standard error (SE). Statistical analyses of the results were performed with one-way analysis of variance (ANOVA) with post hoc Tukey's test using the GraphPad Prism8 (GraphPad Software, Inc., San Diego, CA, USA) software package. $p < 0.05$ was considered to indicate a statistically significant difference.

5. Conclusions

hBD3 inhibited *Pg* LPS-induced oxidative and inflammatory responses in microglia through a direct inhibition of CatB and CatL. Both CatB and CatL might be involved in degradation of IκBα, resulting in a delayed but long-lasting activation of NF-κB pathways.

Supplementary Materials: The supporting information can be downloaded at: https://www.mdpi.com/article/10.3390/ijms232315099/s1.

Author Contributions: Conceptualization, H.N.; methodology, S.N. and H.N.; software, S.N; validation, S.N. and H.N.; formal analysis, E.I., S.M., Y.K. and S.N.; investigation, E.I., S.M., Y.K. and S.N.; resources, H.N.; data curation, E.I., S.M., Y.K. and S.N.; writing—original draft preparation, H.N.; writing—review and editing, E.I., S.M., Y.K., S.N. and H.N.; visualization, S.N. and H.N.; supervision, H.N.; project administration, H.N.; funding acquisition, S.N. and H.N. All authors have read and agreed to the published version of the manuscript.

Funding: This research was funded by the Science Research Promotion Fund from Promotion and Mutual Aid Cooperation for Private Schools of Japan (H.N.), JSPS KAKENHI Grant Numbers JP21K06383 (H.N.) and JP22K15287 (S.N.).

Institutional Review Board Statement: Not applicable.

Informed Consent Statement: Not applicable.

Data Availability Statement: Not applicable.

Acknowledgments: The authors would like to express their sincere thanks to Rin Omoto, Shiori Kanetsuki, Yuzuka Horio (Faculty of Pharmacy, Yasuda Women's University, Hiroshima) for their technical assistance.

Conflicts of Interest: The authors declare no conflict of interest.

Abbreviations

AFC: 7-amino-4-trifluoro-methylcpumarin; AD: Alzheimer's disease; ANOVA: one-way analysis of variance; CatB: cathepsin B; CatL: cathepsin L; CLMS: confocal laser-scanning microscope; ELISA: enzyme-linked immunosorbent assay; G-CSF: granulocyte growth factor; hBDs: human β-defensins; HMC3: human embryonic microglia clone 3; IGF: insulin growth factor; IL-6: interleukin-6; iNOS: inducible nitric oxide synthase; LPS: lipopolysaccharide; MCP-1: monocyte chemoattractant protein-1; MIP-1α: macrophage inflammatory protein-1α; NF-κB: nuclear factor-κB; NO: nitric oxide; *Pg*: *Porphyromonas gingivalis;* qRT-PCR: quantitative real-time polymerase chain reaction; SE: standard error; TLR: Toll-like receptor; VEGF-A: vascular endothelial growth factor-A.

References

1. Nakanishi, H.; Nonaka, S.; Wu, Z. Microglial cathepsin B and *Porphyromonas gingivalis* gingipains as potential therapeutic targets for sporadic Alzheimer's disease. *CNS Neurol. Disord. Drug Targets* **2020**, *19*, 495–502. [CrossRef] [PubMed]
2. Liu, Y.; Wu, Z.; Nakanishi, Y.; Ni, J.; Hayashi, Y.; Takayama, F.; Zhou, Y.; Kadowaki, T.; Nakanishi, H. Infection of microglia with *Porphyromonas gingivalis* promotes cell migration and an inflammatory response through the gingipain-mediated activation of protease-activated receptor-2 in mice. *Sci. Rep.* **2017**, *7*, 11759. [CrossRef] [PubMed]
3. Nonaka, S.; Nakanishi, H. Secreted gingipains from *Porphyromonas gingivalis* induce microglia migration through endosomal signaling by protease-activated receptor 2. *Neurochem. Int.* **2020**, *140*, 104840. [CrossRef]
4. Nonaka, S.; Kadowaki, T.; Nakanishi, H. Secreted gingipains from *Porphyromonas gingivalis* increase permeability in human cerebral microvascular endothelial cells through intracellular degradation of tight junction proteins. *Neurochem. Int.* **2022**, *154*, 105282. [CrossRef] [PubMed]
5. Dominy, S.S.; Lynch, C.; Ermini, F.; Benedyk, M.; Marczyk, A.; Konradi, A.; Nguyen, M.; Haditsch, U.; Raha, D.; Griffin, C.; et al. *Porphyromonas gingivalis* in Alzheimer's disease brains: Evidence for disease causation and treatment with small-molecule inhibitors. *Sci. Adv.* **2019**, *5*, eaau3333. [CrossRef]

6. Poole, S.; Singhrao, S.K.; Kesavalu, L.; Curtis, M.A.; Crean, S. Determining the presence of periodontopathic virulence factors in short-term postmorem Alzheimer's disease brain tissue. *J. Alzheimers Dis.* **2013**, *36*, 665–677. [CrossRef]
7. Ide, M.; Harris, M.; Stevens, A.; Sussams, R.; Hopkins, V.; Culliford, D.; Fuller, J.; Ibbett, P.; Raybould, R.; Thomas, R.; et al. Periodontitis and cognitive decline in Alzheimer's disease. *PLoS ONE* **2016**, *11*, e0151081. [CrossRef]
8. Holmer, J.; Eriksdotter, M.; Schultzberg, M.; Pussinen, P.J.; Buhlin, K. Association between periodontitis and risk of Alzheimer's disease, mild cognitive impairment and subjective cognitive decline: A case-control study. *J. Clin. Periodontol.* **2018**, *45*, 1287–1298. [CrossRef]
9. Kanagasingam, S.; Chukkapalli, S.S.; Welbury, R.; Singhrao, S.K. *Porphyromonas gingivalis* is a strong factor for Alzheimer's disease. *J. Alzheimers Dis. Rep.* **2020**, *4*, 501–511. [CrossRef]
10. Olsen, I.; Singhrao, S.K. Interaction between genetic factors, *Porphyromonas gingivalis* and microglia to promote Alzheimer's disease. *J. Oral. Microbiol.* **2020**, *12*, 1820834. [CrossRef]
11. Seymour, T.; Zhang, J. *Porphyromonas gingivalis* in the pathogenesis of Alzheimer's disease and its therapeutic target. *J. Explor. Res. Pharmacol.* **2022**, *7*, 45–53. [CrossRef]
12. Wu, Z.; Ni, J.; Liu, Y.; Teeling, J.L.; Takayama, F.; Collcutt, A.; Ibbett, P.; Nakanishi, H. Cathepsin B plays a critical role in inducing Alzheimer's disease-like phenotypes following chronic systemic exposure to lipopolysaccharide from *Porphyromonas gingivalis* in mice. *Brain Behav. Immun.* **2017**, *65*, 350–361. [CrossRef] [PubMed]
13. Zhang, J.; Yu, C.; Zhang, X.; Chen, H.; Dong, J.; Lu, W.; Song, Z.; Zhou, W. *Porphyromonas gingivalis* lipopolysaccharide induces cognitive dysfunction, mediated by neuronal inflammation via activation of the TLR4 signaling pathway in C57BL/6 mice. *J. Neuroinflam.* **2018**, *15*, 37. [CrossRef] [PubMed]
14. Hu, Y.; Li, H.; Zhang, J.; Zhang, X.; Xia, X.; Qiu, C.; Liao, Y.; Chen, H.; Song, Z.; Zhou, W. Periodontitis induced by *P. gingivalis*-LPS is associated with neuroinflammation and learning and memory impairment in sprague-dawley rats. *Front. Neurosci.* **2020**, *14*, 658. [CrossRef] [PubMed]
15. Colleran, A.; Ryan, A.; O'Gorman, A.; Mureau, C.; Liptrot, C.; Dockery, P.; Fearnhead, H.; Egan, L.J. Autophagosomal IκBα degradation plays a role in the long term control of tumor necrosis factor-α-induced nuclear factor-κB (NF-κB) activity. *J. Biol. Chem.* **2011**, *286*, 22886–22893. [CrossRef] [PubMed]
16. Criollo, A.; Chereau, F.; Malik, S.A.; Niso-Santano, M.; Mariño, G.; Galluzzi, L.; Maiuri, M.C.; Baud, V.; Kroemer, G. Autophagy is required for the activation of NFκB. *Cell Cycle* **2012**, *11*, 194–199. [CrossRef]
17. Wang, Y.R.; Qin, S.; Han, R.; Wu, J.C.; Liang, Z.Q.; Qin, Z.H.; Wang, Y. Cathepsin L play a role in quinolinic acid-induced NF-κB activation and excitotoxicity in rat striatal neurons. *PLoS ONE* **2013**, *8*, e75702. [CrossRef]
18. Ni, J.; Wu, Z.; Peterts, C.; Yamamoto, K.; Qing, H.; Nakanishi, H. The critical role of proteolytic relay through cathepsins B and E in the phenotypic change of microglia/macrophage. *J. Neurosci.* **2015**, *35*, 12488–12501. [CrossRef]
19. Nakanishi, H. Microglial cathepsin B as a key driver of inflammatory brain diseases and brain aging. *Neural. Regen. Res.* **2020**, *15*, 25–29. [CrossRef]
20. Nakanishi, H. Cathepsin regulation on microglial function. *Biochim. Biophys. Acta. Proteins Proteom.* **2020**, *1868*, 140465. [CrossRef]
21. Kamer, A.R.; Dasanayake, A.P.; Craig, R.G.; Glodzik-Sobanska, L.; Bry, M.; de Leon, M.J. Alzheimer's disease and peripheral infections: The possible contribution from periodontal infections, model and hypothesis. *J. Alzheimers Dis.* **2008**, *13*, 437–449. [CrossRef] [PubMed]
22. Welling, M.M.; Nabuurs, R.J.; van der Weerd, L. Potential role of antimicrobial peptides in the early onset of Alzheimer's disease. *Alzheimer's Dement.* **2015**, *11*, 51–57. [CrossRef] [PubMed]
23. Hao, H.N.; Zhao, J.; Lotoczky, G.; Grever, W.E.; Lyman, W.D. Induction of human β-defensin-2 expression in human astrocytes by lipopolysaccharide and cytokines. *J. Neurochem.* **2001**, *77*, 1027–1035. [CrossRef]
24. Kazakos, E.I.; Kountouras, J.; Polyzos, S.A.; Deretzi, G. Novel aspects of defensins' involvement in virus-induced autoimmunity in the central nervous system. *Med. Hypotheses* **2017**, *102*, 33–36. [CrossRef] [PubMed]
25. Williams, W.M.; Castellani, R.J.; Weinberg, A.; Perry, G.; Smith, M.K. Do β-defensins and other aminomicrobial peptides play a role in neuroimmune function and neurodegeneration? *Sci. World J.* **2012**, 905785. [CrossRef]
26. Nakayama, K.; Okamura, N.; Arai, H.; Sekizawa, K.; Sasaki, H. Expression of human β-defensin-1 in the choroid plexus. *Ann. Neurol.* **1999**, *45*, 685. [CrossRef]
27. Williams, W.M.; Torres, S.; Siedlak, S.L.; Castellani, R.J.; Perry, G.; Smith, M.A.; Zhu, X. Antimicrobial peptide β-defensin-1 expression is upregulated in Alzheimer's brain. *J. Neuroinflam.* **2013**, *10*, 127. [CrossRef] [PubMed]
28. Szekeres, M.; Ivitz, E.; Datki, Z.; Kálmán, J.; Pákáski, M.; Várhelyi, Z.P.; Klivényi, P.; Zadori, D.; Somogyvári, F.; Szolnoki, Z.; et al. Relevance of defensin β-2 and α defensins (HNP1–3) in Alzheimer's disease. *Psychiatry Res.* **2016**, *239*, 342–345. [CrossRef] [PubMed]
29. Semple, F.; Webb, S.; Li, H.-N.; Patel, H.B.; Perretti, M.; Jackson, I.J.; Gray, M.; Davidson, D.J.; Dorin, J.R. Human β-defensin 3 has immunosuppressive activity in vivo and in vitro. *Eur. J. Immunol.* **2010**, *40*, 1073–1078. [CrossRef] [PubMed]
30. Cui, D.; Lyu, J.; Li, H.; Lei, L.; Bian, T.; Li, L.; Yan, F. Human β-defensin 3 inhibits periodontitis development by suppressing inflammatory responses in macrophages. *Mol. Immunol.* **2017**, *91*, 65–74. [CrossRef]
31. Hirano, T. Interleukin 6 and its receptor: Ten years later. *Int. Rev. Immunol.* **1998**, *16*, 249–284. [CrossRef]
32. Chang, S.F.; Lin, S.S.; Yang, H.C.; Chou, Y.Y.; Gao, J.I.; Lu, S.C. LPS-induced G-CSF expression in macrophages is mediated by ERK2, but not ERK1. *PLoS ONE* **2015**, *10*, 0129685. [CrossRef] [PubMed]

33. Libermann, T.A.; Baltimore, D. Activation of interleukin-6 gene expression through the NF-κB transcription factor. *Mol. Cell Biol.* **1990**, *10*, 2327–2334. [CrossRef]
34. Xie, Q.W.; Kashiwabara, Y.; Nathan, C. Role of transcription factor NF-κB/ Rel in induction of nitric oxide synthase. *J. Biol. Chem.* **1994**, *269*, 4705–4708. [CrossRef]
35. Semple, F.; MacPherson, H.; Webb, S.; Cox, S.L.; Mallin, L.J.; Tyrrell, C.; Grimes, G.R.; Semple, C.A.; Nix, M.A.; Millhauser, G.L.; et al. Human β-defensin 3 affects the activity of pro-inflammatory pathways associated with MyD88 and TRIF. *Eur. J. Immunol.* **2011**, *41*, 3291–3300. [CrossRef] [PubMed]
36. Lee, J.Y.; Suh, J.S.; Kim, J.M.; Kim, J.H.; Park, H.J.; Park, Y.J.; Chung, C.P. Identification of a cell-penetrating peptide domain from human β-defensin 3 and characterization of its anti-inflammatory activity. *Int. J. Nanomed.* **2015**, *10*, 5423–5434. [CrossRef]
37. Nativel, B.; Couret, D.; Giraud, P.; Meilhac, O.; Lefebvre d'Hellencourt, C.; Viranaïcken, W.; Da Silva, C.R. *Porphyromonas gingivalis* lipopolysaccharides act exclusively through TLR4 with a resilience between mouse and human. *Sci. Rep.* **2017**, *7*, 15789. [CrossRef] [PubMed]
38. Qui, C.; Yuan, Z.; He, Z.; Chen, H.; Liao, Y.; Li, S.; Zhou, W.; Song, Z. Lipopolysaccharide preparation derived from *Porphyromonas gingivalis* induces a weaker immuno-inflammatory response in BV-2 microglial cells than *Escherichia coli* by differentially activating TLR2/4-mediated NF-κB/STAT3 signaling pathways. *Front. Cell. Infect. Microbiol.* **2021**, *11*, 606986. [CrossRef]
39. Pingel, L.C.; Kohlgraf, K.G.; Hansen, C.J.; Eastman, C.G.; Dietrich, D.E.; Burnell, K.K.; Srikantha, R.N.; Xiao, X.; Bélanger, M.; Progulske-Fox, A.; et al. Human β-defensin 3 binds to hemagglutinin B (rHagB), a non-fimbrial adhesin from *Porphyromonas gingivalis*, and attenuates a pro-inflammatory cytokine response. *Immunol. Cell Biol.* **2008**, *86*, 643–649. [CrossRef]
40. Taggart, C.C.; Greene, C.M.; Smith, S.G.; Levine, R.L.; McCray, P.B., Jr.; O'Neill, S.; McElvaney, N.G. Inactivation of human β-defensins 2 and 3 by elastolytic cathepsins. *J. Immunol.* **2003**, *171*, 931–937. [CrossRef]
41. Nie, R.; Wu, Z.; Ni, J.; Zeng, F.; Yu, W.; Zhang, Y.; Kadowaki, T.; Kashiwazaki, H.; Teeling, J.L.; Zhou, Y. *Porphyromonas gingivalis* infection induces amyloid-β accumulation in monocytes/macrophages. *J. Alzheimers Dis.* **2019**, *72*, 479–494. [CrossRef] [PubMed]
42. Bogyo, M.; Verhelst, S.; Bellingard-Dubouchaud, V.; Toba, S.; Greenbaum, D. Selective targeting of lysosomal cysteine proteases with radiolabeled electrophilic substrate analogs. *Chem. Biol.* **2000**, *7*, 27–38. [CrossRef] [PubMed]
43. Montaser, M.; Lalmanach, G.; Mach, L. CA-074, but not its methyl ester CA-074Me, is a selective inhibitor of cathepsin B within living cells. *Biol. Chem.* **2002**, *383*, 1305–1308. [CrossRef] [PubMed]
44. Hayashi, T.; To, M.; Saruta, J.; Sato, C.; Yamamoto, Y.; Kondo, Y.; Shimizu, T.; Kamata, Y.; Tsukinoki, K. Salivary lactoferrin is transferred into the brain via the sublingual route. *Biosci. Biotech. Biochem.* **2017**, *81*, 1300–1304. [CrossRef]
45. Contini, C.; Olianas, A.; Serrao, S.; Deriu, C.; Iavarone, F.; Boroumand, M.; Bizzarro, A.; Lauria, A.; Faa, G.; Castagnola, M.; et al. Top-down proteomics of human saliva highlights anti-inflammatory, antioxidant, and antimicrobial defense responses in Alzheimer disease. *Front. Neurosci.* **2021**, *15*, 743596. [CrossRef]

International Journal of *Molecular Sciences*

MDPI

Review

Chronic Ulcerative Stomatitis (CUS) as an Interdisciplinary Diagnostic Challenge: A Literature Review

Dominika Cichońska [1,*], Dominika Komandera [2], Magda Mazuś [2] and Aida Kusiak [1]

1 Department of Periodontology and Oral Mucosa Diseases, Medical University of Gdansk, 80-200 Gdansk, Poland
2 Student Research Group at the Department of Periodontology and Oral Mucosa Diseases, Medical University of Gdansk, 80-200 Gdansk, Poland
* Correspondence: dominika.cichonska@gumed.edu.pl

Abstract: Chronic ulcerative stomatitis (CUS) is a rarely reported disease affecting the oral cavity, most often affecting middle-aged Caucasian females. The aim of the present study is to present the diagnosis, differentiation, and interdisciplinary treatment of this rare disease. CUS is characterized by the presence of an oral erosive or ulcerative lesion. The autoimmune pathogenesis of CUS includes affecting the antigen's activity by DNA-breaking and protein-hydrolyzing enzymes. The stratified epithelium-specific antinuclear antibodies (SES-ANA) are associated with CUS development. Clinically, the lesions presented in oral mucosa might resemble an erosive form of oral lichen planus, whereas gingival lesions seem to be similar to desquamative gingivitis related to dermatological diseases manifested in the oral cavity. Patients often report subjective symptoms related to oral mucosa and general symptoms. Histopathological presentation of CUS is often non-specific and includes sub-epithelial separation from underlying connective tissue, atrophic epithelium, and inflammatory infiltrate with an increased number of plasma cells and lymphocytes. Direct immunofluorescence (DIF) might be used in CUS diagnostics. CUS generally remains nonsusceptible to corticosteroid treatments; however, antimalarial drugs and calcineurin inhibitors are more effective. Further research should be conducted in order to implement a diagnostic protocol and observe the long-term results of CUS management.

Keywords: CUS; chronic ulcerative stomatitis; oral mucosa disease

Citation: Cichońska, D.; Komandera, D.; Mazuś, M.; Kusiak, A. Chronic Ulcerative Stomatitis (CUS) as an Interdisciplinary Diagnostic Challenge: A Literature Review. *Int. J. Mol. Sci.* **2022**, *23*, 13772. https://doi.org/10.3390/ijms232213772

Academic Editor: Kenichi Kumagai

Received: 10 October 2022
Accepted: 7 November 2022
Published: 9 November 2022

1. Introduction

Chronic ulcerative stomatitis (CUS) is a rarely reported disease affecting oral cavities [1]. The first to describe CUS as a distinct entity were Parodi et al. in 1990 [1,2], although, in some reports, preliminary investigative statements were published [3–5]. CUS is the disease most often affecting middle-aged Caucasian females, but isolated cases of black and Asian females have also been reported [6]. The majority of the reported cases involved females; however, 10% of CUS cases appeared among male patients [7–9]. People aged 35–81 were most often affected by the disease, mainly in the fifth and sixth decades of life [7,8,10]. The average age of the patients diagnosed with CUS was 59 years [7,9–11].

CUS is characterized by the presence of oral erosive or ulcerative lesions that display distinctively unique direct and indirect immunofluorescence patterns [7,12]. Not only oral mucosa but also the skin can be affected [6,9]. Presumably, 22.5% of cases involve cutaneous tissues [6]. Chronic ulcerative stomatitis is a debilitating condition, its definition consists of chronic oral ulcers and erosions, which can be surrounded by subtle white reticular striations [1,8]. The clinical similarity of CUS to other oral diseases might be the cause of frequent misdiagnosis and such a small number of described cases [2,10]. After more than 29 years, there are still less than 100 patients reportedly with diagnosed CUS [1]. Azzi et al. highlighted that the mean age of the disease development usually differs from the age of diagnosis. Considering this, the mean diagnostic delay was 30 years, but the most extreme delay was even 30 years, which might be a result of the fact that chronic ulcerative stomatitis still remains a poorly understood disease [1,6,8]. CUS is rarely reported; however this does not enable the consideration of CUS as a rare disease [1].

The aim of the present study is to present the diagnosis, differentiation, and interdisciplinary treatment of this rarely reported disease. Therefore, a descriptive review of the literature on the pathogenesis, diagnosis and treatment strategies has been performed.

2. The Autoimmune Pathogenesis

The pathogenesis of chronic ulcerative stomatitis has been investigated since 1990 when Parodi et al. analyzed sera from patients and found circulating antibodies there that targeted a mammalian epithelial antigen. Due to affecting the antigen's activity by DNA-breaking and protein-hydrolyzing-enzymes, it was assumed that the antigen was a multimolecular, non-histonic DNA-protein complex [1,6]. In the meantime, Jaremko et al. described antinuclear antibodies (ANA) associated with the CUS and identified them as stratified epithelium-specific ANA (SES-ANA) [13]. These antibodies were found in both in vivo and in vitro studies [1,13].

Lee et al. were the first to identify the main antigen which was involved in CUS, a multimolecular 70 kDa epithelial nuclear protein, which they called "chronic ulcerative stomatitis protein" (CUSP) [1,14]. Sequencing the cDNA for CUSP autoantigen revealed that CUSP was homologously similar to the p53 tumor suppressor gene and p73 putative tumor suppressor gene and was a splicing variant of the p53-like rat KET gene [7,14,15]. Therefore, p53 was considered a unique protein for a long time, but in 1997 p63 and p73 were identified as the same family members [16,17]. The p63 gene is located on chromosome 3q26-28 and encodes six p53-homologous proteins [6,16,17]. The structural similarity of the p63 and p73 proteins to the p53 proteins applies to the N-terminal transactivation domain, a central DNA-binding protein, and an oligomerization domain close to the C-terminus [6,16–18]. Moreover, each of the p53, p63, and p73 genes use the molecular splicing promoter in different ways, which consequently leads to the formation of various isoforms possessing distinct functional activities. The similarity in the DNA sequences, specifically in the DNA binding domain, enables p53, p63, and p73 to transactivate some of one other's target genes and to regulate their expression. All of these proteins can interact between themselves [18,19]. The three p53-homologous proteins (TAp63α, TAp63β, and TAp63γ) contain a transactivation domain in N-terminus, such as the p53 protein [6,15–17,20]. On the contrary, the other three proteins (ΔNp63α, ΔNp63β, and ΔNp63γ) do not have the N-terminal transactivation domain; thus, they are restricted to the epithelium [6,15,17].

Ebrahimi et al. named CUSP as a ΔNp63α protein because it appeared to be an isoform of p63 [1,16]. The p63 exerts a multifaceted effect on embryogenesis, limb morphogenesis, and also the phenotype and development of stratified squamous epithelia, adnexa, teeth, and glands. It is essential for the differentiation, maintenance of proliferative potential, integrity, and apoptotic epithelial injury through the p53 pathway, which may indicate the importance of the presence of the p63 family protein in the oral mucosa due to the constant exfoliation of epithelial cells and a tendency to minor injuries [4,11,18]. ΔNp63α is normally present in the nuclei of the basal and parabasal cells in the progenitor cell compartment of the stratified squamous epithelium and remains the target to autoantibodies with a stratified epithelial specific-antinuclear antibody (SES-ANA) pattern [18–20]. ΔNp63α poses an impact on the maintenance of epithelial integrity and homeostasis by regulating the expression of cell-to-cell and cell-to-basement membrane adhesion molecules, which affect epithelial development and regeneration [12,19]. Furthermore, ΔNp63α can block the function of the p53 protein [21].

The very first to describe SES-ANA were Jaremko et al. [13]. These antinuclear antibodies were found circulating in the sera of patients with CUS and were observed only in stratified epithelial substrates and indirect immunofluorescence studies [22]. Furthermore, ANA was not detectable in the muscle or fibrous connective tissue nuclei, conventional ANA substrates, such as the human neoplastic Hep-2 cell line, rodent liver, or kidney substrates. The ANA pattern appears not only on the perilesional mucous membrane but also on normal mucous membranes and skin [7,13]. Solomon et al. described the autoimmune response in CUS patients and showed the presence of IgG antibodies, which are bound to the ΔNp63α antigen. Therefore, among 52% of the patients, circulating IgA antibodies were found [1,6,18]. No positive correlation was observed between patients having both circulating IgG and IgA antibodies and more severe courses of the disease, although patients with mucous membrane pemphigoid and dual circulating IgG and IgA antibodies presented a more severe response to the disease. Future studies are necessary to verify if dual antibodies can also pose an impact on the severity of CUS [6,18].

The highly likely patomechanism of chronic ulcerative stomatitis is that SES-ANA interfere with the normal function of the CUSP protein (ΔNp63α) of keratinocytes in the basal and parabasal cell layers of the oral stratified epithelium [4,19]. Due to an intracellular and intranuclear penetration, IgG SES autoantibodies attach to the ΔNp63α of keratinocytes, which cause the detachment from the basal membrane and from each other. Since CUSP is an anti-apoptotic protein, the inhibition of the action of the CUSP by auto-antibody binding leads to apoptotic epithelial injury through the p53 pathway. Clinically, we can observe this as erosions and ulcerations, which are the hallmark of CUS [1,4,10,11,19]. Azzi et al. noted that we do not have a certainty whether the CUS is caused by pathogenic hyperactive IgG autoantibodies binding to ΔNp63α, or if the autoimmune response is handled by physiological IgG reacting to an ΔNp63α overexpression, secondary to the T cell-induced damage to the basal cell layer of the epithelium and to an increase in pro-apoptotic processes [6].

Therefore, some authors have assumed that CUS is not a distinct disease but is recognized as an oral lichen planus variant. This assumption was made due to the fact that autoimmunity directed in ΔNp63α might also be a mechanism involved in LP-epithelial cell damage, which can define CUS as a variant of LP [1,19,23]. Antibodies characteristic of CUS can also be found in LP patients, which was confirmed by Cozzani et al., but some authors claimed that it seems to be an epiphenomenon, which should not be the base of the diagnosis of oral lichen planus (OLP) [1,23]. Contrarily, the other authors believe that chronic ulcerative stomatitis is a hyper-reactive form of OLP, which consists of cytotoxic damage within the basal cells of the epithelium caused by T lymphocytes and the B-cell response to the ΔNp63α antigen. Azzi et al. even proposed a change in the CUS name due to uncertainty if chronic ulcerative stomatitis was a distinct disease [6].

3. Clinical Symptoms

The condition generally manifests as non-healing, erosive, or as an ulcerative lesion with subtle white reticular striations [6,13,24]. The most frequent clinical presentations of CUS in the oral cavity are erosions, white lesions, erythema, and ulcerations. Lesions appear on the tongue, which is usually the most common location, followed by the buccal mucosa and the gingiva [7,9,10,12]. The gingival lesions often appear in the form of desquamative gingivitis, which arises from epithelial sloughing due to even minor manipulation of tissue. This can also resemble erosive oral lichen planus (OLP), mucous membrane pemphigoid (MMP), or pemphigus vulgaris, with non-specific lesions or the presence of lichenoid white plaques or striae [6,10,12]. Rarely the hard palate, lingual and labial mucosa, and lower lip might also remain a place of appearance for CUS lesions [7,9,12]. Regardless of its definition, CUS can also be manifested extra-orally, affecting other mucous membranes, skin, hair, and nails [25]. There were also a few reports describing ocular manifestations, such as cicatricial conjunctivitis and ectropion. It was reported that oral lesions can also be accompanied by gluten-dependent enteropathies and genital lesions [6,21].

Regarding the appearance of the oral cavity lesions of CUS, they are mostly symmetrical and might resemble lichenoid lesions, which appear as shallow, irregular ulcerations with abbreviated or vaguely formed peripheral keratotic striae; however, lesions can also present a non-specific clinical picture [6,25,26]. The healing of those lesions does not involve scarring [11]. A lot of authors emphasize that progressive painful erythematous gingival lesions, with large, tender erosions in CUS, which can be indistinguishable from OLP, lichenoid stomatitis, MMP, dermatitis herpetiformis, linear IgA disease, pemphigus vulgaris, erythema multiforme, pyostomatitis vegetans, and epidermolysis bullosa acquisita [7–9,11,27–29].

Additional reported signs of CUS were the varying severity of desquamation, xerostomia, vesiculation, or positive gingival Nikolsky's sign. Periodic symptom exacerbations and remissions are often observed [7]. Clinically, the lesions presented in oral mucosa might resemble an erosive form of the oral lichen planus, whereas gingival lesions seem to be similar to desquamative gingivitis related to dermatological diseases manifested in the oral cavity [4,6,30]. Clinical symptoms in CUS are mostly symmetrical and might present an OLP appearance, including white striae departing from the borders of the ulcers observed in 60% of cases [31]. According to conducted research, all patients presented at least one clinically evident CUS symptom, and in most of the cases, more than one clinical sign was observed [3,6,7,10,12,20,25,30–39]. The lesions' localization in the oral cavity involved mainly buccal mucosa (nearly 70%) and gingiva (over 50%). Less frequently, lesions were observed on the tongue, hard palate, and labial mucosa. The buccal mucosa and gingiva lesions were more likely to be independently existing lesions. Therefore, lesions in less frequent locations were mostly associated with supplementary lesions in other distinct locations [4,6].

Patients suffering from CUS are affected by painful remitting ulcerations, which are characterized by episodes of remission and exacerbation [8,12]. Patients often report subjective symptoms related to the oral mucosa, such as discomfort or pain and a burning or stinging sensation. General symptoms include nervousness, fatigue, malaise, depression, apathy, and sleeplessness [7,8,11]. Some of the patients reported difficulties in food intake, especially sweet or salty, and drinking hot or cold drinks, which can lead to weight loss in a group of patients [1,7,8]. The inability to maintain proper oral hygiene due to strong pain was also reported by some patients diagnosed with CUS [40].

4. Histopathological Presentation

The histopathological presentation of CUS is often non-specific. According to conducted research, over half of histopathological results were classified as "non-specific mucositis", and nearly half of them were misdiagnosed with lichenoid features [6,41].

The histopathological presentation of CUS was observed as a sub-epithelial separation from the underlying connective tissue, atrophic epithelium, and inflammatory infiltrate with an increased number of plasma cells and lymphocytes. A mixed infiltrate of T-lymphocytes and plasma cells was more specific to chronic ulcerative stomatitis. OLPs' infiltrate is typically exclusive out of the T-lymphocytes, and its location is limited within a superficial layer of lamina propria [6,24,41–43]. Some lesions may present the classic intense, 'band-like' inflammatory infiltrate, that is limited to the superficial lamina propria at the interface with the overlying epithelium and a sharply defined deep edge. However, in some CUS cases, a uniform infiltrate extending in some areas into the deeper lamina propria and producing an irregular or hazy deep edge was also observed [6]. The only histological feature that was observed in all CUS cases was the hydropic degeneration of the basal cell layer [6,37]. CUS and erosive oral lichen planus (OLP) manifest in histopathology as an immunological reaction with lichenoid features and a 'band-like' inflammatory infiltrate. In CUS predominates, an admixture of T lymphocytes and plasma cells, and in OLP, a predomination of T lymphocytes was observed. However, the lymphocytic infiltrate is not a reliable method for distinguishing CUS from OLP due to the fact that an overlap of the lymphocytic subset is commonly observed and also might be related to other oral mucosa diseases [6,9]. In some CUS cases, the deposition of fibrinogen was reported and described as a fluorescence outlining the basement membrane zone with irregular extensions into the superficial lamina propria, yielding a shaggy appearance. However, further investigation is required to evaluate if a fibrinogen deposition might be perceived as a diagnostic criterion for CUS [6,35].

CUS diagnostics should also include immunofluorescence (IF) microscopy, which is a well-established technique used for the detection of a wide variety of antigens in tissues or on cells in suspension and remains a helpful supplement for the accurate diagnosis of immune-mediated dermatological disorders [44,45].

The direct immunofluorescence (DIF) test for tissue-bound autoantibodies provides a verified adjunct for the diagnosis of dermatological bullous autoimmune disorders, enabling the classification of histologically similar conditions which differ in their treatment protocols and prognosis [45]. In addition, DIF combined with histopathology might complete the clinical and histological examination in the diagnosis of a variety of other dermatological diseases, which include connective tissue disorders, vasculitides, and conditions, such as lichen planus or others. DIF is a one-step procedure that involves the application of fluoresceinated antibodies to a frozen section of the skin or mucous membrane and determines the deposition of the immunoreactants in the tissue [45]. So far, DIF remains a golden standard in CUS diagnosis [6,9]. Reviewed DIF tests yielded a positive result—presenting the characteristic SES-ANA speckled pattern located in the basal layer and the bottom three layers of the cells in nearly all cases [6,46,47]. Among less than half of the reviewed cases, fibrinogen deposition was observed. It was located along the basal membrane zone. In some cases, the adjunctive DIF signals were also observed for the IgA, C3, and IgM components [6,43].

Indirect immunofluorescence (IIF) is a method that requires two incubations and detections of the circulating antibodies in the serum. The patient's serum is layered on the substrate, followed by the application of fluoresceinated antibodies. An advantage of the IIF is its increased sensitivity (10–15 times). A modified IIF technique using the patient's own skin as a substrate, known as immunomapping (antigen mapping), is performed to determine the exact site of the cleavage or abnormalities in the distribution of the mutated structural proteins [45]. IIF in CUS diagnosis is the analysis of the SES-ANA autoantibodies located in the basal layer of the epithelium [19]. IIF performed on the remaining negative DIF-analyzed specimens yielded positive. There were also some cases in which DIF has not been performed—the IIF test was always performed on specific epithelial substrates (such as human esophagus/guinea pig, esophagus/monkey esophagus, and esophagus/normal human skin), and all of the IIF test results yielded positive, confirming a CUS presence. Positive IIF results may be used in CUS diagnostics; however, the result is not conclusive, and serum SES-ANA antibodies could also be observed among patients with OLP [6,23,28,48].

The CUS diagnostic protocol might also include an enzyme-linked immunosorbent assay (ELISA) test, which detects the presence of IgG antibodies in the CUS sera. The positive result of an ELISA test for anti-ΔNp63α antibodies was observed in patients with clinical symptoms of CUS and played a significant role in distinguishing CUS from other ulcerative diseases and establishing a relationship with OLP [35].

5. Diagnostics and Differentiation

The CUS diagnostics should include collaboration between dermatologists, pathologists, and oral clinicians. Among patients with clinically observed long-lasting oral erosions and ulcerations, the DIF analysis should be conducted in order to diagnose CUS [6]. Azzi et al. have proposed a diagnostic criterion for CUS that is presented in Table 1. The major criteria include clinical features, such as chronic painful erosions and/or ulcerations and IgG SES-ANA deposition in the lower third of the epithelium with a speckled pattern in DIF analysis. Minor criteria include CUS symptoms that are often observed; however, these were not reported in all cases and concerning clinical features, histopathology, IIF analysis, laboratory findings, and therapy results. Azzi et al. suggested that for CUS diagnosis, two major criteria should be positive. In cases when DIF analysis is not available, four minor criteria should be observed, including one clinical feature, two among laboratory or histopathological findings, and one therapeutic criterion [6].

Table 1. Diagnostic criteria for CUS proposed by Azzi et al. [6].

Major Criteria	Minor Criteria
Clinical features • Chronic painful erosions and/ or ulcerations	**Clinical features** • Middle aged or older women • Chronic course with relapses • Buccal mucosa, tongue (ventral aspect and/or lateral borders), desquamative gingivitis • Lichenoid appearance with white striae departing from lesions borders • Symetrical distribution • Association between diffuse intra-oral distribution and cutaneous lichenoid lesions
DIF analysis • IgG SES-ANA deposition in the lower third of epithelium with a speckled pattern	**Histopathology** • Lichenoid stomatitis, mainly associated with a band-like mixed infiltrate made of lymphocytes and plasma cells **IIF analysis** • IgG SES-ANA deposition in the basal layer of epithelial substrates with speckled pattern • Negative results when using HEp-2 or non-epithelial substrates **Laboratory findings** • 70 kDa protein detected as autoantigen by immunobinding or other techniques • Positive result at ELISA test for anti-ΔNp63α antibodies **Therapy** • Failure or only partial response with corticosteroids • Response to hydroxychloroquine (at least 200 mg/day) alone or combined with low doses of corticosteroids

Clinical and histological similarities to OLP might be the reason for the misdiagnosis of CUS [6,46]. Oral ulcerations could also be caused by mechanical trauma, oral dysplasia, oral squamous cell carcinoma (OSCC), or hematologic abnormalities [49].

OLP, one of the clinical forms of lichen planus (LP), is a common chronic disorder that generally occurs in patients in the fifth to sixth decades of life and is observed twice more often in females than in males. Oral involvement in LP is very common, and it is assessed that even 15–35% of LP patients might be the only clinical manifestation of the disease. OLP exclusively affects the stratified squamous epithelium that presents as a muco-cutaneous inflammatory disease [42]. Oral mucosa lesions tend to occur as one of three general types: 1. Reticular, including white lines, plaques, and papules which is the most common clinical manifestation; 2. atrophic or erythematous, and 3. erosive, including ulcerations and bullae that resemble CUS lesions [35,50,51]. OLP lesions are mostly symmetrical and are often observed in trauma-prone areas, such as the buccal mucosa and lateral surface of the tongue; however, they might also be present on the gingiva, labial mucosa, and vermilion of the lower lip [2,52]. Erythematous lesions that affect the gingiva cause desquamative gingivitis. Uncommon areas of OLP manifestation are the upper lip, palate, and the floor of the mouth. In the majority of OLP cases, lesions were observed in multiple areas; however, a single patient with lesions isolated to only the lop or tongue has been described. The clinical manifestation of OPL is diversified and might resemble other oral mucosa diseases. Striated white lesions, with or without erosions, might be similar to lupus erythematosus lesions, and the plaque-like OLP lesions may resemble leukoplakia, which is a white keratosis and a precancerous lesion mostly related to tobacco smoking [49,52,53]. The ulcerative form of OLP might be difficult to distinguish from vesiculoerosive dermatological diseases, such as pemphigus and pemphigoid, or can resemble OSCC [49,52]. OLP is a chronic disease that is characterized by periods of exacerbation and remission; however, spontaneous remissions are rarely observed. The pathogenesis of the lichen planus is defined as a lymphocytic immunologic reaction to the epithelial basal cells. The histopathological examination presents basal layer degeneration and apoptotic bodies. In early lesions, the predomination of CD4+ T cells is observed, and in chronic lesions, CD8+ T cells occur more often [35]. A DIF examination in OPL presents a characteristic fibrillar pattern of the fibrin deposition at the basement membrane zone; however, this result in not pathognomonic and can be interpreted only as suggestive OPL. The treatment for OLP can include both local and systemic corticosteroid implementation, calcineurin inhibitors, or retinoids. Taking into consideration only clinical and histopathological presentation, CUS might be indistinguishable from erosive OLP [35,54]. The differentiation of CUS and erosive OLP, including clinical symptoms, histopathology, DIF, and treatment, is presented in Table 2.

Table 2. Differentiation of CUS and erosive OLP.

	Chronic Ulcerative Stomatitis (CUS)	Erosive Oral Lichen Planus (OLP)
Clinical symptoms	Oral non-healing ulcerative lesions with subtle white reticular striations located on the tongue, the buccal mucosa, and the gingival tissues (desquamative gingivitis), mostly symmetrical.	Oral mucosa lesions manifested as reticular, including white lines, plaques, and papules; atrophic or erythematous; erosions and ulcerations; mostly symmetrical, located on the buccal mucosa and lateral surface of the tongue, gingiva (desquamative gingivitis) and labial mucosa.
Histopathology	Sub-epithelial separation from underlying connective tissue, atrophic epithelium, and inflammatory infiltrate with increased number of plasma cells and lymphocytes (non-specific).	Basal layer degeneration and apoptotic bodies; CD4+ T and CD8+ T cells (non-specific).
Direct immunofluorescence (DIF)	SES-ANA speckled pattern, located in the basal layer and the bottom three layers of cells.	Fibrillar pattern of fibrin deposition at the basement membrane zone.
Treatment	Chloroquine and hydroxychloroquine combined with corticosteroids or a single drug treatment.	Reticular OLP—observation; erosive OLP—pharmacological treatment (local and systemic corticosteroids implementation, calcineurin inhibitors or retinoids).

Ulcerations in the oral cavity might also result from either acute or chronic trauma. Oral ulcers resulting from acute trauma are generally self-resolving within 14 days without complications; however, chronic ulcerations that are not related to a clear source of trauma require a biopsy to excuse neoplasia or other oral mucosa conditions. The majority of traumatic lesions have nonspecific histologic findings, and treatment should include the removal of the etiologic source of the trauma, promoting healing, and preventing infection [49].

A significant aspect regarding oral ulcers is the diagnosis of ulcerated malignant lesions, such as oral dysplasia and OSCC. All non-healing oral ulcerations require histopathology, especially in a group of patients reporting tobacco and alcohol use. The most malignant suspected are non-symmetrical lesions, and those located on the lateral and ventral surfaces of the tongue and floor of the mouth tend to present a higher risk for malignant transformation. Histopathology depends on the stage of progression and may range from mild to severe dysplasia, carcinoma-in-situ, to invasive carcinoma [49].

There are multiple hematologic abnormalities that may manifest in the oral cavity, including malignant and non-malignant lesions of the B or T-cell origin. Leukemia and neutropenia are the most commonly observed hematological reasons for oral ulcerations. Those conditions also commonly involve gingival bleeding and hypertrophic gingivitis [49,55].

6. Treatment Methods

The CUS treatment should promote the healing of erosions and ulcers, relieving the symptoms and preventing secondary infections. CUS generally remains nonsusceptible to both topical and general corticosteroid treatments, on the contrary to other immune-mediated diseases [42,46]. There have been several CUS patient cases treated with corticosteroids—only 11% of them presented therapeutic success (including one combined therapy with dapsone administration). Patients who did not develop a successful response to corticosteroids were passed to antimalarial drugs, mostly chloroquine and hydroxychloroquine combined with corticosteroids or a single drug treatment. Almost half of the treatments resulted in a general improvement or complete clearance, and in over half of the reported cases, this therapy resulted in benefits relapsing when tapering the antimalarial dose [1,8,11]. The improvement or complete healing of the oral lesions was observed after the administration of low doses of hydroxychloroquine (200 mg/day) [35]; however, some authors suggested a higher dosage, even 400 mg/day and 800 mg/day [7]. Hydroxychloroquine interferes with the antigen-processing mechanisms of macrophages and other antigen-presenting cells, which result in the downregulation of the immune response against antigenic peptides. However, the hydroxychloroquine treatment may result in side effects, such as aplastic anemia, agranulocytosis, irreversible retinopathy, toxic psychosis, or neuromyopathy, which leads to the necessity of constant monitoring of patients and a collaboration with the patient's physician [35]. The therapeutic protocol of CUS may include low doses of antimalarial drugs combined with corticosteroids administrated for a prolonged time [6].

Another approach to CUS management might include tacrolimus, a topical calcineurin inhibitor involved in the production of interleukin-2, which promotes T-lymphocyte proliferation and recruitment. Therefore, it is used in T-cells and mediated diseases, such as eczema, psoriasis, and, potentially, CUS, for its beneficial immunosuppressive effects. According to Stoopler et al., patients undergoing a combined treatment with the antimalarial drug (390 mg hydroxychloroquine) and tacrolimus (0,1%) presented a positive outcome [31]. Cyclosporine, another calcineurin inhibitor, was administered when combined with chloroquine as a CUS treatment method, which also resulted in a successful response [40].

The management of oral lesions among patients with CUS can be challenging, and a multidisciplinary approach is required. The prevention of local irritation by avoiding spicy and hard food with the elimination of alcohol consumption and cessation of smoking plays an essential role. The patient should also be instructed on how to properly care for oral hygiene. It is recommended to use soft toothbrushes and antiseptic mouthwashes, such as chlorhexidine gluconate 0.2%. Topical analgesics, such as benzydamine hydrochloride 0.15% (rinse or spray), might be applied to relieve pain and discomfort, especially prior to eating or tooth brushing. It is also recommended to regularly visit a periodontist in order to remove dental calculus and control periodontal diseases. Sanitation of the oral cavity consisting of the treatment of dental cavities and the removal of non-prognostic teeth allows the elimination of inflammation in the oral cavity. By smoothing the sharp edges of the teeth and parafunction treatment, it is possible to reduce oral injuries that exacerbate CUS-related symptoms. [9,56–61].

Summary data on the clinical symptoms, diagnosis, and treatment methods of CUS are presented in Table 3.

Table 3. Clinical symptoms, diagnostic tests, and treatment methods of CUS.

	Clinical symptoms	Diagnostic tests	Treatment
Azzi et al. [6]	Painful oral erosions and/or ulcers most often located on the buccal mucosa, the gingiva (desquamative gingivitis), and the tongue.	DIF: SES-ANA deposition, mainly composed of IgGs, in cells of the basal layer and the bottom three layers of cells.	Low doses of antimalarial drugs combined with corticosteroids administrated for a prolonged time.
Islam et al. [7]	Oral erosive or ulcerative lesions located on the tongue, the buccal mucosa, and the gingival tissues (desquamative gingivitis).	DIF: a speckled or finely granular pattern of IgG limited to the basal and parabasal layers of the epithelium, often perinuclear distribution.	Steroidal combination therapy and/or dose regulation of hydroxychloroquine.
Mustafa et al. [9]	Persistent or recurrent painful erosive, ulcerative, vesicular lesions, predominately affecting the tongue, the buccal mucosa, and the gingiva.	DIF: a speckled, finely granular pattern of IgG deposition in the nuclei of keratinocytes. SES-ANA signal is confined to the basal cells and the lower third of the spinous layers.	The same as other oral mucosa erosions and ulcers (no specific treatment was described).
Ko et al. [10]	Oral erosions or ulcerations with periods of exacerbation and remission; the tongue, buccal mucosa, and gingiva are the most commonly affected.	DIF: a speck-led pattern of IgG deposition in the nuclei of keratinocytes limited to the lower layers of the oral squamous epithelium. The presence of SES-ANA distinguishes CUS from oral LP.	No response to corticosteroids.
Solomon et al. [12]	Oral erosive or ulcerative lesions that ale most often present on the tongue, then on the buccal mucosa and gingiva (desquamative gingivitis).	DIF: a speckled, finely granular pattern of IgG deposition in the nuclei of keratinocytes. The SES-ANA signal is confined to the basal cells and lower third of the Malphigian layers.	A combination of small doses of steroids and hydroxychloroquine.
Stoopler et al. [24]	Symptomatic chronic oral ulcers: Wickham's striae, erythema, and ulceration which commonly affect the buccal mucosa, the tongue, and the gingiva (desquamative gingivitis).	DIF: a speckled pattern of IgG deposition in keratinocyte nuclei limited to the lower layers of the oral squamous epithelium; the presence of SES-ANA antibodies.	Promoting healing, symptom relief, mitigating risks of secondary infection, hydroxychloroquine. No response to corticosteroids.

7. Conclusions

Chronic ulcerative stomatitis (CUS) manifests as non-healing, erosive, or ulcerative lesions in the oral cavity and is an often-misdiagnosed disease due to both its clinical and histological resemblance to other oral mucosa conditions. A proper diagnosis is essential for a successful treatment administration. Further research should be conducted in order to implement a diagnostic protocol and observe the long-term results of CUS management. Taking the presented data into consideration, clinicians should consider the diagnosis of CUS for all erosive or ulcerative lesions appearing cyclically in the oral cavity at the same site, with moderate pain and a slightly specific histopathological picture, after previously excluding traumatic factors.

Author Contributions: Conceptualization, A.K. and D.C.; methodology, A.K. and D.C.; software, D.C.; validation, A.K. and D.C.; formal analysis, A.K; investigation, D.K., M.M. and D.C.; resources, D.K., M.M. and D.C.; data curation, A.K. and D.C.; writing—original draft preparation, D.C., D.K. and M.M.; writing—review and editing, A.K. and D.C.; visualization, D.C.; supervision, A.K.; project administration, A.K. and D.C.; funding acquisition, A.K. All authors have read and agreed to the published version of the manuscript.

Funding: This research received no external funding.

Institutional Review Board Statement: Not applicable.

Informed Consent Statement: Not applicable.

Data Availability Statement: Not applicable.

Conflicts of Interest: The authors declare no conflict of interest.

References

1. Herzum, A.; Burlando, M.; Cozzani, E.; Parodi, A. The 30th birthday of chronic ulcerative stomatitis: A systematic review. *Int. J. Immunopathol. Pharmacol.* **2021**, *35*, 20587384211052437. [CrossRef] [PubMed]
2. Müller, S. Oral lichenoid lesions: Distinguishing the benign from the deadly. *Mod. Pathol.* **2017**, *30*, S54–S67. [CrossRef] [PubMed]
3. Wörle, B.; Wollenberg, A.; Schaller, M.; Kunzelmann, K.H.; Plewig, G.; Meurer, M. Chronic ulcerative stomatitis. *Br. J. Dermatol.* **1997**, *137*, 262–265. [CrossRef] [PubMed]
4. Solomon, L.W.; Stark, P.C.; Winter, L.; Kumar, V.; Sinha, S. ELISA test for p63 antibodies in chronic ulcerative stomatitis. *Oral Dis.* **2010**, *16*, 151–155. [CrossRef] [PubMed]
5. Parodi, A.; Cozzani, E.; Chorzelski, T.P.; Beutner, E.H.; Rebora, A. A molecule of about 70 kd is the immunologic marker of chronic ulcerative stomatitis. *J. Am. Acad. Dermatol.* **1998**, *38 Pt 1*, 1005–1006. [CrossRef]
6. Azzi, L.; Cerati, M.; Lombardo, M. Chronic ulcerative stomatitis: A comprehensive review and proposal for diagnostic criteria. *Oral Dis.* **2019**, *25*, 1465–1491. [CrossRef]
7. Islam, M.N.; Cohen, D.M.; Ojha, J.; Stewart, C.M.; Katz, J.; Bhattacharyya, I. Chronic ulcerative stomatitis: Diagnostic and management challenges–four new cases and review of literature. *Oral Surg. Oral Med. Oral Pathol. Oral Radiol. Endod.* **2007**, *104*, 194–203. [CrossRef]
8. Ferrisse, T.M.; Travassos, D.C.; Rocha, A.F.; Massucato, E.M.; Bufalino, A. Chronic ulcerative stomatitis: A systematic review of the clinical and microscopic features. *Med. Oral Patol. Oral Cir. Bucal.* **2019**, *24*, e698–e703. [CrossRef]
9. Mustafa, M.B.; Porter, S.R.; Smoller, B.R.; Sitaru, C. Oral mucosal manifestations of autoimmune skin diseases. *Autoimmun. Rev.* **2015**, *14*, 930–951. [CrossRef]
10. Ko, E.M.; Danciu, T.E.; Fullen, D.R.; Chan, M.P. Chronic ulcerative stomatitis: Case series of an under-recognized entity. *J. Cutan. Pathol.* **2018**, *45*, 927–932. [CrossRef]
11. Sirisha, V. A rare entity of chronic ulcerative stomatitis—A case report. *Int. J. Oral Health Dent.* **2019**, *5*, 233–235. [CrossRef]
12. Solomon, L.W.; Aguirre, A.; Neiders, M.; Costales-Spindler, A.; Jivide, G.J.; Zwick, M.G.; Kumar, V. Chronic ulcerative stomatitis: Clinical, histopathologic, and immunopathologic findings. *Oral Surg. Oral Med. Oral Pathol. Oral Radiol. Endod.* **2003**, *96*, 718–726. [CrossRef]
13. Jaremko, W.M.; Beutner, E.H.; Kumar, V.; Kipping, H.; Condry, P.; Zeid, M.; Kauffmann, C.; Tatakis, D.N.; Chorzelski, T.P. Chronic ulcerative stomatitis associated with a specific immunologic marker. *J. Am. Acad. Dermatol.* **1990**, *22 Pt 1*, 215–220. [CrossRef]
14. Lee, L.A.; Walsh, P.; Prater, C.A.; Marchbank, A.; Egbert, T.B.; Dellavalle, R.P.; Targoff, I.N.; Kaufman, K.M.; Chorzelski, T.P.; Jablonska, S. Characterization of an autoantigen associated with chronic ulcerative stomatitis: The CUSP autoantigen is a member of the p53 family. *J. Investig. Dermatol.* **1999**, *113*, 146–151. [CrossRef] [PubMed]
15. Dellavalle, R.P.; Egbert, T.B.; Marchbank, A.; Su, L.J.; Lee, L.A.; Walsh, P. CUSP/p63 expression in rat and human tissues. *J. Dermatol. Sci.* **2001**, *27*, 82–87. [CrossRef]
16. Ebrahimi, M.; Wahlin, Y.B.; Coates, P.J.; Wiik, A.; Roos, G.; Nylander, K. Detection of antibodies against p63 and p73 isoforms in sera from patients diagnosed with oral lichen planus. *J. Oral Pathol. Med.* **2007**, *36*, 93–98. [CrossRef]
17. Yang, A.; Kaghad, M.; Wang, Y.; Gillett, E.; Fleming, M.D.; Dötsch, V.; Andrews, N.C.; Caput, D.; McKeon, F. p63, a p53 homolog at 3q27-29, encodes multiple products with transactivating, death-inducing, and dominant-negative activities. *Mol. Cell* **1998**, *2*, 305–316. [CrossRef]
18. Solomon, L.W.; Neiders, M.E.; Zwick, M.G.; Kirkwood, K.L.; Kumar, V. Autoimmunity to deltaNp63alpha in chronic ulcerative stomatitis. *J. Dent. Res.* **2007**, *86*, 826–831. [CrossRef]
19. Feller, L.; Khammissa, R.A.G.; Lemmer, J. Is chronic ulcerative stomatitis a variant of lichen planus, or a distinct disease? *J. Oral Pathol. Med.* **2017**, *46*, 859–863. [CrossRef]
20. Carlson, M.W.; Garlick, J.A.; Solomon, L.W. Chronic ulcerative stomatitis: Evidence of autoimmune pathogenesis. *Oral Surg. Oral Med. Oral Pathol. Oral Radiol. Endod.* **2011**, *111*, 742–748. [CrossRef]
21. Mrowiecka, M.K.; Macugowska, M.C.; Michcik, A.; Chomik, P.; Wlodarkiewicz, A. Chronic ulcerative stomatitis (CUS). Lichen planus disseminatus, unguium pedis utriusque and pseudopelade. *Am. J. Case Rep.* **2010**, *11*, 252–257. [CrossRef]
22. Parodi, A.; Cozzani, E.; Cacciapuoti, M.; Rebora, A. Chronic ulcerative stomatitis: Antibodies reacting with the 70-kDa molecule react with epithelial nuclei. *Br. J. Dermatol.* **2000**, *143*, 671–672. [CrossRef] [PubMed]
23. Cozzani, E.; Cacciapuoti, M.; di Marco, E.; Zerega, B.; Descalzi Cancedda, F.; Parodi, A. Patients with oral erosive and cutaneous lichen planus may have antibodies directed against the chronic ulcerative stomatitis protein antigen of 70-kDa. *Acta Derm. Alp Pannonica Adriat* **2008**, *17*, 120–124.
24. Stoopler, E.T.; Kulkarni, R.; Alawi, F.; Sollecito, T.P. Novel combination therapy of hydroxychloroquine and topical tacrolimus for chronic ulcerative stomatitis. *Int. J. Dermatol.* **2021**, *60*, e162–e163. [CrossRef] [PubMed]

25. Chorzelski, T.P.; Olszewska, M.; Jarzabek-Chorzelska, M.; Jablonska, S. Is chronic ulcerative stomatitis an entity? Clinical and immunological findings in 18 cases. *Eur. J. Dermatol.* **1998**, *8*, 261–265. [PubMed]
26. Rech, B.O.; Tenório, J.R.; Braz-Silva, P.H.; Ortega, K.L. Chronic ulcerative stomatitis: A case report. *Oral Surg. Oral Med. Oral Pathol. Oral Radiol.* **2020**, *130*, e152. [CrossRef]
27. Gururaj, N.; Hasinidevi, P.; Janani, V.; Divynadaniel, T. Diagnosis and management of oral lichen planus-Review. *J. Oral Maxillofac. Pathol.* **2021**, *25*, 383–393. [CrossRef]
28. Parodi, A.; Cozzani, E.; Massone, C.; Rebora, A.; Priano, L.; Ghigliotti, G.; Balbi, P.; Rongioletti, F.; Micalizzi, C.; Cestari, R.; et al. Prevalence of stratified epithelium-specific antinuclear antibodies in 138 patients with lichen planus. *J. Am. Acad. Dermatol.* **2007**, *56*, 974–978. [CrossRef]
29. Lorenzana, E.R.; Rees, T.D.; Glass, M.; Detweiler, J.G. Chronic ulcerative stomatitis: A case report. *J. Periodontol.* **2000**, *71*, 104–111. [CrossRef]
30. Stewart, C.; Bhattacharyya, I. Chronic ulcerative stomatitis in primary Sjögren's syndrome. *Oral Surg. Oral Med. Oral Pathol. Oral Radiol. Endod.* **2007**, *102*, 335–336. [CrossRef]
31. Fourie, J.; van Heerden, W.F.; McEachen, S.C.; van Zyl, A. Chronic ulcerative stomatitis: A distinct clinical entity? *SADJ* **2011**, *66*, 119–121. [PubMed]
32. Jacyk, W.K.; Fourie, J.; Van Heerden, W.F. Chronic ulcerative stomatitis and lichen planus: Just a coincidence or a direct link between the two diseases? *Dermatol. Klin.* **2012**, *14*, 127–129.
33. Beutner, E.H.; Chorzelski, T.P.; Parodi, A.; Schosser, R.; Guin, J.; Cardo, P.P.; Maciejowska, E.; Valeski, J.E.; Kumar, V. Ten cases of chronic ulcerative stomatitis with stratified epithelium-specific antinuclear antibody. *J. Am. Acad. Dermatol.* **1991**, *24 Pt 1*, 781–782. [CrossRef]
34. Lewis, J.E.; Beutner, E.H.; Rostami, R.; Chorzelski, T.P. Chronic ulcerative stomatitis with stratified epithelium-specific antinuclear antibodies. *Int. J. Dermatol.* **1996**, *35*, 272–275. [CrossRef] [PubMed]
35. Solomon, L.W. Chronic ulcerative stomatitis. *Oral Dis.* **2008**, *14*, 383–389. [CrossRef]
36. Rinaggio, J.; Crossland, D.M.; Zeid, M.Y. A determination of the range of oral conditions submitted for microscopic and direct immunofluorescence analysis. *J. Periodontol.* **2007**, *78*, 1904–1910. [CrossRef]
37. Qari, H.; Villasante, C.; Richert, J.; Rees, T.; Kessler, H. The diagnostic challenges of separating chronic ulcerative stomatitis from oral lichen planus. *Oral Surg. Oral Med. Oral Pathol. Oral Radiol.* **2015**, *120*, 622–627. [CrossRef]
38. Alshagroud, R.; Neiders, M.; Kramer, J.M.; Suresh, L. Clinicopathologic significance of in vivo antinuclear autoantibodies in oral mucosal biopsies. *Oral Surg. Oral Med. Oral Pathol. Oral Radiol.* **2017**, *124*, 475–482. [CrossRef]
39. Reddy, R.; Fitzpatrick, S.G.; Bhattacharyya, I.; Cohen, D.M.; Islam, M.N. Seventeen New Cases of Chronic Ulcerative Stomatitis with Literature Review. *Head Neck Pathol.* **2019**, *13*, 386–396. [CrossRef]
40. Church, L.F., Jr.; Schosser, R.H. Chronic ulcerative stomatitis associated with stratified epithelial specific antinuclear antibodies. A case report of a newly described disease entity. *Oral Surg. Oral Med. Oral Pathol.* **1992**, *73*, 579–582. [CrossRef]
41. Carrozzo, M.; Porter, S.; Mercadante, V.; Fedele, S. Oral lichen planus: A disease or a spectrum of tissue reactions? Types, causes, diagnostic algorhythms, prognosis, management strategies. *Periodontology 2000* **2019**, *80*, 105–125. [CrossRef] [PubMed]
42. Carrozzo, M.; Thorpe, R. Oral lichen planus: A review. *Minerva Stomatol.* **2009**, *58*, 519–537.
43. Khudhur, A.S.; Di Zenzo, G.; Carrozzo, M. Oral lichenoid tissue reactions: Diagnosis and classification. *Expert. Rev. Mol. Diagn.* **2014**, *14*, 169–184. [CrossRef] [PubMed]
44. Mysorekar, V.V.; Sumathy, T.K.; Shyam Prasad, A.L. Role of direct immunofluorescence in dermatological disorders. *Indian Dermatol. Online J.* **2015**, *6*, 172–180. [CrossRef] [PubMed]
45. Chhabra, S.; Minz, R.W.; Saikia, B. Immunofluorescence in dermatology. *Indian J. Dermatol. Venereol. Leprol* **2012**, *78*, 677–691. [CrossRef] [PubMed]
46. Reddy, D.R.; Fitzpatrick, D.S.; Davidova, D.L.; Bhattacharyya, I.; Cohen, D.; Islam, M. Chronic ulcerative stomatitis: A lichenoid or vesiculobullous disease? *Oral Surg. Oral Med. Oral Pathol. Oral Radiol.* **2019**, *128*, e81. [CrossRef]
47. Ezzatt, O. *Chronic Ulcerative Stomatitis-Case Report*; Ain Shams University: Cairo, Egypt, 2015. [CrossRef]
48. Cacciapuoti, M.; Di Marco, E.; Cozzani, E.; Zerega, B.; Descalzi Cancedda, F.; Parodi, A. The antibody to the 70-kd antigen in chronic ulcerative stomatitis and lichen planus. *J. Am. Acad. Dermatol.* **2004**, *50*, 486. [CrossRef]
49. Fitzpatrick, S.G.; Cohen, D.M.; Clark, A.N. Ulcerated Lesions of the Oral Mucosa: Clinical and Histologic Review. *Head Neck Pathol.* **2019**, *13*, 91–102. [CrossRef]
50. Eisen, D. The clinical features, malignant potential, and systemic associations of oral lichen planus: A study of 723 patients. *J. Am. Acad. Dermatol.* **2002**, *46*, 207–214. [CrossRef]
51. Scully, C.; Carrozzo, M. Oral mucosal disease: Lichen planus. *Br. J. Oral Maxillofac. Surg.* **2008**, *46*, 15–21. [CrossRef]
52. Müller, S. Oral manifestations of dermatologic disease: A focus on lichenoid lesions. *Head Neck Pathol.* **2011**, *5*, 36–40. [CrossRef] [PubMed]
53. Kusiak, A.; Maj, A.; Cichońska, D.; Kochańska, B.; Cydejko, A.; Świetlik, D. The Analysis of the Frequency of Leukoplakia in Reference of Tobacco Smoking among Northern Polish Population. *Int. J. Environ. Res. Public Health* **2020**, *17*, 6919. [CrossRef]
54. Parodi, A.; Cardo, P.P. Patients with erosive lichen planus may have antibodies directed to a nuclear antigen of epithelial cells: A study on the antigen nature. *J. Investig. Dermatol.* **1990**, *94*, 689–693. [CrossRef] [PubMed]

55. Francisconi, C.F.; Caldas, R.J.; Oliveira Martins, L.J.; Fischer Rubira, C.M.; da Silva Santos, P.S. Leukemic oral manifestations and their management. *Asian Pac. J. Cancer Prev.* **2016**, *17*, 911–915. [CrossRef]
56. Kneisel, A.; Hertl, M. Autoimmune bullous skin diseases. Part 2: Diagnosis and therapy. *J. Dtsch. Dermatol. Ges.* **2011**, *9*, 927–947. [CrossRef]
57. Tziotzios, C.; Brier, T.; Lee, J.Y.W.; Saito, R.; Hsu, C.K.; Bhargava, K.; Stefanato, C.M.; Fenton, D.A.; McGrath, J.A. Lichen planus and lichenoid dermatoses: Conventional and emerging therapeutic strategies. *J. Am. Acad. Dermatol.* **2018**, *79*, 807–818. [CrossRef] [PubMed]
58. Diebold, S.; Overbeck, M. Soft Tissue Disorders of the Mouth. *Emerg. Med. Clin. N. Am.* **2019**, *37*, 55–68. [CrossRef]
59. Price, S.M.; Murrah, V.A. Why the general dentist needs to know how to manage oral lichen planus. *Gen. Dent.* **2015**, *63*, 16–22.
60. Murrah, V.A.; Perez, L.M. Oral lichen planus: Parameters affecting accurate diagnosis and effective management. *Pract. Periodontics Aesthet Dent.* **1997**, *9*, 613–620; quiz 622.
61. Edwards, P.C.; Kelsch, R. Oral lichen planus: Clinical presentation and management. *J. Can. Dent. Assoc.* **2002**, *68*, 494–499.

International Journal of
Molecular Sciences

Article

The Role of Polymorphisms at the Interleukin-1, Interleukin-4, GATA-3 and Cyclooxygenase-2 Genes in Non-Surgical Periodontal Therapy

Kay-Arne Walther [1,*], José Roberto Gonzales [1], Sabine Gröger [1], Benjamin Ehmke [2], Dogan Kaner [3,4], Katrin Lorenz [5], Peter Eickholz [6], Thomas Kocher [7], Ti-Sun Kim [8], Ulrich Schlagenhauf [9], Raphael Koch [10,†] and Jörg Meyle [1,†]

1 Department of Periodontology, University of Giessen, Schlangenzahl 14, 35392 Giessen, Germany; office@prof-gonzales.de (J.R.G.); sabine.e.groeger@dentist.med.uni-giessen.de (S.G.); joerg.meyle@dentist.med.uni-giessen.de (J.M.)
2 Department of Periodontology, University of Münster, Albert-Schweitzer-Campus 1, Waldeyerstraße30, 48149 Münster, Germany; ehmke@uni-muenster.de
3 Departments of Periodontology and Synoptic Dentistry, Charite Centrum 3, Charite-Universitatsmedizin Berlin, Aßmannshauserstraße 4-6, 14197 Berlin, Germany; dogan.kaner@uni-wh.de
4 Department of Periodontology, University of Witten/Herdecke, Alfred-Herrhausen-Straße 44, 58455 Witten, Germany
5 Department of Periodontology, TU Dresden, Fetscherstraße 74, 01307 Dresden, Germany; katrin.lorenz@tu-dresden.de
6 Department of Periodontology, Johann Wolfgang Goethe-University Frankfurt, Theodor-Stern-Kai 7, 60596 Frankfurt, Germany; eickholz@med.uni-frankfurt.de
7 Unit of Periodontology, University of Greifswald, Rotgerberstraße 8, 17475 Greifswald, Germany; kocher@uni-greifswald.de
8 Section of Periodontology, Department of Conservative Dentistry, University of Heidelberg, Im Neuenheimer Feld 400, 69120 Heidelberg, Germany; tisun.kim@med.uni-heidelberg.de
9 Department of Periodontology, University of Würzburg, Pleicherwall 2, 97070 Würzburg, Germany; schlagenha_u@ukw.de
10 Institute of Biostatistics and Clinical Research, University of Münster, Schmeddingstraße 56, 48149 Münster, Germany; raphael.koch@ukmuenster.de
* Correspondence: kay-arne.walther@dentist.med.uni-giessen.de; Tel.: +49-(0)-641-99-46191
† These authors contributed equally to this work.

Citation: Walther, K.-A.; Gonzales, J.R.; Gröger, S.; Ehmke, B.; Kaner, D.; Lorenz, K.; Eickholz, P.; Kocher, T.; Kim, T.-S.; Schlagenhauf, U.; et al. The Role of Polymorphisms at the Interleukin-1, Interleukin-4, GATA-3 and Cyclooxygenase-2 Genes in Non-Surgical Periodontal Therapy. *Int. J. Mol. Sci.* **2022**, *23*, 7266. https://doi.org/10.3390/ijms23137266

Academic Editors: Kenichi Kumagai and Marcella Reale

Received: 13 May 2022
Accepted: 27 June 2022
Published: 30 June 2022

Abstract: Periodontitis is a multifactorial disease. The aim of this explorative study was to investigate the role of Interleukin-(IL)-1, IL-4, GATA-3 and Cyclooxygenase-(COX)-2 polymorphisms after non-surgical periodontal therapy with adjunctive systemic antibiotics (amoxicillin/metronidazole) and subsequent maintenance in a Caucasian population. Analyses were performed using blood samples from periodontitis patients of a multi-center trial (ClinicalTrials.gov NCT00707369=ABPARO-study). Polymorphisms were analyzed using quantitative real-time PCR. Clinical attachment levels (CAL), percentage of sites showing further attachment loss (PSAL) $\geq$1.3 mm, bleeding on probing (BOP) and plaque score were assessed. Exploratory statistical analysis was performed. A total of 209 samples were genotyped. Patients carrying heterozygous genotypes and single-nucleotide-polymorphisms (SNP) on the GATA-3-IVS4 +1468 gene locus showed less CAL loss than patients carrying wild type. Heterozygous genotypes and SNPs on the IL-1A-889, IL-1B +3954, IL-4-34, IL-4-590, GATA-3-IVS4 +1468 and COX-2-1195 gene loci did not influence CAL. In multivariate analysis, CAL was lower in patients carrying GATA-3 heterozygous genotypes and SNPs than those carrying wild-types. For the first time, effects of different genotypes were analyzed in periodontitis progression after periodontal therapy and during supportive treatment using systemic antibiotics demonstrating a slight association of GATA-3 gene locus with CAL. This result suggests that GATA-3 genotypes are a contributory but non-essential risk factor for periodontal disease progression.

Keywords: periodontitis; polymorphisms; risk factor; periodontal therapy; antibiotics; GATA-3; Interleukin-1; Interleukin-4; Cyclooxygenase-2

1. Introduction

Periodontitis is a multifactorial disease accompanied by attachment loss which finally results in tooth loss. It is initiated by a dysbiotic biofilm which elicits a destructive inflammatory response. Periodontitis is clinically classified into staging, grading, extent and distribution. The latest classification of periodontal diseases addresses to future research the identification of specific genetic markers to differentiate between distinct periodontitis phenotypes. This could reflect clinically the initiation and progression of periodontitis [1,2]. It is known that genetic factors are involved in the pathogenesis of periodontitis [3–8]. However, there are limited data on the role of genetic factors in disease progression [9–11]. Genetic factors and environmental modifiers are known to influence disease severity. Analysis of disease expression in twins showed that a considerable variance in clinical phenotype is explained by genetic factors [3,4]. The IL-1 cytokine family has key regulatory roles in innate and adaptive immunity. Variations in IL-1 genes were first associated with chronic periodontitis (CP) in Caucasians in 1997 [12]. Many studies have subsequently explored the role of IL-1 gene polymorphisms in periodontitis with mixed results [13–17]. A couple of studies evaluated the influence of IL-1 SNP on the progression of periodontitis with a small sample size, partially with a short observation period and different results [9,10,18–22].

IL-4 is a multifunctional cytokine which induces polyclonal B-cell proliferation. IL-4 differentiates naive CD4+ T cells to TH2 cells, which in turn produce IL-4 [23,24]. A meta-analysis by Yan et al. [25] showed a significant association between CP patients and the T/T genotype at position IL-4 −590 in Caucasians, whereas a recent meta-analysis by a Bayesian approach found no association between CP and IL-4 −590 [17]. The gene loci IL-4 −590 and IL-4 −34 are in linkage disequilibrium which has been demonstrated in CP [26] and in aggressive periodontitis (AgP) patients [27].

GATA factors are pleiotropic transcription factors of the C4 Zinc finger family. GATA-3 is present in early thymocyte cells. Thereby, it is a key transcription factor for gene expression in the single stages of TH2 cell differentiation [28]. GATA-3 expression is induced by IL-4 in a STAT-6-dependent signaling pathway. For STAT-6 deficiency, GATA-3 can continue the TH2 differentiation. If GATA-3 is absent, T cell differentiation into TH2 cells is disturbed [29].

To date, there are only few GATA-3 SNPs reported. A well-documented locus is on the gene locus rs3802604. This locus was associated with type 1 diabetes mellitus [30], a significant reduction in breast cancer risk [31] and a higher relapse-free survival rate [32].

COX-2 is produced mainly in inflammatory processes [33]. Many studies have highlighted the importance of arachidone derivatives on the progression of periodontal disease. In particular, prostaglandin E2 and leukotriene B4 are involved in periodontal destruction. They are strongly increased in the inflamed periodontal tissue and in the crevicular fluid in gingivitis, periodontitis and peri-implantitis patients [34–38]. Interesting results were published regarding the gene locus COX-2 −1195 (rs689466). SNP on this position was associated with CP in a Chinese population [39]. On the contrary, there was no association in the European population [40].

The present study investigated genetic factors from patient samples with periodontitis that participated in the placebo-controlled, multi-center ABPARO trial (ClinicalTrials.gov NCT00707369). Clinical, microbiological and systemic results have been previously reported [41–45].

The aim of the present exploratory subanalysis was to evaluate the role of the IL-1A −889, IL-1B +3954, IL-4 −34, IL-4 −590, GATA-3 IVS4 +1468 and COX-2 −1195 gene loci in patients with periodontitis after a non-surgical periodontal therapy with or without additional adjunctive systemic antibiotics.

2. Results

In the present study the results of the genetic analyses of the gene loci at positions IL-1A −889 (rs1800587), IL-1B +3954 (rs1143634), IL-4 −34 (rs2070874), IL-4 −590 (rs2243250), GATA-3 IVS4 +1468 (rs3802604) and COX-2 −1195 (rs689466) are shown. The results of the clinical and microbiological parameters have been demonstrated in previous publications [41–45].

A total number of 209 FTA elute micro cards was genotyped. The patient characteristics and the smoking status of all individuals are shown in Table 1. The external and internal validation of the genetic analyses did not show abnormalities. Genotype frequencies of the analyzed collectives are presented in Table 2. All genotypes were presented in a sufficient number of individuals to perform the analyses.

Table 1. Patient characteristics.

	Total Genotyped Patients n = 209	Placebo Group n = 104	Antibiotics Group n = 105	p-Value
Sex; n (%)				
Male	116 (55.5%)	58 (55.8%)	58 (55.2%)	1.000 ^{F}
Female	93 (44.5%)	46 (44.2%)	47 (44.8%)	
Age; years	53.1 ±10.0	53.0 ±10.6	53.3 ±9.5	0.929 ^{U}
Active smokers; n	45 (21.5%)	19 (18.3%)	26 (24.8%)	0.313 ^{F}
CO ≥ 7 ppm (smoker); n	37 (17.7%)	16 (15.7%)	19 (18.1%)	0.712 ^{F}
Diabetes mellitus type II; n	12 (5.7%)	6 (5.8%)	6 (5.7%)	1.000 ^{F}

Categorical variables are reported as absolute and relative frequencies. Continuous variables are shown as mean ± standard deviation. p-values are from Fisher's exact test ^{F} or Mann–Whitney U test ^{U}. Abbreviations: CO: carbon monoxide in exhaled air.

Table 2. Genotype frequencies.

Gene Loci	Genotype	Number of Patients; n (%)	Placebo Group Patients; n (%)	Antibiotics Group Patients; n (%)
rs1800587 IL-1A −889C > T	C/C	102 (48.8%)	49 (47.1%)	53 (50.5%)
	C/T	87 (41.6%)	46 (44.2%)	41 (39%)
	T/T	20 (9.6%)	9 (8.7%)	11 (10.5%)
	Allele T = MAF	30.9%		
rs1143634 IL-1B +3954C > T	C/C	128 (61.2%)	64 (61.5%)	64 (61%)
	C/T	64 (30.6%)	31 (29.8%)	33 (31.4%)
	T/T	17 (8.1%)	9 (8.7%)	8 (7.6%)
	Allele T = MAF	28.5%		
rs2070874 IL-4 −34C > T	C/C	155 (74.2%)	76 (73.1%)	79 (75.2%)
	C/T	48 (23%)	26 (25%)	22 (21%)
	T/T	6 (2.9%)	2 (1.9%)	4 (3.8%)
	Allele T = MAF	16.9%		
rs2243250 IL-4 −590C > T	C/C	154 (73.7%)	76 (73.1%)	78 (74.3%)
	C/T	48 (23%)	26 (25%)	22 (21%)
	T/T	7 (3.4%)	2 (1.9%)	5 (4.8%)
	Allele T = MAF	18.3%		
rs3802604 GATA-3 IVS4 +1468G > A	G/G	35 (16.8%)	17 (16.3%)	18 (17.1%)
	G/A	89 (42.6%)	51 (49%)	38 (36.2%)
	A/A	85 (40.7%)	36 (34.6%)	49 (46.7%)
	Allele G = MAF	40.9%		
rs689466 COX-2 −1195A > G	A/A	138 (66.0%)	69 (66.3%)	69 (65.7%)
	A/G	64 (30.6%)	32 (30.8%)	32 (30.5%)

Table 2. *Cont.*

Gene Loci	Genotype	Number of Patients; *n* (%)	Placebo Group Patients; *n* (%)	Antibiotics Group Patients; *n* (%)
	G/G Allele G = MAF	7 (3.4%) 18.3%	3 (2.9%)	4 (3.8%)

Note: Because of rounding of the values, the percentages do not always add up to 100%. Abbreviations: MAF: minor allele frequency.

Table 3 shows the changes of the clinical attachment level (CAL) from month 27.5 (visit 12)-baseline (visit 2) and from month 27.5 (visit 12)-month 3.5 (visit 4) in the total group of individuals, the placebo group and in the antibiotics group, with their respective *p*-values.

Table 3. Clinical measurements of the change of the CAL.

	Genotype	Total Group Patients *n* CAL Median (25% Quantile, 75% Quantile)	*p*-Value	Placebo Group Patients *n* CAL Median (25% Quantile, 75% Quantile)	*p*-Value	Antibiotics Group Patients *n* CAL Median (25% Quantile, 75% Quantile)	*p*-Value
CAL (mm)				**IL-1A −889 (rs1800587)**			
Change 27.5 months vs. baseline	HG (C/T)	87 −0.5 (−0.8, −0.2)	0.4633	46 −0.4 (−0.8, −0.2)	0.1427	41 −0.6 (−0.9, −0.2)	0.9515
	SNP (T/T)	20 −0.4 (−0.9, 0.1)	0.7852	9 −0.4 (−0.7, 0.2)	0.9319	11 −0.4 (−1.2, 0.0)	0.5832
	WT (C/C)	102 −0.4 (−0.9, 0.0)	ref.	49 −0.2 (−0.7, 0.1)	ref.	53 −0.6 (−1.0, −0.3)	ref.
Change 27.5 months vs. 3.5 months	HG (C/T)	87 0.0 (−0.3, 0.3)	0.1434	46 0.0 (−0.5, 0.2)	**0.0444**	41 0.0 (−0.2, 0.4)	0.9599
	SNP (T/T)	20 0.1 (−0.3, 0.3)	0.9804	9 −0.2 (−0.4, 0.2)	0.4799	11 0.2 (0.0, 0.3)	0.3103
	WT (C/C)	100 0.1 (−0.3, 0.5)	ref.	48 0.2 (−0.1, 0.5)	ref.	52 0.0 (−0.3, 0.4)	ref.
CAL (mm)				**IL−1B +3954 (rs1143634)**			
Change 27.5 months vs. baseline	HG (C/T)	64 −0.4 (−0.8, −0.1)	0.3594	31 −0.3 (−0.6, 0.0)	0.5399	33 −0.6 (−0.9, −0.1)	0.5616
	SNP (T/T)	17 −0.4 (−1.0, −0.2)	0.8660	9 −0.7 (−1.0, −0.4)	0.1375	8 −0.2 (−0.8, 0.1)	0.1920
	WT (C/C)	128 −0.5 (−0.9, 0.0)	ref.	64 −0.4 (−0.8, 0.1)	ref.	64 −0.6 (−1.0, −0.3)	ref.
Change 27.5 months vs. 3.5 months	HG (C/T)	64 0.0 (−0.1, 0.4)	0.3037	31 0.0 (−0.1, 0.5)	0.6826	33 0.0 (−0.1, 0.4)	0.2695
	SNP (T/T)	17 0.1 (−0.3, 0.2)	0.9429	9 −0.3 (−0.4, −0.2)	0.0898	8 0.3 (0.1, 0.6)	**0.0427**
	WT (C/C)	126 0.0 (−0.4, 0.4)	ref.	63 0.1 (−0.5, 0.5)	ref.	63 0.0 (−0.4, 0.4)	ref.
CAL (mm)				**IL−4 −590 (rs2243250)**			
Change 27.5 months vs. baseline	HG (C/T)	48 −0.4 (−0.9, 0.0)	0.7252	26 −0.3 (−0.7, 0.1)	0.6487	22 −0.5 (−1.1, −0.2)	0.9109
	SNP (T/T)	7 −0.6 (−0.8, −0.5)	0.5598	2 −0.7 (−0.8, −0.5)	0.3874	5 −0.6 (−0.7, −0.6)	0.9316

Table 3. *Cont.*

	Genotype	Total Group Patients n CAL Median (25% Quantile, 75% Quantile)	p-Value	Placebo Group Patients n CAL Median (25% Quantile, 75% Quantile)	p-Value	Antibiotics Group Patients n CAL Median (25% Quantile, 75% Quantile)	p-Value
	WT (C/C)	154 −0.4 (−0.9, 0.0)	ref.	76 −0.4 (−0.8, 0.0)	ref.	78 −0.6 (−0.9, −0.1)	ref.
Change 27.5 months vs. 3.5 months	HG (C/T)	47 0.2 (−0.1, 0.4)	0.0864	26 0.2 (−0.1, 0.5)	0.3095	21 0.2 (−0.1, 0.3)	0.1156
	SNP (T/T)	7 0.3 (−0.1, 0.5)	0.3824	2 0.2 (0.1, 0.3)	0.5553	5 0.4 (−0.1, 0.5)	0.5483
	WT (C/C)	153 0.0 (−0.4; 0.4)	ref.	75 0.0 (−0.4, 0.5)	ref.	78 0.0 (−0.3, 0.4)	ref.
CAL (mm)				**GATA−3 IVS4 +1468 (rs3802604)**			
Change 27.5 months vs. baseline	HG (C/T)	89 −0.5 (−0.9, 0.0)	0.1322	51 −0.4 (−0.9, 0.2)	0.1308	38 −0.6 (−0.9, −0.2)	0.6191
	SNP (T/T)	85 −0.5 (−0.9, −0.2)	**0.0184**	36 −0.5 (−0.8, −0.2)	**0.0085**	49 −0.6 (−1.1, −0.1)	0.4352
	WT (C/C)	35 −0.3 (−0.7, 0.0)	ref.	17 −0.1 (−0.3, 0.0)	ref.	18 −0.4 (−0.9, −0.1)	ref.
Change 27.5 months vs. 3.5 months	HG (C/T)	88 0.0 (−0.3, 0.4)	**0.0473**	51 0.0 (−0.4, 0.5)	0.1344	37 0.0 (−0.3, 0.2)	0.3691
	SNP (T/T)	84 0.0 (−0.4, 0.4)	**0.0172**	35 −0.1 (−0.5, 0.4)	**0.0306**	49 0.0 (−0.3, 0.4)	0.2255
	WT (C/C)	35 0.2 (0.0, 0.5)	ref.	17 0.2 (0.1, 0.5)	ref.	18 0.0 (−0.1, 0.5)	ref.
CAL (mm)				**COX−2 −1195 (rs689466)**			
Change 27.5 months vs. baseline	HG (C/T)	64 −0.4 (−0.7, −0.1)	0.1231	32 −0.3 (−0.7, 0.0)	0.4053	32 −0.4 (−0.8, −0.2)	0.1609
	SNP (T/T)	7 −0.5 (−0.7, 0.0)	0.7789	3 −0.5 (−2.0, 0.0)	0.5373	4 −0.4 (−0.7, 0.1)	0.2534
	WT (C/C)	138 −0.5 (−0.9, 0.0)	ref.	69 −0.4 (−0.8, 0.0)	ref.	69 −0.7 (−1.1, −0.1)	ref.
Change 27.5 months vs. 3.5 months	HG (C/T)	64 0.1 (−0.3, 0.5)	0.2653	32 0.0 (−0.3, 0.6)	0.5883	32 0.1 (−0.3, 0.5)	0.2533
	SNP (T/T)	7 0.2 (−0.2, 0.8)	0.4115	3 0.4 (−0.9, 0.8)	0.5986	4 0.1 (−0.1, 0.6)	0.5819
	WT (C/C)	136 0.0 (−0.3, 0.3)	ref.	68 0.1 (−0.5, 0.4)	ref.	68 0.0 (−0.2, 0.3)	ref.

Results are from n = 209 patients (placebo n = 104, antibiotics n = 105) and reported as median (25% quantile, 75% quantile) for continuous variables. p-values are from Mann–Whitney U-tests for the pairwise comparison with the WT group. Negative values represent an improvement and positive values represent a deterioration. Abbreviations: CAL: mean clinical attachment level per patient, HG: heterozygote, SNP: single nucleotide polymorphism, WT: wild type, vs.: versus, mm: millimeter, ref.: reference category for pairwise comparison.

With regard to the GATA-3 genotypes, the results presented in Table 3 are summarized as follow: after 27.5 months, changes in CAL (between visit 12 and visit 2) and changes between visit 12 and visit 4 were larger in individuals presenting the GATA-3 SNP (A/A) than GATA-3 wild-type (G/G), which means that between baseline and 27.5 months there was a greater CAL gain in GATA-3 SNP (A/A) individuals than in wild-type individuals, while wild-type patients showed more CAL loss after the 3.5 months visit than GATA-3

SNP (A/A) patients. In the total group and in the placebo group, there was a higher attachment loss for the wild-type patients ($p < 0.05$). In addition, in the total group, the GATA-3 heterozygous patients showed less CAL loss after 27.5 months (between visit 12 and visit 4) than wild-type patients ($p = 0.0473$). With regard to the IL-1A genotypes, there was higher gain of CAL after 6 months (between visit 6 and visit 2) in the total group of IL-1A −889 wild-type patients compared to heterozygous patients ($p = 0.0170$). In the placebo group, there was a higher CAL gain after 27.5 months (between visit 12 and visit 4) in the heterozygous patients versus the wild-type patients ($p = 0.0444$). In contrast, CAL loss was higher after 27.5 months (between visit 12 and visit 2 and between visit 12 and visit 4, respectively) in the groups of patients carrying the IL-1B +3954 wild type compared to patients with heterozygous and SNP genotypes, but not statistically noticeably different ($p > 0.05$).

In addition, there was a higher CAL loss after 27.5 months (between visit 12 and visit 4) between the patients carrying the SNP versus the wild-type patients in the antibiotics group ($p = 0.0427$). In the placebo group, differences between the genotypes were smaller.

After 27.5 months, changes in BOP (Table 4) and plaque scores (Table 5) (between visit 12 and visit 2 as well as between visit 12 and visit 4) were not influenced by genotypes of the analyzed gene loci.

Table 4. Clinical measurements of the change of the bleeding on probing (BOP).

	Genotype	Total Group Patients *n* BOP Median (25% Quantile, 75% Quantile)	*p*-Value	Placebo Group Patients *n* BOP Median (25% Quantile, 75% Quantile)	*p*-Value	Antibiotics Group Patients *n* BOP Median (25% Quantile, 75% Quantile)	*p*-Value
BOP (%)				IL-1A −889 (rs1800587)			
Change 27.5 months vs. baseline	HG (C/T)	87 −21.0 (−33.3, −7.5)	0.5562	46 −13.3 (−29.2, −3.9)	0.5553	41 −24.2 (−42.3, −14.2)	0.7237
	SNP (T/T)	20 −17.1 (−31.6, −3.0)	0.9176	9 −11.9 (−2649.4, −3.0)	0.9489	11 −22.2 (−35.5, −1.2)	0.7972
	WT (C/C)	102 −19.8 (−33.9, −8.3)	ref.	49 −10.8 (−26.8, −7.9)	ref.	53 −23.5 (−39.9, −9.7)	ref.
Change 27.5 months vs. 3.5 months	HG (C/T)	87 0.6 (−7.3, 6.1)	0.0914	46 0.8 (−6.7, 7.2)	0.2574	41 0.6 (−7.3, 5.0)	0.1697
	SNP (T/T)	20 0.7 (−7.2, 5.6)	0.2476	9 0.8 (−7.1, 5.6)	0.2424	11 0.0 (−7.2, 5.6)	0.6326
	WT (C/C)	102 3.0 (−3.5, 10.4)	ref.	49 3.3 (−2.7, 13.3)	ref.	53 2.3 (−3.6, 10.0)	ref.
BOP (%)				IL-1B +3954 (rs1143634)			
Change 27.5 months vs. baseline	HG (C/T)	64 −17.7 (−28.6, −3.4)	0.3445	31 −10.9 (−26.4, −2.2)	0.3066	33 −22.7 (−35.5, −14.2)	0.5240
	SNP (T/T)	17 −11.9 (−33.9, −3.0)	0.4081	9 −11.9 (−29.2, −4.9)	0.9267	8 −12.7 (−38.6, 0.6)	0.3191
	WT (C/C)	128 −20.4 (−34.1, −9.3)	ref.	64 −12.2 (−29.4, −8.3)	ref.	64 −24.5 (−42.6, −13.5)	ref.
Change 27.5 months vs. 3.5 months	HG (C/T)	64 1.8 (−6.8, 6.9)	0.6668	31 5.6 (−4.8, 16.0)	0.3805	33 −0.7 (−7.3, 5.0)	0.1349
	SNP (T/T)	17 0.0 (−7.2, 5.6)	0.2950	9 0.7 (−15.4, 5.6)	0.3056	8 −0.6 (−5.7, 10.5)	0.6364
	WT (C/C)	128 1.9 (−5.0, 10.2)	ref.	64 2.2 (−5.2, 10.0)	ref.	64 1.6 (−4.0, 10.2)	ref.

Table 4. *Cont.*

	Genotype	Total Group Patients *n* BOP Median (25% Quantile, 75% Quantile)	*p*-Value	Placebo Group Patients *n* BOP Median (25% Quantile, 75% Quantile)	*p*-Value	Antibiotics Group Patients *n* BOP Median (25% Quantile, 75% Quantile)	*p*-Value
BOP (%)				IL-4 −590 (rs2243250)			
Change 27.5 months vs. baseline	HG (C/T)	48 −22.2 (−28.9, −8.6)	1.0	26 −13.3 (−27.0, −6.5)	0.9359	22 −25.3 (−44.9, −14.9)	0.9009
	SNP (T/T)	7 −20.8 (−36.4, 6.9)	0.8848	2 −27.0 (−33.1, −20.8)	0.3380	5 −9.9 (−36.4, 6.9)	0.3366
	WT (C/C)	154 −19.0 (−34.0, −8.2)	ref.	76 −11.5 (−28.6, −3.0)	ref.	78 −22.8 (−39.9, −12.6)	ref.
Change 27.5 months vs. 3.5 months	HG (C/T)	48 2.3 (−2.6, 14.3)	0.2807	26 3.1 (−2.7, 16.0)	0.4220	22 0.6 (−2.6, 13.5)	0.4607
	SNP (T/T)	7 4.2 (−3.7, 14.7)	0.4419	2 8.7 (2.8, 14.7)	0.3704	5 4.2 (−3.7, 5.6)	0.7171
	WT (C/C)	154 1.5 (−6.5, 9.3)	ref.	76 1.8 (−6.9, 9.6)	ref.	78 0.9 (−6.3, 6.7)	ref.
BOP (%)				GATA-3 IVS4 +1468 (rs3802604)			
Change 27.5 months vs. baseline	HG (C/T)	89 −20.0 (−33.3, −9.7)	0.1889	51 −14.1 (−33.3, −7.2)	0.0912	38 −21.1 (−34.4, −14.3)	0.8545
	SNP (T/T)	85 −20.2 (−31.3, −8.7)	0.2245	36 −10.7 (−25.7, −4.6)	0.5764	49 −25.0 (−40.2, −14.2)	0.5266
	WT (C/C)	35 −10.0 (−35.5, 3.2)	ref.	17 −9.8 (−24.1, 0.5)	ref.	18 −26.3 (−43.1, 3.2)	ref.
Change 27.5 months vs. 3.5 months	HG (C/T)	89 2.5 (−4.5, 9.7)	0.7126	51 1.8 (−4.5, 9.3)	0.2068	38 4.2 (−5.1, 10.0)	0.4486
	SNP (T/T)	85 0.0 (−5.9, 5.6)	0.3605	36 0.6 (−6.3, 6.6)	0.1617	49 0.0 (−4.1, 5.0)	0.8602
	WT (C/C)	35 4.2 (−7.3, 13.5)	ref.	17 11.1 (−7.3, 18.8)	ref.	18 −0.7 (−6.3, 9.0)	ref.
BOP (%)				COX-2 −1195 (rs689466)			
Change 27.5 months vs. baseline	HG (C/T)	64 −18.0 (−33.0, −4.1)	0.5439	32 −10.1 (−27.2, −2.1)	0.4474	32 −26.1 (−40.0, −7.3)	0.9188
	SNP (T/T)	7 −24.5 (−36.3, −14.2)	0.5714	3 −36.3 (−36.8, −24.1)	0.0670	4 −19.4 (−25.9, 1.6)	0.3538
	WT (C/C)	138 −19.4 (−33.9, −9.5)	ref.	69 −12.2 (−26.9, −6.2)	ref.	69 −22.2 (−41.0, −12.8)	ref.
Change 27.5 months vs. 3.5 months	HG (C/T)	64 −0.3 (−7.1, 5.3)	0.1567	32 0.3 (−4.2, 5.7)	0.5034	32 −0.6 (−13.8, 4.9)	0.1588
	SNP (T/T)	7 0.6 (−8.8, 16.7)	0.6159	3 −8.6 (−15.5, 2.5)	0.1472	4 8.7 (−4.1, 17.9)	0.4621
	WT (C/C)	138 2.4 (−4.3, 9.7)	ref.	69 3.3 (−5.4, 11.1)	ref.	69 1.5 (−3.5, 9.0)	ref.

Results are from n = 209 patients (placebo n = 104, antibiotics n = 105) and reported as median (25% quantile, 75% quantile) for continuous variables. p-value from Mann–Whitney U-tests. Abbreviations: BOP: percentage of sites showing bleeding on probing per patient, HG: heterozygote, SNP: single nucleotide polymorphism, WT: wild type, vs.: versus.

The multivariable linear regression model confirmed the univariate analysis to explain CAL gain or loss after 27.5 months (Table 6). Individuals presenting with GATA-3 SNPs and heterozygous patients had statistically noticeably lower CAL loss over 27.5 months than wild-type patients. Patients carrying the GATA-3 SNP had −0.32 mm (95% CI: −0.57 mm, −0.07 mm) and heterozygous patients had −0.18 mm (95% CI: −0.43 mm, 0.07 mm) less change in the mean CAL than wild-type patients. In comparison, systemic antibiotic therapy leads to 0.22 mm (95% CI: 0.04 mm, 0.39 mm) more CAL improvement between 27.5 months and baseline than placebo.

Table 5. Clinical measurements of the change of the percentage of sites with plaque per patient (PS).

	Genotype	Total Group Patients *n* PS Median (25% Quantile, 75% Quantile)	*p*-Value	Placebo Group Patients *n* PS Median (25% Quantile, 75% Quantile)	*p*-Value	Antibiotics Group Patients *n* PS Median (25% Quantile, 75% Quantile)	*p*-Value
PS (%)				**IL-1A −889 (rs1800587)**			
Change 27.5 months vs. baseline	HG (C/T)	87 −2.9 (−19.2, 15.0)	0.8510	46 −3.3 (−18.5, 9.2)	0.4561	41 0.0 (−20.4, 17.6)	0.3430
	SNP (T/T)	20 5.3 (−16.3, 20.1)	0.2619	9 3.6 (−16.9, 17.7)	0.9489	11 12.6 (−15.6, 22.5)	0.1755
	WT (C/C)	102 −1.3 (−20.8, 10.6)	ref.	49 1.8 (−18.6, 12.2)	ref.	53 −10.2 (−21.4, 9.2)	ref.
Change 27.5 months vs. 3.5 months	HG (C/T)	87 9.0 (−2.9, 22.5)	0.4696	46 8.3 (−2.9, 22.5)	0.5073	41 11.5 (−5.4, 21.6)	0.7198
	SNP (T/T)	20 5.5 (−6.8, 27.7)	0.6811	9 −3.4 (−10.7, 6.0)	0.3910	11 23.3 (0.9, 46.3)	0.1357
	WT (C/C)	100 8.4 (−5.7, 19.2)	ref.	48 8.4 (−9.6, 19.2)	ref.	52 8.4 (−2.5, 19.1)	ref.
PS (%)				**IL-1B +3954 (rs1143634)**			
Change 27.5 months vs. baseline	HG (C/T)	64 0.4 (−16.0, 15.4)	0.5137	31 −2.9 (−17.6, 9.2)	0.4768	33 7.1 (−15.6, 17.6)	0.1002
	SNP (T/T)	17 0.0 (−19.2, 17.7)	0.7731	9 −5.0 (−19.2, 17.7)	0.9002	8 2.2 (−22.4, 17.5)	0.5926
	WT (C/C)	128 −3.0 (−20.6, 10.3)	ref.	64 0.7 (−18.5 17.3)	ref.	64 −7.8 (−24.0, 8.2)	ref.
Change 27.5 months vs. 3.5 months	HG (C/T)	64 9.3 (−3.8, 19.9)	0.8409	31 9.0 (−5.9, 21.2)	0.9584	33 9.5 (0.8, 17.9)	0.7996
	SNP (T/T)	17 6.0 (−5.2, 28.3)	0.7627	9 5.0 (−3.4, 10.0)	0.7473	8 25.8 (−8.5, 37.8)	0.3078
	WT (C/C)	126 7.3 (−3.9, 20.0)	ref.	63 7.1 (−5.4, 22.5)	ref.	63 7.9 (−2.6, 20.0)	ref.
PS (%)				**IL-4 −590 (rs2243250)**			
Change 27.5 months vs. baseline	HG (C/T)	48 −4.9 (−20.6, 12.6)	0.3825	26 −4.9 (−19.2, 9.0)	0.1833	22 −7.0 (−21.4, 17.2)	0.9603
	SNP (T/T)	7 −21.4 (−50.0, 3.0)	0.0773	2 −35.7 (−50.0, −21.4)	0.0889	5 −6.9 (−34.3, 3.0)	0.4075
	WT (C/C)	154 0.4 (−18.5, 13.0)	ref.	76 1.7 (−17.2, 22.4)	ref.	78 −1.2 (−19.2, 11.1)	ref.
Change 27.5 months vs. 3.5 months	HG (C/T)	47 7.0 (−6.3, 21.2)	0.7473	26 9.1 (−5.2, 21.2)	0.8102	21 6.5 (−7.1, 19.6)	0.5087
	SNP (T/T)	7 −1.0 (−15.3, 20.0)	0.4539	2 −8.1 (−15.3, −1.0)	0.2335	5 11.1 (−2.8, 20.0)	0.8263
	WT (C/C)	153 8.8 (−3.6, 21.6)	ref.	75 7.4 (−5.4, 19.7)	ref.	78 10.1 (0.0, 22.0)	ref.

Table 5. *Cont.*

	Genotype	Total Group Patients *n* PS Median (25% Quantile, 75% Quantile)	*p*-Value	Placebo Group Patients *n* PS Median (25% Quantile, 75% Quantile)	*p*-Value	Antibiotics Group Patients *n* PS Median (25% Quantile, 75% Quantile)	*p*-Value
PS (%)				**GATA-3 IVS4 +1468 (rs3802604)**			
Change 27.5 months vs. baseline	HG (C/T)	89 0.8 (−21.4, 17.7)	0.9669	51 −3.2 (−23.4, 22.4)	0.2672	38 3.4 (−15.3, 17.6)	0.2150
	SNP (T/T)	85 −1.9 (−19.1, 8.3)	0.4561	36 1.0 (−13.5, 9.4)	0.4665	49 −10.3 (−20.4, 7.1)	0.8768
	WT (C/C)	35 −2.9 (−17.6, 15.5)	ref.	17 0.0 (−7.8, 22.6)	ref.	18 −5.7 (−30.2, 15.2)	ref.
Change 27.5 months vs. 3.5 months	HG (C/T)	88 9.7 (−3.8, 20.4)	0.5387	51 6.0 (−5.9, 19.7)	0.6878	37 13.6 (5.6, 21.6)	0.1354
	SNP (T/T)	84 6.1 (−5.4, 22.5)	0.7468	35 9.4 (−1.5, 18.1)	0.9690	49 4.5 (−5.6, 23.3)	0.6778
	WT (C/C)	35 6.5 (−6.9, 18.5)	ref.	17 4.6 (−2.9, 24.0)	ref.	18 6.8 (−8.7, 15.1)	ref.
PS (%)				**COX-2 −1195 (rs689466)**			
Change 27.5 months vs. baseline	HG (C/T)	64 −6.5 (−21.1, 12.8)	0.2872	32 −8.3 (−25.0, 13.7)	0.0985	32 −1.2 (−17.1, 12.8)	0.9333
	SNP (T/T)	7 −17.6 (−20.4, 9.2)	0.5683	3 4.5 (−17.6, 9.2)	1.0	4 −19.7 (−37.1, 22.2)	0.5072
	WT (C/C)	138 0.6 (−19.2, 14.3)	ref.	69 1.5 (17.6, 15.0)	ref.	69 −0.5 (−22.0, 13.0)	ref.
Change 27.5 months vs. 3.5 months	HG (C/T)	64 7.7 (−6.0, 22.3)	0.8578	32 10.2 (−4.4, 22.7)	0.9061	32 6.8 (−7.7, 20.5)	0.6826
	SNP (T/T)	7 8.8 (−5.6, 46.3)	0.6440	3 4.6 (−20.4, 18.8)	0.7650	4 27.5 (1.6, 49.4)	0.3471
	WT (C/C)	136 8.0 (−3.3, 19.8)	ref.	68 4.4 (−5.3, 18.3)	ref.	68 10.2 (−0.2, 21.0)	ref.

Results are from $n = 209$ patients (placebo $n = 104$, antibiotics $n = 105$) and reported as median (25% quantile, 75% quantile) for continuous variables. *p*-value from Mann–Whitney U-tests. Abbreviations: PS: percentage of sites with plaque per patient, HG: heterozygote, SNP: single nucleotide polymorphism, WT: wild type, vs.: versus.

Table 6. Multivariable linear model estimates for the change of mean CAL from 27.5 months versus baseline and from 27.5 months versus 3.5 month.

		Dependent Variable							
		Change of Mean CAL (27.5 Month Baseline) Per Patient				**Change of Mean (27.5 Month - 3.5 Month) Per Patient**			
Independent Variables		**β**	**95% CI**		***p*-Value**	**β**	**95% CI**		***p*-Value**
Intercept		−0.39	−0.64	−0.13	0.0027	0.22	0.01	0.43	0.0393
Therapy	Placebo vs. AB	0.22	0.04	0.39	**0.0132**	−0.03	−0.18	0.11	0.6509
IL-1A -889	Global				0.1800				0.0878
	HG vs. WT	−0.17	−0.39	0.04	0.1281	−0.17	−0.36	0.0	0.0574
	SNP vs. WT	0.06	−0.36	0.49	0.7801	0.03	−0.32	0.38	0.8654
IL-1B +3954	Global				0.1117				0.3163
	HG vs. WT	0.20	−0.03	0.44	0.1010	0.13	−0.06	0.33	0.1833

Table 6. *Cont.*

		Dependent Variable							
		Change of Mean CAL (27.5 Month Baseline) Per Patient				Change of Mean (27.5 Month - 3.5 Month) Per Patient			
Independent Variables		β	95% CI		*p*-Value	β	95% CI		*p*-Value
	SNP vs. WT	−0.12	−0.57	0.31	0.5754	−0.02	−0.39	0.33	0.8816
IL-4 −34	Global				0.4058				0.2384
	HG vs. WT	0.86	−0.43	2.15	0.1903	0.17	−0.89	1.24	0.7425
	SNP vs. WT	0.65	−1.21	2.52	0.4915	1.12	−0.41	2.67	0.1518
IL-4 −590	Global				0.4114				0.3364
	HG vs. WT	−0.86	−2.14	0.42	0.1886	−0.07	−1.13	0.99	0.8961
	SNP vs. WT	−0.72	−2.54	1.08	0.4298	−0.86	−2.36	0.64	0.2600
GATA-3 +1468	Global				**0.0355**				0.0605
	HG vs. WT	−0.18	−0.43	0.07	0.1591	−0.21	−0.42	0.0	**0.0424**
	SNP vs. WT	−0.32	−0.57	−0.07	**0.0116**	−0.24	−0.45	−0.03	**0.0218**
COX-2 -1195	Global				0.3306				0.4510
	HG vs. WT	0.14	−0.04	0.33	0.1384	0.09	−0.06	0.25	0.2523
	SNP vs. WT	0.02	−0.47	0.51	0.9261	0.14	−0.26	0.55	0.4935

Results are from n = 209 patients. Intercept: Therapy = Antibiotics. IL-1A −889 = WT, IL-1B +3954 = WT, IL-4 −34 = WT, IL-4 −590 = WT, GATA-3 +1468 = WT, and COX-2-1195 = WT. For example a placebo patient with IL-1A −889 = HG, IL-1B +3954 = SNP, IL-4 −34 = HG, IL-4 −590 = SNP, GATA-3 +1468= SNP, and COX-2-1195= WT has an expected change of the mean CAL between 27.5 months and baseline of: −0.39 + 0.22 − 0.17 − 0.12 + 0.86 − 0.72 − 0.32 + 0 = −0.64 mm. *p*-values for pairwise comparisons are from the Wald Tests and global *p*-values are from the F-Test. Abbreviations: AB = Antibiotic, HG = Heterozygous genotype, SNP = Single nucleotide polymorphism, WT: wild type, β = regression coefficient (least square mean estimate), CI = Confidence interval (lower limit, upper limit).

The IL-1A- 889 gene locus seems to influence CAL change (between visit 12 and visit 4) after 27.5 months. In addition, the figures of CAL change (between visit 12 and visit 2 respectively between visit 12 and visit 4) from the GATA-3 IVS4 +1468 gene loci are shown in Figure 1.

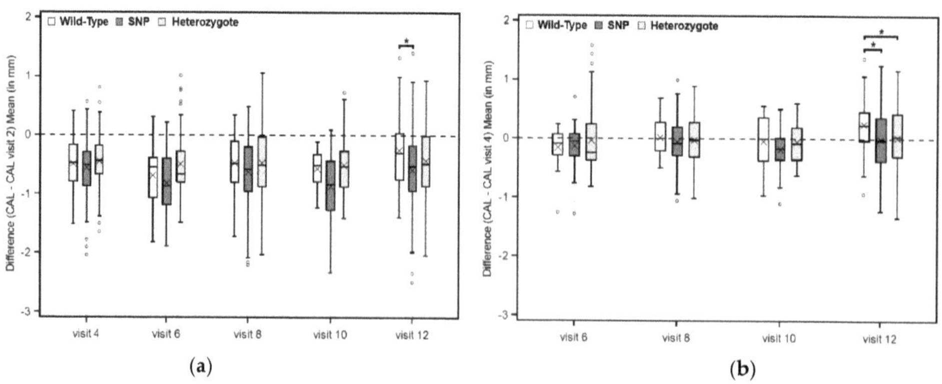

Figure 1. Boxplots of GATA-3: Mean CAL differences to visit 2 (**a**) and to visit 4 (**b**) in the total collective: * *p*-values are from Mann–Whitney U-tests for the pairwise comparison with the wild-type group. Negative values represent an improvement and positive values represent a deterioration. Abbreviations: CAL: mean clinical attachment level per patient, SNP: single nucleotide polymorphism, mm: millimeter, × marks the mean.

The percentage of sites per patient showing further attachment loss (PSAL) ≥1.3 mm showed similar results (results are shown in the supplementary file).

3. Discussion

In the present study, the influence of IL-1, IL-4, GATA-3 and COX-2 polymorphisms on the outcomes of non-surgical therapy with and without antibiotics and 2 years of maintenance was analyzed in Caucasian periodontitis patients. This is an exploratory subanalysis of a subset of the per-protocol collective from the prospective, randomized, double-blind, multi-center ABPARO trial (ClinicalTrials.gov NCT00707369) on the effect of adjunctive systemic amoxicillin 500 mg plus metronidazole 400 mg (3×/day, 7 days) [44,46]. The clinical results reported improvement in all clinical parameters after non-surgical periodontal therapy. Overall, after 27.5 months, the additional use of adjunctive antibiotics led to significantly better clinical results than mechanical debridement alone [44]. In severely diseased and younger patients, adjunctive antibiotics led to clinically relevant better improvements than placebo [42]. Nevertheless, the use of antibiotics should be carefully considered. Recent studies indicate that probiotics, paraprobiotics and postbiotics have a positive clinical effect on periodontitis remission [47,48]. This could be an alternative to reduce antibiotic medication.

In the last decades, some genetic markers which may be associated with the progression of the disease have been identified [12]. It is known that periodontitis is a polygenetic disease [49–51]. Individual genetic factors may modify the host immune response to the dysbiotic biofilm [7,52,53]. Associations between genetic factors, systemic diseases and lifestyle factors were also reported [54]. Meta-analyses of case-control studies demonstrated that SNPs were associated with periodontitis [14–17,25,55,56]. Only few studies demonstrated large effects of SNPs on the stage of disease [39,57,58].

This is the first study evaluating longitudinally the effect of 6 gene loci on changes in CAL, PSAL ≥1.3 mm, BOP and plaque scores following non-surgical periodontal treatment and 2 years of maintenance. The primary outcome variable of this subanalysis was the change of the mean CAL per patient between 27.5 months (visit 12) and baseline (visit 2), and between 27.5 months and 3.5 months (visit 4), respectively. Periodontitis progression was quantified by the paired comparison of the mean CAL of the 3 genotypes in 6 gene loci. In addition, the adjusted influence of the gene loci on disease progression was assessed by a multivariate analysis. A total of 209 patients (104 placebos, 105 antibiotics) were genotyped. Out of these patients, 136 could not be genotyped because the amount of sampled blood was insufficient.

The main results demonstrate differences for the CAL change between the genotypes. Patients carrying the GATA-3 SNP or the heterozygous genotype showed lower CAL loss during supportive periodontal therapy (SPT) in comparison with the patients carrying the wild-type. These data provide new insights since only few studies have investigated this genetic marker [30–32] and actually no periodontal-genetic clinical study was performed yet. Zhao et al. [59] analyzed the expression of Th17/Th1/Th2 cytokines and transcription factors, and Th17 cell vibration in Chinese CP patients. The expression of transcription factors (RORC, T-bet and GATA-3) in peripheral blood was measured by real-time PCR, and the levels of Th17 cells in CD4(+) T cells were analyzed by flow cytometry. In the mouse model, it was demonstrated that the expression of the Th2 differentiation maintaining transcription factor GATA-3 [28] in Th2 cells after treatment compared to the expression before treatment was increased 1.76-fold ($p < 0.05$). Intracellular staining of IL-17 revealed that the quantity of Th17 cells in contrast decreased ($p < 0.05$), especially the IL-17(+) IFN-γ(+) subgroup. The results indicated a protective effect of Th2 cells.

The results of the present study with regard to the IL-1 genotype are similar to the results of the study by Ehmke et al. [9] investigating the prognostic value of the IL-1 haplotype on the progression of periodontal disease following therapy. Forty-eight adult patients with untreated periodontitis harboring *Aggregatibacter actinomycetemcomitans* and/or *Porphyromonas gingivalis* were randomly assigned to receive full-mouth scaling alone (control) or in combination with systemic antibiotic treatment (metronidazole plus amoxicillin) and supragingival irrigation with chlorhexidine digluconate (test). All patients received SPT at 3 to 6-month intervals. In 33 patients, DNA was analyzed for polymorphism in the IL-1A gene at position −889 and IL-1B gene at position +3953. These results indicated that the IL-1 haplotype may be of limited value for the prognosis of periodontal disease progression following non-surgical therapy.

The study of Meisel et al. [21] evaluated the genetic influence of IL-1A, IL-1B and IL-1RN polymorphisms on periodontal variables in relation to environmental factors such as smoking. No differences were found in allele frequencies or combined allotypes between subjects with mild or moderate versus those with severe periodontitis. However, the extent of CAL loss defined as percentage of sites >4 mm was significantly associated with the composite genotype of IL-1A/-1B in smokers. The results provided evidence that the genotypes studied show interaction with smoking, the main exposition-related risk factor of periodontitis.

Eickholz et al. [10] analyzed patient-related factors contributing (1) to tooth loss and (2) to the quality of treatment outcomes 10 years after initiation of non-surgical therapy. All patients who had received active periodontal treatment 10 years ago by the same examiner were recruited until a total of 100 patients was re-examined. The following risk factors for tooth loss were identified: ineffective oral hygiene, irregular SPT, IL-1A/-1B polymorphism, initial diagnosis, smoking, age and sex.

In the present study, there were small or no differences in the change of CAL between the genotypes for the IL-1B +3954 gene locus. Wild-type patients showed higher improvement of CAL after non-surgical periodontal therapy than heterozygote or SNP patients. In the antibiotic group there was a difference in the change of the CAL between 27.5 months and 3.5 months between SNP versus wild-type patients ($p = 0.0427$).

Successful anti-infective therapy as well as SPT in short intervals [60] led to insignificant differences for BOP and plaque score between genotypes. With regard to the IL-1A −889 genotypes the median change in CAL (between 27.5 months and 3.5 months) observed in the placebo group was 0.0 mm (−0.5 mm, 0.2 mm) compared to 0.2 mm (−0.1 mm, 0.5 mm) between heterozygote versus wild-type patients ($p = 0.0444$). Therefore, it may be concluded that heterozygote patients had less CAL loss during 24 months of SPT. The multivariable linear regression models support these results.

Although many case-control studies have been conducted on the association of IL-4 with periodontitis [25], there is no longitudinal study that can be compared with the present analysis. Comparison of the clinical parameters in the IL-4 and COX-2 genotype showed no differences or indifferent results between the genotypes. The role of COX-2 in the progression of periodontitis has been previously reported. Important mediators are the end-products of the COX cycles such as prostaglandin, prostacycline, thromboxane and leukotrienes [61]. Mesa et al. [62] reported an association of higher COX-2 expression levels with CAL loss, BOP and loss of connective tissue in gingivitis and periodontitis patients. Beikler et al. [63] reported a significant reduction of the gene expression of COX-2 after non-surgical periodontal therapy.

Three case-control studies focused on the frequency of the COX-2 −1195 gene locus in periodontitis patients [39,40,64], including a Caucasian population [40]. The ethnologically comparable study of Schäfer et al. [40] with the present study did not identify an association between SNP at position −1195 and CP or AgP patients. In contrast, Xie et al. [39] demonstrated a significantly increased risk of CP in a Chinese population in the presence of SNP at position −1195. Daing et al. [64] confirmed these results in a North Indian population.

The secondary outcomes in the present study were PSAL $\geq$ 1.3 mm, BOP and plaque score. PSAL $\geq$ 1.3 mm demonstrated similar results as CAL. No difference in the change of BOP and plaque score was observed between the genotypes. Our results were in concordance with the data of Ehmke et al. (1999) [9] and Konig et al. (2005) [19] who analyzed both IL-1 gene loci from Caucasian CP patients. PSAL $\geq$ 2 mm and BOP showed no differences between the IL-1 genotypes 2 years after SPT [9]. Furthermore, no differences in pocket probing depths (PPD), tooth loss and plaque score could be demonstrated after 13 years of SPT [19].

The present study investigated the association of several known genetic factors in the pathogenesis of periodontitis with disease progression and the influence of systemic antibiotics. Young patients would benefit from an early diagnosis of severe periodontal disease. If many teeth had to be extracted in the early years of life due to periodontitis, additional complications will arise in the future, including rehabilitation by implants. Peri-implantitis has a more rapid disease progression than periodontitis [65] and today's life expectancy for young people suggests multiple future implant placements at the same site. This creates a high level of physical and financial burden for the affected patients.

4. Materials and Methods

4.1. Study Design

The present study was designed as a subanalysis of the per-protocol collective data from the prospective, randomized, stratified, double-blind, multi-center trial (ClinicalTrials.gov NCT00707369) [46]. The results of the clinical study were previously reported [41–45]. The trial analyzed the effects of adjunctive systemic administration of amoxicillin 500 mg plus metronidazole 400 mg (3×/day, 7 days) on different periodontal parameters in patients suffering from moderate to severe periodontitis. Patients who followed the study timeline according to the protocol were included in the per-protocol collective.

Caucasian patients with untreated moderate to severe CP and AgP were included (periodontitis stage III or IV with grade B or C). Key inclusion criteria were: age (18–75 years), a Community Periodontal Index of Treatments Needs (CPITN) of IV in at least one sextant, at least 10 natural teeth in situ and PPD of ≥6 mm at a minimum of 4 teeth. Key exclusion criteria were: confirmed or assumed allergies or former hypersensitive skin reactions to amoxicillin and/or metronidazole, systemic medications affecting periodontal health and pregnancy. Type 2 diabetes mellitus was diagnosed by taking a blood sample and determining the HbA1c. Smoking status was checked by Bedfont-Smokerlyzer (Bedfont, UK) and the body mass index was calculated. Other systemic diseases were inquired by anamnesis (e.g., Down Syndrome, AIDS/HIV or systemic medication affecting periodontal conditions). For details, refer to Harks et al. [46].

Genetic evaluations were performed after closure of the ABPARO database [44], i.e., the laboratory staff were blinded to all dental measurements. All per-protocol patients in whom genotyping was possible were included.

The institutional review boards (IRB) of the participating centers approved the protocol and all patients had provided written informed consent. An independent data and safety monitoring board reviewed the safety data throughout the trial.

4.2. Periodontal Therapy

The study duration per patient comprised 12 visits over 27.5 months. Within 1.5 months after baseline examination (visit 2), patients received supra- and subgingival debridement in one or two sessions on one/two consecutive days (visit 3). All mechanical therapy was performed with hand instruments and/or machine-driven scalers. After completion of mechanical debridement, the antibiotic group of patients received two antibiotics (amoxicillin $3H_2O$ 574 mg (Amoxicillin-ratiopharm 500 mg®, Ratiopharm, Ulm, Germany); metronidazole 400 mg (Flagyl® 400, Sanofi-Aventis, Frankfurt, Germany)). The placebo group patients received two placebo tablets, each to be taken 3 times a day for 7 days. In addition, patients were instructed to rinse their mouth with Chlorhexidine 0.2% twice per day for 7 days. Furthermore, they were instructed to brush with their toothbrush 2 min twice per day at home applying a technique suitable for the individual. Additional oral hygiene devices, such as dental floss or interdental brushes, were recommended depending on the patient's individual need.

Re-evaluation (visit 4) was performed 3.5 months after baseline and at least 2 months after mechanical debridement. Thereafter, all patients received SPT, including full-mouth supragingival debridement and oral hygiene instructions at 3-month intervals. Sites with PPD $\geq$ 4 mm also received subgingival re-debridement.

4.3. Examinations

Periodontal parameters were assessed at 6 sites of each tooth by blinded examiners not involved in the periodontal treatment. All measurements were performed at baseline (visit 2), 3.5 months (re-evaluation, visit 4) and at 27.5 months (visit 12). CAL was calculated by measurements of PPD and gingival recession that were performed with an electronic pressure-sensitive probe (Standard Florida Probe, Gainesville, FL, USA) in increments of 0.2 mm. The mean CAL (in mm) was calculated for each patient. The difference in the mean CAL between the 27.5 months (visit 12) to visit 2 and visit 4 described the changes of the CAL (gain or loss of tooth supporting tissue). After the X-ray appraisal and the evaluation of the anamnesis, the diagnosis was made according to the 1999s classification [66,67].

The primary outcome variable in this study was the change in mean CAL per patient and was assessed in an exploratory manner. Thus, the presence of attachment loss is representative for disease progression. In addition, the following secondary endpoints were included: PSAL $\geq$1.3 mm, percentage of sites per patient with BOP [68] and percentage of sites per patient showing supragingival plaque score [69]. BOP and Plaque scores were recorded at the baseline investigation (before subgingival instrumentation), at re-evaluation and during 24-month SPT (plaque score and BOP in a 3-month interval) (compare Harks et al. (2015) [44]). Differences in the percentage of BOP and in the percentage of supragingival plaque score per patient were calculated between 27.5 months (visit 12) and baseline (visit 2) or re-evaluation (visit 4).

4.4. Genotyping

Gene loci at positions: IL-1A −889 (rs1800587), IL-1B +3954 (rs1143634), IL-4 −34 (rs2070874), IL-4 −590 (rs2243250), GATA-3 IVS4 +1468 (rs3802604) and COX-2 −1195 (rs689466) were analyzed using qPCR. 345 FTA elute micro cards (GE Healthcare UK, Little Chalfont, UK) from patients were used to obtain genomic DNA. A Ø 1.25 mm Harris micro punch (GE Healthcare UK, Little Chalfont, UK) was used to excise 6 pieces from a FTA card. The pieces were placed into a micro centrifuge tube, added with 500 µL of dH2O and vortexed 3x for 5 s. Thereafter, the pipette was used to remove the water. After 5 s centrifugation, the DNA elution was performed in 50 µL dH2O at 95 °C for 30 min in a heating block. The quality of DNA was analyzed by NanoDrop 2000 (Thermo Fisher Scientific, Watham/MA, USA) UV-Vis Spectrophotometer and monitored for the values of OD260/OD280. Then, 1 µL of the DNA extract was used as template in each qPCR. The qPCR was performed with the TaqMan SNP Assay®, and the TaqMan Genotyping Master Mix® (both Applied Biosystems Inc., Foster City/CA, USA) and analyzed using a Bio-Rad C1000 thermal cycler (Bio-Rad Laboratories Inc., Hercules/CA, USA). For validation of each SNP, one positive and two negative controls were included. Each sample was analyzed three times. The results of the TaqMan qPCR were verified by Sanger sequencing.

4.5. Statistical Analysis

Statistical analyses were performed using SAS software, version 9.4 of the SAS System for Windows (SAS Institute, Cary/NC, USA). Inferential statistics such as p-values and confidence intervals were intended to be exploratory, not confirmatory. The p-values represent a metric measure of evidence against the respective null hypothesis and were used only to generate new hypotheses. Therefore, neither global nor local significance levels were determined, and no adjustment for multiplicity was applied. Consequently, explorative two-sided p-values $\leq$ 0.05 were denoted as statistically noticeable instead of significant.

Sample size calculation was performed for the initial randomized ABPARO trial (compare [44]). In the exploratory analyses in this study, all per-protocol patients in whom genotyping was possible (n = 209) were included.

Standard univariate statistical analyses were performed to describe demographical and clinical characteristics. Categorical variables are shown as absolute and relative frequencies. Normally distributed continuous variables are shown as mean ± standard deviation. Not normally distributed continuous variables are reported as median (25% quantile–75% quantile). The relationship between the categorical variables was verified using Fisher's exact test. Pairwise comparisons of outcome variables between wild type, heterozygous genotype and SNP were performed using two-sided nonparametric Mann–Whitney U-tests. In order to estimate the adjusted influence of the gene loci on the changes of the CAL, a multivariate analysis was performed by fitting a linear regression model. The dependent variables were the difference in the mean CAL per patient between visit 12 and 2 and the difference between visit 12 and 4. The influencing factors were the main effect of treatment group and of each gene loci: IL-1A −889 (rs1800587), IL-1B +3954 (rs1143634), IL-4 −34 (rs2070874), IL-4 −590 (rs2243250), GATA-3 IVS4 +1468 (rs3802604) and COX-2 -1195 (rs689466). Results are reported as regression coefficients and the corresponding 95% confidence interval (CI).

5. Conclusions

Within the limitations of this explorative analysis, the results of the present study demonstrate that specific genetic polymorphisms may have an impact on long-term progression of periodontitis. GATA-3 genotypes might be useful for predicting disease progression in severe periodontitis. Individuals with the GATA-3 IVS4 +1468 SNP showed better CAL reduction and attachment gain than the wild type after non-surgical therapy and 2 years of maintenance. Testing the genotype status in the future might be valuable for individualizing SPT. The results for IL-1, IL-4 and COX-2 polymorphisms were inconsistent as regards the risk for further CAL loss in a well-maintained patient group. Further prospective studies are required to confirm these results.

Supplementary Materials: The following supporting information can be downloaded at: https://www.mdpi.com/article/10.3390/ijms23137266/s1.

Author Contributions: K.-A.W. performed most of the experiments, analyzed the data and wrote the original draft of the manuscript. J.R.G. and S.G. were responsible for all experimental design and critically revising the entire manuscript. B.E., D.K., K.L., P.E., T.K., T.-S.K. and U.S. contributed to the conception, provided clinical patient data and were involved in the development of this article. R.K. performed the statistical analyses and editing the manuscript. B.E. is the Principal Investigator of the ABPARO study. J.M. is the Principal Investigator of this study and was involved in the final revision of the manuscript. All authors have read and agreed to the published version of the manuscript.

Funding: This research was funded by a grant from the German Research Foundation (Deutsche Forschungsgemeinschaft (DFG) grant number: EH 365 1-1), the Deutsche Gesellschaft für Zahn-, Mund- und Kieferheilkunde Research Grant (grant number: 62880362), and from the authors' institutions.

Institutional Review Board Statement: The study was conducted in accordance with the Declaration of Helsinki, and approved by the University of Giessen, Faculty of Medicine Ethics Committee under protocol number AZ 53/13.

Informed Consent Statement: Informed consent was obtained from all subjects involved in the study.

Data Availability Statement: The raw data and other related data in the manuscript are available from the corresponding author, K.-A.W., upon reasonable request.

Acknowledgments: The authors are greatly indebted to the collaborators and staff members who represent the ABPARO-Group for their successful work on this project. The members of the ABPARO-Group are listed below: Study center, University Hospital Muenster: Christina Elberg, Heike Frieling-Braithwaite, Anna-Maria Marx, Marie Christin Ohlmeier, Martin Sachs and Thomas Weniger. University Hospital Berlin: Peter Purucker, Marta Czownicka, Kathleen Kraatz, Nicole Pischon and Bernd-Michael Kleber. University Hospital Dresden: Gerlinde Bruhn and Ihssan Khallili. University Hospital Frankfurt: Bettina Dannewitz, Katrin Nickles, Lasse Röllke, Susanne Scharf and Martin Wohlfeil. University Hospital Giessen: Heidi Fastnacht and Tomas Cabrera-Chica. University Hospital Greifswald: Jutta Daus and Jutta Fanghänel. University Hospital Heidelberg: Raluca Cosggarea, Amelie Meyer-Bäumer, Nihad El Sayed, Sven Zehaczek and Nils Zimmermann. University Hospital Wuerzburg: Markus Bechtold, Yvonne Jockel-Schneider and Simone Veihelmann. Institute of Biostatistics and Clinical Research, Medical Faculty Muenster: Andreas Faldum, Joachim Gerß and Achim Heinecke. Clinical Pharmacy, University Hospital Dresden: Ina-Maria Klut and Madeleine Schubert. Institute of Clinical Chemistry and Laboratory Medicine, University Hospital Greifswald: Matthias Nauck, Astrid Petersmann and Helma Preez. Data Monitoring and Safety Board: Guido Knapp, Gregor Petersilka and Anne Sonntag.

Conflicts of Interest: The authors declare no conflict of interest. The funders had no role in the design of the study; in the collection, analyses, or interpretation of data; in the writing of the manuscript, or in the decision to publish the results.

References

1. Chapple, I.L.C.; Mealey, B.L.; van Dyke, T.E.; Bartold, P.M.; Dommisch, H.; Eickholz, P.; Geisinger, M.L.; Genco, R.J.; Glogauer, M.; Goldstein, M.; et al. Periodontal health and gingival diseases and conditions on an intact and a reduced periodontium: Consensus report of workgroup 1 of the 2017 world workshop on the classification of periodontal and peri-implant diseases and conditions. *J. Clin. Periodontol.* **2018**, *45* (Suppl. 20), S68–S77. [CrossRef] [PubMed]
2. Papapanou, P.N.; Sanz, M.; Buduneli, N.; Dietrich, T.; Feres, M.; Fine, D.H.; Flemmig, T.F.; Garcia, R.; Giannobile, W.V.; Graziani, F.; et al. Periodontitis: Consensus report of workgroup 2 of the 2017 world workshop on the classification of periodontal and peri-implant diseases and conditions. *J. Clin. Periodontol.* **2018**, *45* (Suppl. 20), S162–S170. [CrossRef] [PubMed]
3. Michalowicz, B.S.; Aeppli, D.; Virag, J.G.; Klump, D.G.; Hinrichs, J.E.; Segal, N.L.; Bouchard, T.J., Jr.; Pihlstrom, B.L. Periodontal findings in adult twins. *J. Periodontol.* **1991**, *62*, 293–299. [CrossRef] [PubMed]
4. Michalowicz, B.S.; Diehl, S.R.; Gunsolley, J.C.; Sparks, B.S.; Brooks, C.N.; Koertge, T.E.; Califano, J.V.; Burmeister, J.A.; Schenkein, H.A. Evidence of a substantial genetic basis for risk of adult periodontitis. *J. Periodontol.* **2000**, *71*, 1699–1707. [CrossRef]
5. Mucci, L.A.; Bjorkman, L.; Douglass, C.W.; Pedersen, N.L. Environmental and heritable factors in the etiology of oral diseases—A population-based study of Swedish twins. *J. Dent. Res.* **2005**, *84*, 800–805. [CrossRef]
6. Torres de Heens, G.L.; Loos, B.G.; van der Velden, U. Monozygotic twins are discordant for chronic periodontitis: Clinical and bacteriological findings. *J. Clin. Periodontol.* **2010**, *37*, 120–128. [CrossRef]
7. Meyle, J.; Chapple, I. Molecular aspects of the pathogenesis of periodontitis. *Periodontology 2000* **2015**, *69*, 7–17. [CrossRef]
8. Brodzikowska, A.; Gorska, R.; Kowalski, J. Interleukin-1 genotype in periodontitis. *Arch. Immunol. Ther. Exp.* **2019**, *67*, 367–373. [CrossRef]
9. Ehmke, B.; Kress, W.; Karch, H.; Grimm, T.; Klaiber, B.; Flemmig, T.F. Interleukin-1 haplotype and periodontal disease progression following therapy. *J. Clin. Periodontol.* **1999**, *26*, 810–813. [CrossRef]
10. Eickholz, P.; Kaltschmitt, J.; Berbig, J.; Reitmeir, P.; Pretzl, B. Tooth loss after active periodontal therapy. 1: Patient-related factors for risk, prognosis, and quality of outcome. *J. Clin. Periodontol.* **2008**, *35*, 165–174. [CrossRef]
11. Kinane, D.F.; Shiba, H.; Hart, T.C. The genetic basis of periodontitis. *Periodontology 2000* **2005**, *39*, 91–117. [CrossRef]
12. Kornman, K.S.; Crane, A.; Wang, H.Y.; di Giovine, F.S.; Newman, M.G.; Pirk, F.W.; Wilson, T.G., Jr.; Higginbottom, F.L.; Duff, G.W. The interleukin-1 genotype as a severity factor in adult periodontal disease. *J. Clin. Periodontol.* **1997**, *24*, 72–77. [CrossRef] [PubMed]
13. Gonzales, J.R.; Michel, J.; Rodriguez, E.L.; Herrmann, J.M.; Bodeker, R.H.; Meyle, J. Comparison of interleukin-1 genotypes in two populations with aggressive periodontitis. *Eur. J. Oral Sci.* **2003**, *111*, 395–399. [CrossRef] [PubMed]
14. Deng, J.S.; Qin, P.; Li, X.X.; Du, Y.H. Association between interleukin-1beta C (3953/4)T polymorphism and chronic periodontitis: Evidence from a meta-analysis. *Hum. Immunol.* **2013**, *74*, 371–378. [CrossRef] [PubMed]
15. Karimbux, N.Y.; Saraiya, V.M.; Elangovan, S.; Allareddy, V.; Kinnunen, T.; Kornman, K.S.; Duff, G.W. Interleukin-1 gene polymorphisms and chronic periodontitis in adult whites: A systematic review and meta-analysis. *J. Periodontol.* **2012**, *83*, 1407–1419. [CrossRef]
16. Mao, M.; Zeng, X.T.; Ma, T.; He, W.; Zhang, C.; Zhou, J. Interleukin-1alpha −899 (+4845) C–>T polymorphism increases the risk of chronic periodontitis: Evidence from a meta-analysis of 23 case-control studies. *Gene* **2013**, *532*, 114–119. [CrossRef]
17. Da Silva, F.R.P.; Pessoa, L.D.S.; Shin, J.I.; Alves, E.H.P.; Koga, R.S.; Smith, C.V.; Vasconcelos, D.F.P.; Pereira, A. Polymorphisms in the interleukin genes and chronic periodontitis: A field synopsis and revaluation by Bayesian approaches. *Cytokine* **2021**, *138*, 155361. [CrossRef]
18. Cattabriga, M.; Rotundo, R.; Muzzi, L.; Nieri, M.; Verrocchi, G.; Cairo, F.; Pini Prato, G. Retrospective evaluation of the influence of the interleukin-1 genotype on radiographic bone levels in treated periodontal patients over 10 years. *J. Periodontol.* **2001**, *72*, 767–773. [CrossRef]
19. Konig, J.; Ruhling, A.; Plagmann, H.C.; Meisel, P.; Kocher, T. Influence of Interleukin (IL)-1 Composite Genotype on Clinical Variables in Non-Smoking, Well-Maintained Compliant Patients with Chronic Periodontitis. *Swed. Dent. J.* **2005**, *29*, 11–16. Available online: https://www.ncbi.nlm.nih.gov/pubmed/15898359 (accessed on 26 June 2022).

20. McGuire, M.K.; Nunn, M.E. Prognosis versus actual outcome. IV. The effectiveness of clinical parameters and IL-1 genotype in accurately predicting prognoses and tooth survival. *J. Periodontol.* **1999**, *70*, 49–56. [CrossRef]
21. Meisel, P.; Siegemund, A.; Dombrowa, S.; Sawaf, H.; Fanghaenel, J.; Kocher, T. Smoking and polymorphisms of the interleukin-1 gene cluster (IL-1alpha, IL-1beta, and IL-1RN) in patients with periodontal disease. *J. Periodontol.* **2002**, *73*, 27–32. [CrossRef] [PubMed]
22. Mazurek-Mochol, M.; Dembowska, E.; Malinowski, D.; Safranow, K.; Pawlik, A. IL-1ss rs1143634 and rs16944 polymorphisms in patients with periodontal disease. *Arch. Oral Biol.* **2019**, *98*, 47–51. [CrossRef] [PubMed]
23. Maggi, E.; Parronchi, P.; Manetti, R.; Simonelli, C.; Piccinni, M.P.; Rugiu, F.S.; de Carli, M.; Ricci, M.; Romagnani, S. Reciprocal Regulatory Effects of IFN-Gamma and IL-4 on the In Vitro Development of Human Th1 and Th2 Clones. *J. Immunol.* **1992**, *148*, 2142–2147. Available online: https://www.ncbi.nlm.nih.gov/pubmed/1532000 (accessed on 26 June 2022).
24. Swain, S.L. Regulation of the development of distinct subsets of CD4+ T cells. *Res. Immunol.* **1991**, *142*, 14–18. [CrossRef]
25. Yan, Y.; Weng, H.; Shen, Z.H.; Wu, L.; Zeng, X.T. Association between interleukin-4 gene -590 c/t, -33 c/t, and 70-base-pair polymorphisms and periodontitis susceptibility: A meta-analysis. *J. Periodontol.* **2014**, *85*, e354–e362. [CrossRef]
26. Takabayashi, A.; Ihara, K.; Sasaki, Y.; Kusuhara, K.; Nishima, S.; Hara, T. Novel polymorphism in the 5′-untranslated region of the interleukin-4 gene. *J. Hum. Genet.* **1999**, *44*, 352–353. [CrossRef] [PubMed]
27. Gonzales, J.R.; Mann, M.; Stelzig, J.; Bodeker, R.H.; Meyle, J. Single-nucleotide polymorphisms in the IL-4 and IL-13 promoter region in aggressive periodontitis. *J. Clin. Periodontol.* **2007**, *34*, 473–479. [CrossRef]
28. Zheng, W.; Flavell, R.A. The transcription factor GATA-3 is necessary and sufficient for Th2 cytokine gene expression in CD4 T cells. *Cell* **1997**, *89*, 587–596. [CrossRef]
29. Zhu, J.; Min, B.; Hu-Li, J.; Watson, C.J.; Grinberg, A.; Wang, Q.; Killeen, N.; Urban, J.F., Jr.; Guo, L.; Paul, W.E. Conditional deletion of Gata3 shows its essential function in T(H)1-T(H)2 responses. *Nat. Immunol.* **2004**, *5*, 1157–1165. [CrossRef]
30. Zhu, M.; Xu, K.; Chen, Y.; Gu, Y.; Zhang, M.; Luo, F.; Liu, Y.; Gu, W.; Hu, J.; Xu, H.; et al. Identification of novel T1D risk loci and their association with age and islet function at diagnosis in autoantibody-positive T1D individuals: Based on a two-stage genome-wide association study. *Diabetes Care* **2019**, *42*, 1414–1421. [CrossRef]
31. Garcia-Closas, M.; Troester, M.A.; Qi, Y.; Langerod, A.; Yeager, M.; Lissowska, J.; Brinton, L.; Welch, R.; Peplonska, B.; Gerhard, D.S.; et al. Common genetic variation in GATA-binding protein 3 and differential susceptibility to breast cancer by estrogen receptor alpha tumor status. *Cancer Epidemiol. Biomark. Prev.* **2007**, *16*, 2269–2275. [CrossRef] [PubMed]
32. Larsen, V.; Barlow, W.E.; Yang, J.J.; Zhu, Q.; Liu, S.; Kwan, M.L.; Ergas, I.J.; Roh, J.M.; Hutchins, L.F.; Kadlubar, S.A.; et al. Germline genetic variants in GATA3 and breast cancer treatment outcomes in SWOG S8897 trial and the pathways study. *Clin. Breast Cancer* **2019**, *19*, 225–235.e2. [CrossRef] [PubMed]
33. Dubois, R.N.; Abramson, S.B.; Crofford, L.; Gupta, R.A.; Simon, L.S.; van de Putte, L.B., Lipsky, P.E. Cyclooxygenase in Biology and Disease. *FASEB J.* **1998**, *12*, 1063–1073. Available online: https://www.ncbi.nlm.nih.gov/pubmed/9737710 (accessed on 26 June 2022). [CrossRef]
34. Goodson, J.M.; Dewhirst, F.E.; Brunetti, A. Prostaglandin E2 levels and human periodontal disease. *Prostaglandins* **1974**, *6*, 81–85. [CrossRef]
35. Inada, M.; Matsumoto, C.; Uematsu, S.; Akira, S.; Miyaura, C. Membrane-bound prostaglandin E synthase 1 mediated prostaglandin E2 production by osteoblast plays a critical role in lipopolysaccharide-induced bone loss associated with inflammation. *J. Immunol.* **2006**, *177*, 1879–1885. [CrossRef]
36. Offenbacher, S.; Odle, B.M.; Braswell, L.D.; Johnson, H.G.; Hall, C.M.; McClure, H.; Orkin, J.L.; Strobert, E.A.; Green, M.D. Changes in cyclooxygenase metabolites in experimental periodontitis in Macaca mulatta. *J. Periodontal. Res.* **1989**, *24*, 63–74. [CrossRef]
37. Offenbacher, S.; Heasman, P.A.; Collins, J.G. Modulation of host PGE_2 secretion as a determinant of periodontal disease expression. *J. Periodontol.* **1993**, *64*, 432–444. [CrossRef]
38. Tsai, C.C.; Hong, Y.C.; Chen, C.C.; Wu, Y.M. Measurement of prostaglandin E2 and leukotriene B4 in the gingival crevicular fluid. *J. Dent.* **1998**, *26*, 97–103. [CrossRef]
39. Xie, C.J.; Xiao, L.M.; Fan, W.H.; Xuan, D.Y.; Zhang, J.C. Common single nucleotide polymorphisms in cyclooxygenase-2 and risk of severe chronic periodontitis in a Chinese population. *J. Clin. Periodontol.* **2009**, *36*, 198–203. [CrossRef]
40. Schaefer, A.S.; Richter, G.M.; Nothnagel, M.; Laine, M.L.; Noack, B.; Glas, J.; Schrezenmeir, J.; Groessner-Schreiber, B.; Jepsen, S.; Loos, B.G.; et al. COX-2 is associated with periodontitis in Europeans. *J. Dent. Res.* **2010**, *89*, 384–388. [CrossRef]
41. Eickholz, P.; Nickles, K.; Koch, R.; Harks, I.; Hoffmann, T.; Kim, T.S.; Kocher, T.; Meyle, J.; Kaner, D.; Schlagenhauf, U.; et al. Is furcation involvement affected by adjunctive systemic amoxicillin plus metronidazole? A clinical trials exploratory subanalysis. *J. Clin. Periodontol.* **2016**, *43*, 839–848. [CrossRef] [PubMed]
42. Eickholz, P.; Koch, R.; Kocher, T.; Hoffmann, T.; Kim, T.S.; Meyle, J.; Kaner, D.; Schlagenhauf, U.; Harmsen, D.; Harks, I.; et al. Clinical benefits of systemic amoxicillin/metronidazole may depend on periodontitis severity and patients' age: An exploratory sub-analysis of the ABPARO trial. *J. Clin. Periodontol.* **2019**, *46*, 491–501. [CrossRef] [PubMed]
43. Hagenfeld, D.; Koch, R.; Junemann, S.; Prior, K.; Harks, I.; Eickholz, P.; Hoffmann, T.; Kim, T.S.; Kocher, T.; Meyle, J.; et al. Do we treat our patients or rather periodontal microbes with adjunctive antibiotics in periodontal therapy? A 16S rDNA microbial community analysis. *PLoS ONE* **2018**, *13*, e0195534. [CrossRef] [PubMed]

44. Harks, I.; Koch, R.; Eickholz, P.; Hoffmann, T.; Kim, T.S.; Kocher, T.; Meyle, J.; Kaner, D.; Schlagenhauf, U.; Doering, S.; et al. Is progression of periodontitis relevantly influenced by systemic antibiotics? A clinical randomized trial. *J. Clin. Periodontol.* **2015**, *42*, 832–842. [CrossRef] [PubMed]
45. Kocher, T.; Holtfreter, B.; Petersmann, A.; Eickholz, P.; Hoffmann, T.; Kaner, D.; Kim, T.S.; Meyle, J.; Schlagenhauf, U.; Doering, S.; et al. Effect of periodontal treatment on HbA1c among patients with prediabetes. *J. Dent. Res.* **2019**, *98*, 171–179. [CrossRef]
46. Harks, I.; Harmsen, D.; Gravemeier, M.; Prior, K.; Koch, R.; Doering, S.; Petersilka, G.; Weniger, T.; Eickholz, P.; Hoffmann, T.; et al. A Concept for Clinical Research Triggered by Suggestions from Systematic Reviews about Adjunctive Antibiotics. *Appl. Clin. Res. Clin. Trials Regul. Aff.* **2014**, *1*, 43–50. Available online: http://www.eurekaselect.com/node/121188/article (accessed on 26 June 2022). [CrossRef]
47. Butera, A.; Gallo, S.; Maiorani, C.; Molino, D.; Chiesa, A.; Preda, C.; Esposito, F.; Scribante, A. Probiotic alternative to chlorhexidine in periodontal therapy: Evaluation of clinical and microbiological parameters. *Microorganisms* **2020**, *9*, 69. [CrossRef]
48. Butera, A.; Gallo, S.; Pascadopoli, M.; Maiorani, C.; Milone, A.; Alovisi, M.; Scribante, A. Paraprobiotics in non-surgical periodontal therapy: Clinical and microbiological aspects in a 6-month follow-up domiciliary protocol for oral hygiene. *Microorganisms* **2022**, *10*, 337. [CrossRef]
49. De Carvalho, F.M.; Tinoco, E.M.; Govil, M.; Marazita, M.L.; Vieira, A.R. Aggressive periodontitis is likely influenced by a few small effect genes. *J. Clin. Periodontol.* **2009**, *36*, 468–473. [CrossRef]
50. Fraser, D.A.; Loos, B.G.; Boman, U.; van Winkelhoff, A.J.; van der Velden, U.; Schenck, K.; Dembic, Z. Polymorphisms in an interferon-gamma receptor-1 gene marker and susceptibility to periodontitis. *Acta Odontol. Scand.* **2003**, *61*, 297–302. [CrossRef]
51. Schafer, A.S.; Jepsen, S.; Loos, B.G. Periodontal genetics: A decade of genetic association studies mandates better study designs. *J. Clin. Periodontol.* **2011**, *38*, 103–107. [CrossRef] [PubMed]
52. Genco, R.J.; Slots, J. Host responses in periodontal diseases. *J. Dent. Res.* **1984**, *63*, 441–451. [CrossRef] [PubMed]
53. Hajishengallis, G.; Darveau, R.P.; Curtis, M.A. The keystone-pathogen hypothesis. *Nat. Rev. Microbiol.* **2012**, *10*, 717–725. [CrossRef] [PubMed]
54. Schaefer, A.S.; Dommisch, H.; Jepsen, S. Parodontitis: Gene als Risikofaktor. *ZM Online* **2015**, *10*, 1–6. Available online: https://www.zm-online.de/archiv/2015/10/titel/parodontitis-gene-als-risikofaktor/seite/alle/ (accessed on 26 June 2022).
55. Jiang, L.; Weng, H.; Chen, M.Y.; Zhang, C.; Zeng, X.T. Association between cyclooxygenase-2 gene polymorphisms and risk of periodontitis: A meta-analysis involving 5653 individuals. *Mol. Biol. Rep.* **2014**, *41*, 4795–4801. [CrossRef] [PubMed]
56. Nikolopoulos, G.K.; Dimou, N.L.; Hamodrakas, S.J.; Bagos, P.G. Cytokine gene polymorphisms in periodontal disease: A meta-analysis of 53 studies including 4178 cases and 4590 controls. *J. Clin. Periodontol.* **2008**, *35*, 754–767. [CrossRef] [PubMed]
57. Schaefer, A.S.; Richter, G.M.; Groessner-Schreiber, B.; Noack, B.; Nothnagel, M.; El Mokhtari, N.E.; Loos, B.G.; Jepsen, S.; Schreiber, S. Identification of a shared genetic susceptibility locus for coronary heart disease and periodontitis. *PLoS Genet.* **2009**, *5*, e1000378. [CrossRef]
58. Schaefer, A.S.; Jochens, A.; Dommisch, H.; Graetz, C.; Jockel-Schneider, Y.; Harks, I.; Staufenbiel, I.; Meyle, J.; Eickholz, P.; Folwaczny, M.; et al. A large candidate-gene association study suggests genetic variants at IRF5 and PRDM1 to be associated with aggressive periodontitis. *J. Clin. Periodontol.* **2014**, *41*, 1122–1131. [CrossRef]
59. Zhao, L.; Zhou, Y.; Xu, Y.; Sun, Y.; Li, L.; Chen, W. Effect of non-surgical periodontal therapy on the levels of Th17/Th1/Th2 cytokines and their transcription factors in Chinese chronic periodontitis patients. *J. Clin. Periodontol.* **2011**, *38*, 509–516. [CrossRef]
60. Becker, W.; Becker, B.E.; Berg, L.E. Periodontal treatment without maintenance. A retrospective study in 44 patients. *J. Periodontol.* **1984**, *55*, 505–509. [CrossRef]
61. Patrono, C.; Ciabattoni, G.; Pugliese, F.; Pierucci, A.; Blair, I.A.; FitzGerald, G.A. Estimated rate of thromboxane secretion into the circulation of normal humans. *J. Clin. Investig.* **1986**, *77*, 590–594. [CrossRef] [PubMed]
62. Mesa, F.; Aguilar, M.; Galindo-Moreno, P.; Bravo, M.; O'Valle, F. Cyclooxygenase-2 expression in gingival biopsies from periodontal patients is correlated with connective tissue loss. *J. Periodontol.* **2012**, *83*, 1538–1545. [CrossRef] [PubMed]
63. Beikler, T.; Peters, U.; Prior, K.; Eisenacher, M.; Flemmig, T.F. Gene expression in periodontal tissues following treatment. *BMC Med. Genom.* **2008**, *1*, 30. [CrossRef] [PubMed]
64. Daing, A.; Singh, S.V.; Saimbi, C.S.; Khan, M.A.; Rath, S.K. Cyclooxygenase 2 gene polymorphisms and chronic periodontitis in a North Indian population: A pilot study. *J. Periodontal Implant Sci.* **2012**, *42*, 151–157. [CrossRef]
65. Lang, N.P.; Berglundh, T.; Working Group 4 of Seventh European Workshop on Periodontology. Periimplant diseases: Where are we now?—Consensus of the seventh European workshop on periodontology. *J. Clin. Periodontol.* **2011**, *38* (Suppl. 11), 178–181. [CrossRef] [PubMed]
66. Lindhe, J.; Ranney, R.; Lamster, I.; Charles, A.; Chung, C.-P.; Flemmig, T.; Kinane, D.; Listgarten, M.; Löe, H.; Schoor, R.; et al. Consensus report: Chronic periodontitis. *Ann. Periodontol.* **1999**, *4*, 38. [CrossRef]
67. Lang, N.P.; Bartold, P.M.; Cullinan, M.; Jeffcoat, M.; Mombelli, A.; Murakami, S.; Page, R.; Papapanou, P.; Tonetti, M.; van Dyke, T. Consensus report: Aggressive periodontitis. *Ann. Periodontol.* **1999**, *4*, 53. [CrossRef]
68. Lang, N.P.; Adler, R.; Joss, A.; Nyman, S. Absence of bleeding on probing. An indicator of periodontal stability. *J. Clin. Periodontol.* **1990**, *17*, 714–721. [CrossRef]
69. O'Leary, T.J.; Drake, R.B.; Naylor, J.E. The plaque control record. *J. Periodontol.* **1972**, *43*, 38. [CrossRef]

International Journal of *Molecular Sciences*

MDPI

Article

Shear Stress Enhances the Paracrine-Mediated Immunoregulatory Function of Human Periodontal Ligament Stem Cells via the ERK Signalling Pathway

Ravipha Suwittayarak [1], Nuttha Klincumhom [1,*], Utapin Ngaokrajang [1], Worachat Namangkalakul [2], João N. Ferreira [3], Prasit Pavasant [1] and Thanaphum Osathanon [2]

1 Center of Excellence for Regenerative Dentistry, Department of Anatomy, Faculty of Dentistry, Chulalongkorn University, Bangkok 10330, Thailand; 6378011232@student.chula.ac.th (R.S.); utapin.ng@gmail.com (U.N.); rumtdent@gmail.com (P.P.)

2 Dental Stem Cell Biology Research Unit, Department of Anatomy, Faculty of Dentistry, Chulalongkorn University, Bangkok 10330, Thailand; worachat.n@chula.ac.th (W.N.); thanaphum.o@chula.ac.th (T.O.)

3 Avatar Biotechnologies for Oral Health and Healthy Longevity Research Unit, Faculty of Dentistry, Chulalongkorn University, Bangkok 10330, Thailand; joao.f@chula.ac.th

* Correspondence: nuttha.k@chula.ac.th

Citation: Suwittayarak, R.; Klincumhom, N.; Ngaokrajang, U.; Namangkalakul, W.; Ferreira J.N.; Pavasant, P.; Osathanon, T. Shear Stress Enhances the Paracrine-Mediated Immunoregulatory Function of Human Periodontal Ligament Stem Cells via the ERK Signalling Pathway. *Int. J. Mol. Sci.* **2022**, *23*, 7119. https://doi.org/10.3390/ijms23137119

Academic Editor: Kenichi Kumagai

Received: 27 May 2022
Accepted: 23 June 2022
Published: 27 June 2022

Publisher's Note: MDPI stays neutral with regard to jurisdictional claims in published maps and institutional affiliations.

Abstract: Relevant immunomodulatory effects have been proposed following allogeneic cell-based therapy with human periodontal ligament stem cells (hPDLSCs). This study aimed to examine the influence of shear stress on the immunosuppressive capacity of hPDLSCs. Cells were subjected to shear stress at different magnitudes (0.5, 5 and 10 dyn/cm^2). The expression of immunosuppressive markers was evaluated in shear stress-induced hPDLSCs using qRT-PCR, western blot, enzyme activity and enzyme-linked immunosorbent assays. The effects of a shear stress-derived condition medium (SS-CM) on T cell proliferation were examined using a resazurin assay. Treg differentiation was investigated using qRT PCR and flow cytometry analysis. Our results revealed that shear stress increased mRNA expression of *IDO* and *COX2* but not *TGF-β1* and *IFN-γ*. IDO activity, kynurenine and active TGF-β1 increased in SS-CM when compared to the non-shear stress-derived conditioned medium (CTL-CM). The amount of kynurenine in SS-CM was reduced in the presence of cycloheximide and ERK inhibitor. Subsequently, T cell proliferation decreased in SS-CM compared to CTL-CM. Treg differentiation was promoted in SS-CM, indicated by *FOXP3*, *IL-10* expression and CD4$^+$CD25^{hi}CD127$^{lo/-}$ subpopulation. In conclusion, shear stress promotes kynurenine production through ERK signalling in hPDLSC, leading to the inhibition of T cell proliferation and the promotion of Treg cell differentiation.

Keywords: immunosuppression; IDO; TGF-β1; periodontal ligament stem cells; shear stress; T cells; mechanotransduction

1. Introduction

Human periodontal ligament stem cells (hPDLSCs) can be isolated from the periodontal ligament, the connecting fibrous tissue between the tooth root and adjacent alveolar bone. To date, hPDLSCs have exhibited mesenchymal stem cell (MSC) characteristics owing to their potential for self-renewal and differentiation [1,2]. These cells have been proposed as a potential cell source for periodontal tissue regeneration due to their ability to differentiate and generate all periodontal tissue components after in vivo transplantation [3–5]. The therapeutic potential of hPDLSCs is indicated not only in the periodontal tissues but also in other tissues, such as the alveolar bone [6].

From an immunomodulatory standpoint, hPDLSCs possess striking features allowing them to escape immune recognition, inhibit activated immune cell function and regulate inflammatory cytokines during tissue regeneration [1,7–13]. These properties were initially recognised in MSCs that possess the ability to regulate the proliferation and differentiation of immune cells during an inflammatory response in order to promote tissue healing [14–16]. The significant immunomodulatory effect of paracrine immunosuppressive cytokines, including indoleamine 2,3-dioxygenase (IDO), hepatocyte growth factor (HGF), interleukin-10 (IL-10) and transforming growth factor beta1 (TGF-β1), has been reported to modulate immune cell behaviour.

Transforming growth factor beta 1 (TGF-β1) is constitutively expressed in MSCs and periodontal ligament (PDL) cells [10,17]. It is secreted as a latent form bounded to the cell membrane and extracellular matrix (ECM) proteins prior to activation [18,19]. In addition to regulation of tissue homeostasis, TGF-β1 can suppress T cell proliferation and activation [17,20,21] as well as promote the differentiation of regulatory CD4$^+$ T (Treg) cell [22,23].

Indoleamin-2,3-dioxygenase (IDO) is an enzyme catabolising L-tryptophan into kynurenine. The depletion of tryptophan through the IDO–kynurenine pathway modulates immune response by inhibiting T cell proliferation and promoting Treg differentiation [24–27]. The role of IDO as an immune suppressor was shown to be activated in the presence of pro-inflammatory cytokines, for instance, IL-1β [28], IL-12 [29], Toll-like receptors (TLRs) [7,30], cyclooxygenase 2 (COX2)/prostaglandin E2 (PGE2) [31,32], interferon gamma (IFN-γ) [1,30,33] as well as TGF-β1 [34], leading to suppression of inflammation. Recently, our group showed that TLR3 induced the expression IFN-γ-independent IDO, resulting in inhibition of peripheral blood mononuclear cell (PBMC) proliferation and upregulation of a Treg-specific gene marker, *forkhead box P3* or *FOXP3* [7]. Additionally, mechanical stimuli have also been applied in vitro to regulate immunomodulatory properties of both PDL cells [35,36] and bone marrow-derived MSCs (BM-MSCs) [37]

PDL cells are considered as mechanosensory cells that can perceive and respond to mechanical stress driven by mastication, speech or orthodontic movement [38]. Several studies suggested that mechanical stimuli are important factors in regulating periodontal tissue homeostasis [39–41]. Previously, our group showed that intermittent compressive force regulated SOST/POSTN expression via the TGF-β1 signalling pathway in PDL cells [42].

In periodontal tissue, PDL cells are exposed to interstitial fluid shear stress during tooth movement [43,44]. Shear stress has been shown to play an essential role in cell behaviour in several cell types including embryonic stem cells [45], osteocytes [46], endothelial cells [47], MSCs [48] as well as PDL cells [43,49–53]. Previously, in vitro shear stress stimulation (3–15 dyn/cm^2) was shown to modulate PDL cell properties such as osteogenic differentiation, cell proliferation and ECM remodelling [49–52]. Moreover, shear stress has the potential to enhance the immunosuppression of MSCs via COX2/PGE2 expression [37].

However, the influence of shear stress in triggering immunomodulatory properties of hPDLSCs has not been elucidated yet. Herein, we assess the effects of shear stress on the immunosuppressive properties of hPDLSCs. The influence of shear stress-activated immunosuppressive molecules in regulating CD4$^+$ T cell proliferation and Treg differentiation was investigated. Our study suggests the use of mechanical stimuli of hPDLSCs as a promising approach to induce immunosuppression while targeting tissue regeneration with allogeneic therapies.

2. Materials and Methods

2.1. Isolation and Culture of hPDLSCs

Human PDL tissues were scraped from the middle third of the root surface obtained from normal healthy teeth that were scheduled to be extracted according to the treatment plan at the Department of Oral and Maxillofacial Surgery, Faculty of Dentistry, Chulalongkorn University. The human cell isolation protocol was approved by the Ethical Committee for Human Research (HREC-DCU 2022-010). The scraped PDL tissues from at least 3 patients were used for cell explant culture. The outgrown cells were maintained in Dulbecco's modified Eagle medium (DMEM) (cat. No. 11885-084, Thermo Scientific,

Waltham, MA, USA) containing 10% (*v*/*v*) fetal bovine serum (FBS) (cat. No. SV30160.03, Thermo Scientific), 1% (*v*/*v*) L-glutamine (Glutamax TM-1) (cat. No. 35050-061, Thermo Scientific), 1% (*v*/*v*) antibiotic–antimycotic (cat. No. 15240-062, Thermo Scientific) at 37 °C in a humidified 5% CO_2 atmosphere. The culture medium was removed and changed every 2 days. Human periodontal stem cells (hPDLSCs) were subcultured using 0.25% trypsin/EDTA (cat. No. 25200-072, Thermo Scientific) until cells reached confluency. Cells in passages 3 through 7 were characterised and used in this study.

2.2. Flow Cytometry Analysis

For characterisation of hPDLSCs, cells were stained with PERCP-conjugated anti-human CD45 (cat. No. 21810455, ImmunoTools, Friesoythe, Germany), FITC-conjugated anti-human CD73 (cat. No. 212270733, ImmunoTools), FITC-conjugated anti-human CD90 (cat. No. ab11155, Abcam, Cambridge, UK) and PE-conjugated anti-human CD105 antibodies (cat. No. 21271054, ImmunoTools). For the Treg population, after indirect co-culture, all conditions of T cell culture were stained with VioBlue-conjugated anti-human CD45, VioGreen-conjugated anti-human CD4, VioBright 515-conjugated anti-human CD25 and PE-conjugated anti-human CD127 antibodies from the Treg Detection Kit (cat. No.130-122-994, MACS, San Diego, CA, USA). Mouse IgG1 PERCP, PE and FITC antibodies (MACS) were used as an isotype control. The stained cells were further detected by a flow cytometer (BD FACSCalibur and Cell Quest software, BD Bioscience, San Jose, CA, USA).

2.3. Induction of Osteogenic Differentiation

The cells were seeded into 24-well plates (cat. No. 142475, Thermo Scientific) at a density of 2×10^4 cells/cm^2 and cultured for 10–14 days in an osteogenic medium containing 10% FBS-DMEM supplemented with 50 μg/mL L-ascorbic acid (cat. No. A-4034, Sigma-Aldrich, St. Louis, MO, USA), 100 nM dexamethasone (cat. No. D8893, Sigma-Aldrich) and 5 mM β-glycerophosphate (cat. No. G9422, Sigma-Aldrich). The medium was changed every 2 days.

2.4. Induction of Adipogenic Differentiation

The cells were seeded into 24-well plates (Thermo Scientific) at a density of 2×10^4 cells/cm^2 and cultured for 16–20 days with an adipogenic medium containing 10% FBS-DMEM supplemented with 500 μmol 3-isobutyl-1-methylxanthine (IBMX) (cat. No. PI121124, Thermo Scientific), 1 μg/mL insulin from bovine pancreas, 100 μM indomethacin (cat. No. I7378, Sigma-Aldrich) that was kept at room temperature before use. The medium was changed every 2 days.

2.5. Alizarin Red S and Von Kossa Staining

For alizarin red S staining, the cells were fixed with cold methanol (cat. No. 100230, honey, Ulsan, Korea) for 10 min and washed with deionised (DI) water. Then, cells were stained with 1% alizarin red S (cat. No. A5533, Sigma-Aldrich) solution for 5 min at room temperature (RT). The samples were then rinsed with DI water gently and left to dry.

For von Kossa staining, cells were fixed with 4% (*v*/*v*) cold paraformaldehyde (AR1068, BOSTER, Pleasanton, CA, USA) in for 20 min, then rinsed in DI water. Silver nitrate (cat. No. 21572.188, VWR International GmbH, Vienna, Austria) at 5% (*w*/*v*) in DI water was added to the fixed cells for 30 min. The cells were exposed to a 100 w UV lamp for 10 min and left to dry. All samples were observed by inverted microscopy (Nikon ECLIPSE Ts2, Nikon, Melville, NY, USA).

2.6. Oil Red O Staining

The cells were fixed with 4% (*v*/*v*) formalin (cat. No. F-1268, Sigma-Aldrich) in DI water and gently washed with DI water. The samples were fixed with oil red O (cat. No. O0625, Sigma-Aldrich) solution in methanol (cat. No. 1.06009.2500, Merk, Germany) for

15 min. The samples were rinsed and DI water was replaced. All samples were then observed by inverted microscopy (Nikon ECLIPSE Ts2).

2.7. Isolation and Activation of CD4$^+$ T Cells from PBMCs

Human PBMCs were isolated from the human buffy coat with permission from the Thai Red Cross Society. The protocols were approved by the Ethical Committee for Human Research (HREC-DCU-2022-010). The CD4$^+$ T cell isolation was performed using Sepmate-50 (Stem Cell Technologies, Singapore). The isolated CD4$^+$ T cells were maintained in RPMI 1640 (cat. No. 11835-030, Thermo Scientific) supplemented with 10% (v/v) FBS (Thermo Scientific), 1% (v/v) L-glutamine (Glutamax TM-1) (Thermo Scientific) and 1% (v/v) antibiotic–antimycotic (Thermo Scientific). For the T cell activation, the isolated CD4$^+$ T cells were seed into 24-well plates (Thermo Scientific) at a density of 10^6 cells/well and activated with 10% FBS-RPMI medium supplemented with 1 μg/mL CD3 (coated on 24-well plate overnight) (cat. No. 21850030, Immuno Tool), 1 mg/mL CD28 (cat. No. 302933, Biolegend) and 50 U/mL IL-2 (cat. No. 200-02, Peprotech) before the start of the conditioned medium treatment.

2.8. Shear Stress Stimulation

The hPDLSCs were seeded into 35 mm culture dishes (cat. No. 430165, Costar® Corning) at a density of 4×10^5 cells/well overnight. The culture medium was then changed to fresh medium. The hPDLSCs were subjected to different magnitudes of shear stress (0.5, 5 and 10 dyn/cm^2) using a cone-shaped rotating disk for 3 h [54]. Thereafter, the hPDLSCs were cultured for 24 h in normal culture conditions before sample collection for mRNA, protein analysis and conditioned medium. For inhibitory experiments, the hPDLSCs were pretreated with 100 nM cycloheximide (CHX) (cat. No. C-0934, Sigma-Aldrich) or 1.5 nM ERK inhibitor (cat. No. 328006, Calbiochem, San Diego, CA, USA) for 1 h prior to shear stress stimulation. The hPDLSCs without shear stress stimulation and non-shear stress-induced hPDLSC-derived conditioned medium were used as the control and CTL-CM, respectively.

2.9. Cell Viability Assay

The Cell Counting Kit-8 (CCK-8) (cat. No. ab228554, Abcam, Cambridge®, UK) was used to detect the cell viability of hPDLSCs. After shear stress stimulation, the hPDLSCs were incubated with CCK-8 solution for 30 min. The 100 μL of secreted soluble formazan in the culture medium was put into 96-well plates and the absorbance was measured at 460 nm using a microplate reader (Synergy H1, Biotek multi-mode reader, Winooski, VT, USA) according to the manufacturer's protocol. The data were calculated as the percentage of viable cells.

2.10. Immunofluorescent Staining

After shear stress stimulation, hPDLSCs were fixed with 4% (v/v) paraformaldehyde (BOSTER) in DI water and incubated with rhodamine–phalloidin (dilution 1:1000) (cat. No. ab235138, Abcam, Cambridge, UK). Stained cells were then counterstained with DAPI (dilution 1:2000) (TOCRIS bioscience, Bristol, UK). Then, 50% glycerol was added to the plate and stored at 4 °C. Apotome microscopy (Axio Observer Z1 and ZEN pro, ZEISS International, Oberkochen, Germany) was performed.

2.11. Conditioned Medium Preparation and Treatment

The conditioned medium was collected from hPDLSC cultures and kept at −80 °C before use. Prior to T cell culture experiments, frozen conditioned medium was thawed and lyophilised with a Tabletop Freeze Dryer (Medfuture Biotech, Jinan, China). Lyophilised conditioned medium was suspended with 10% FBS-RPMI medium and used for CD4$^+$ T cell culture that was activated as previously described. All conditioned medium-treated T cells were collected for the T cell proliferation assay and investigation of Treg cell differentiation on day 3 and day 5 of the culture period, respectively.

2.12. T Cell Proliferation Analysis Using Resazurin Assay

For investigation of T cell proliferation, the activated $CD4^+$ T cells were co-cultured with the conditioned medium for 3 days [7]. The $CD4^+$ T cells were incubated with 7-hydroxy-10-oxidophenoxazin-10-ium-3-one, sodium (resazurin) (cat. No. R7017, Sigma-Aldrich) for 2 h. The fluorescence signal of secreted pink resorufin in culture medium converted from blue resazurin solution was measured using a microplate reader at excitation of 560 nm and emission of 590 nm and gain = 50. The data were presented as the percentage of proliferative cells.

2.13. Development of Regulatory T Cell (Treg) Cells

For induction of Treg cell differentiation, activated $CD4^+$ T cells were exogenously treated with kynurenine (1–100 μM) (cat. No. K8625, Sigma-Aldrich), or cultured with hPDLSC-derived conditioned medium for 5 days before sample collection for qRT-PCR and flow cytometry analysis. The activated $CD4^+$ T cells treated with 100 μM kynurenine were used as a positive control (or induced Treg).

2.14. RNA Isolation and Real-Time RT-PCR Analysis

Total RNA was extracted by using RiboExTM solution (cat. No. 301-902, GeneAll®, Seoul, Korea). The quality of RNA concentration was measured using a NanoDrop (Thermo Scientific). The amount of RNA was converted into complementary DNA (cDNA) using an ImProm-IITM Reverse Transcription System (cat. No. A3800, Promega, Madison, WI, USA). A FastStart Essential DNA Green Master kit was used for the real-time polymerase chain reaction process. The reaction was performed on a Bio-Rad PCR system (CFX Connect Real-Time System, Bio-Rad, Hercules, CA, USA). Cycling conditions were set at 95 °C for 30 s, followed by 45 cycles of 95 °C for 3 s and 60 °C for 30 s. Relative gene expression was calculated using the $2^{-\Delta\Delta CT}$ method [55]. The expression value was normalised to the GAPDH expression value and the control. Oligonucleotide primers used in this study are shown in Table S1.

2.15. Enzyme-Linked Immunosorbent Assay (ELISA)

The supernatant and cell lysate were collected and extracted by RIPA buffer containing protease inhibitor buffer (cat. No. P8340, Sigma-Aldrich) after the shear stress experiment. Protein levels were measured using PGE2 (cat No. KGE004B, R&D system, Minneapolis, MN, USA), IFN-γ (cat. No. DY285-05, R&D system) and TGF β1 (cat No. DY240 05, R&D system) following the manufacturer's instructions.

2.16. IDO Activity Assay and Kynurenine Measurement

All the conditioned media were collected and prepared with IDO buffer containing 40 mM L-ascorbic acid (Sigma-Aldrich), 20 μM methylene blue (cat. No. MB-1, Sigma-Aldrich), 200 μg/mL catalase (cat. No. C9322, Sigma-Aldrich) and 800 μM of L-tryptophan solution (cat. T-8659, Sigma-Aldrich). In short, the samples were mixed with IDO buffer in a 1:1 ratio for 1 h at 37 °C in a humidified 5% CO_2 atmosphere to convert tryptophan into kynurenine. To stop the reaction, 30% (v/v) trichloroacetic acid (TCA) in DI water (cat. No. T6399, Sigma-Aldrich) was added and incubated at 56 °C for 30 min. The sample was centrifuged at 13,000× g and added to 2% (w/v) Ehrlich reagent (cat. No. 39070, Sigma-Aldrich) in glacial acetic acid in a 1:1 ratio. For measurement of kynurenine product, 100 μL of all samples were mixed with 50 μL of 30% v/v TCA, then centrifuged at 8000× g for 5 min. Then, 75 μL of supernatant was equally mixed with 2.5% (w/v) Ehrlich reagent (Thermo Scientific) in glacial acetic acid. All samples were then read at 492 nm using a microplate reader [1]. The recombinant kynurenine (Sigma-Aldrich) was used as a standard in this experiment.

2.17. Western Blot Assay

The protein from shear stress-induced hPDLSCs was extracted by RIPA buffer containing protease inhibitor buffer (cat. No. P8340, Sigma-Aldrich). The concentration of lysate proteins was measured with a BCA protein assay (PierceTM BCA detection reagent, cat No. 23228 and 23224, Thermo Scientific). Lysate proteins were separated on 12% SDS-polyacrylamide gel and transferred to a nitrocellulose membrane. The membrane was placed in a solution containing a monoclonal antibody to rabbit-anti actin (cat. No. A2066, Sigma-Aldrich), rabbit anti-TGF-β1 (cat No. ab92486, Abcam), rabbit anti-ERK1/2 (T202/Y204, Cell Signaling) or rabbit anti-phosphorylated ERK1/2 (cat. No. 137F5, Cell Signaling Technology) antibodies. Then, the membranes were developed with the horseradish peroxidase-linked antibody (cat No. 7074S, Cell signaling Technology). The membranes were visualised and exposed to chemiluminescence (SuperSignal® West Pico Chemiluminescent Substrate, cat No. 34577, Thermo Scientific) and an image analyser (GE Healthcare, Pittsburgh, PA, USA), respectively. The band density was measured using ImageJ software. The band density was normalised to band density of actin and to the control.

2.18. Statistical Analysis

All data are presented as mean ± standard deviation (SD). One way-ANOVA and the Mann–Whitney U test were using for comparisons between groups. A p-value below 0.05 was considered significant. The analysis was performed by using the statistical software GraphPad Prism version 8 (GraphPad software, San Diego, CA, USA). At least three replicates from different donors were performed for each experiment.

3. Results

3.1. Shear Stress Enhanced the Expression of Immunosuppressive Regulators

Isolated hPDLSCs were characterised and positive for MSC-specific markers and the ability to differentiate into osteoblastic and adipogenic lineages (Figure S1). The effect of shear stress on viability of hPDLSCs was determined using the CCK-8 assay. We found that all magnitudes of shear stress (0.5–10 dyn/cm^2) had no significant effect on cell viability, which was indicated by the mitochondrial activity of hPDLSCs (Figure S2).

To determine whether shear stress stimulates the expression of immunomodulatory regulators in hPDLSCs, cells were subjected to shear stress at 0.5, 5, 10 dyn/cm^2 for 3 h and continuously cultured afterwards up to 24 h. Cells were harvested for mRNA expression analysis of immunomodulatory regulators, including *IDO*, *TGF-β1*, *COX2* and *IFN-γ*. The stress at 5 dyn/cm^2 significantly increased the gene expression of IDO (Figure 1A) and *COX2* (Figure 1D) in hPDLSCs, while there is no significant difference in the expression of *IFN-γ* (Figure 1C) and *TGF-β1* (Figure 1B).

We further investigated the activity of IDO and kynurenine products, which were measured in all conditioned media after shear stress stimulation (Figure 2A,B). The IDO activity increased in shear stress-derived conditioned medium (SS-CM) at 5 and 10 dyn/cm^2 compared to the non-shear stress-derived conditioned medium, also referred to as a control (CTL-CM) (Figure 2A). Subsequently, the amount of kynurenine product in SS-CM at 5 dyn/cm^2 was significantly higher than CTL-CM (Figure 2B). With regard to *TGF-β1*, the protein expression of active TGF-β1 was determined in the conditioned medium. Although the secretion of total TGFβ1 was increased in 5 dyn/cm^2 shear stress, active TGF-β1 in SS-CM was increased at all shear stress magnitudes (0.5, 5 and 10 dyn/cm^2) compared to the CTL-CM (Figure 2C). In contrast, compared to control, shear stress decreased the active form of TGF-β1 in cell lysates at 5 dyn/cm^2 yet had no effect on the latent form of TGF-β1 (Figure 2D–F). The amount of IFN-γ in cell lysate and conditioned medium was measured using an ELISA assay. Interestingly, this assay showed that the amount of cell-bound IFN-γ in all magnitudes of shear stress was not different compared to control (Figure 2G). In contrast, the amount of secreted IFN-γ decreased at 0.5 and 10 dyn/cm^2 (Figure 2H). As for the *COX2* expression, we analysed COX2-dependent PGE2 synthesis in hPDLSCs with different magnitudes of shear stress. Outcomes showed no difference in the PGE2 product of all groups of SS-CM compared to CTL-CM (Figure 2I). Our results are supported by previous literature [49–52] and indicated that a 5 dyn/cm^2 magnitude of shear stress can optimally enhance IDO, kynurenine and TGF-β1 gene/protein expression.

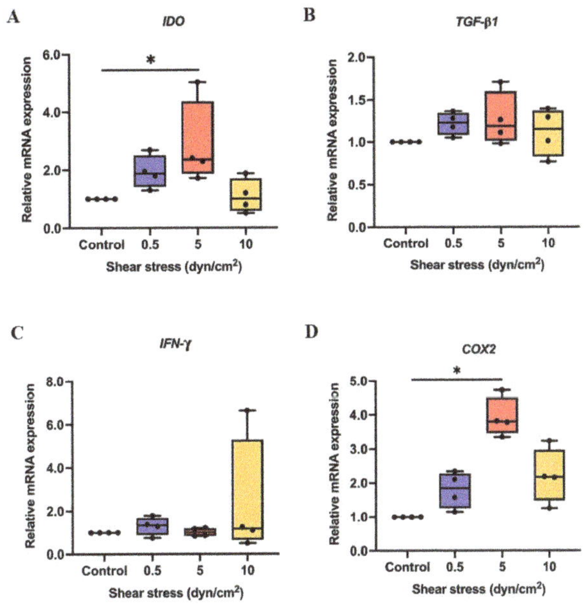

Figure 1. The mRNA expression of immunosuppressive regulators in shear stress-induced hPDLSCs. After shear stress stimulation, the relative mRNA expression of *IDO* (**A**), *TGF-β1* (**B**), *IFN-γ* (**C**) and *COX2* (**D**) was detected using qRT-PCR. * $p < 0.05$ vs. control group.

3.2. Shear Stress Enhanced the Product of IDO Activity via ERK1/2 Signalling Pathway

To investigate the regulatory mechanism by which shear stress enhances IDO expression and the amount of kynurenine in hPDLSC-derived conditioned medium, cells were pretreated with protein synthesis inhibitor cycloheximide (CHX) for 1 h before shear stress stimulation (5 dyn/cm^2). As a result, CHX attenuated the expression of *IDO* and the amount of kynurenine when compared to control (Figure 3A,B). This result suggested the involvement of intermediate molecules in the regulatory mechanism of shear stress-induced *IDO* gene expression in hPDLSCs.

Next, the role of extracellular signal-regulated kinase 1/2 (ERK1/2) was demonstrated to be associated with the activation of immunosuppressive properties of human dendritic cells [51]. To investigate whether shear stress regulates the IDO–kynurenine and TGF-β1 secretion via ERK1/2 activation, hPDLSCs were pretreated with ERK inhibitor for 1 h and subsequently were placed under shear stress at 5 dyn/cm^2 for 3 h. In the control condition, cells received shear stress for 3 h without pretreatment with an ERK inhibitor. The effect of shear stress on protein expressions of ERK1/2 and phosphorylated ERK1/2 (P-ERK1/2) was determined by Western blot. Shear stress induced the phosphorylation of ERK1/2 of hPDLSCs, which was abolished by the ERK inhibitor (Figure 3C–E). The effect of shear stress on the amount of kynurenine in hPDLSCs was also attenuated in the presence of ERK inhibitor (Figure 3F), but not TGF-β1 and its active form (Figure 3G). Therefore, shear stress enhanced the kynurenine secretion in hPDLSCs via the activation of the ERK1/2 signalling pathway.

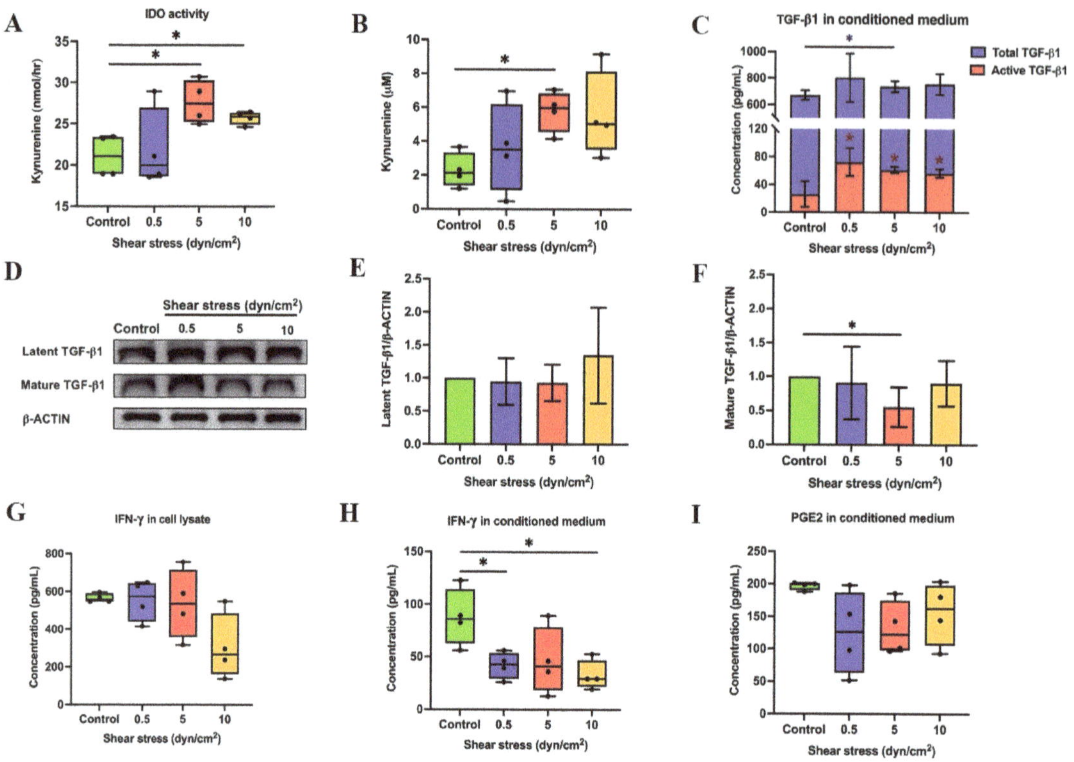

Figure 2. Protein expression of immunosuppressive regulators in shear stress-induced hPDLSCs. Shear stress promoted IDO activity (**A**) and the kynurenine product (**B**) in hPDLSCs. Shear stress increased the secretion of active TGF-β1 in PDLSC-derived conditioned medium (**C**). TGF-β1 protein expression in hPDLSCs was determined by Western blot analysis (**D**). The quantitative analysis of Western blot band intensity of latent TGF-β1(**E**) and active TGF-β1 (**F**). The concentration of IFN-γ in cell lysate (**G**) and conditioned medium (**H**) and the amount of COX2-independent PGE2 were determined by ELISA assay (**I**). * $p < 0.05$ vs. control group.

3.3. Shear Stress-Derived Conditioned Medium (SS-CM) Suppressed T Cell Proliferation

To evaluate the inhibitory effect of SS-CM on the proliferation of CD4$^+$ T cells from PBMCs, the conditioned medium was lyophilised and suspended with a fresh 10% FBS-RPMI medium. Isolated T cells were activated with CD3, CD28 and IL-2, which are co-stimulatory signals from antigen-presenting cells (APCs) for activation and expansion of T cells [52]. The activated T cells were treated with SS-CM for 3 days as the optimal timing [7]. The CTL-CM was used as a control. Suspended cells in each condition were observed and collected for a T cell proliferation assay. After T cell activation, the activated T cells remarkably formed clusters, increased cell size and increased in number (Figure S3A–D). T cell proliferation was significantly decreased after treatment with CTL-CM and SS-CM compared to activated T cells without conditioned medium treatment. In addition, the proliferation of T cells was significantly lower in SS-CM than in CTL-CM (Figure 4A). Our results suggested that a conditioned medium derived from hPDLSCs inhibited the proliferation of T cells, and this effect can be enhanced by shear stress stimulation.

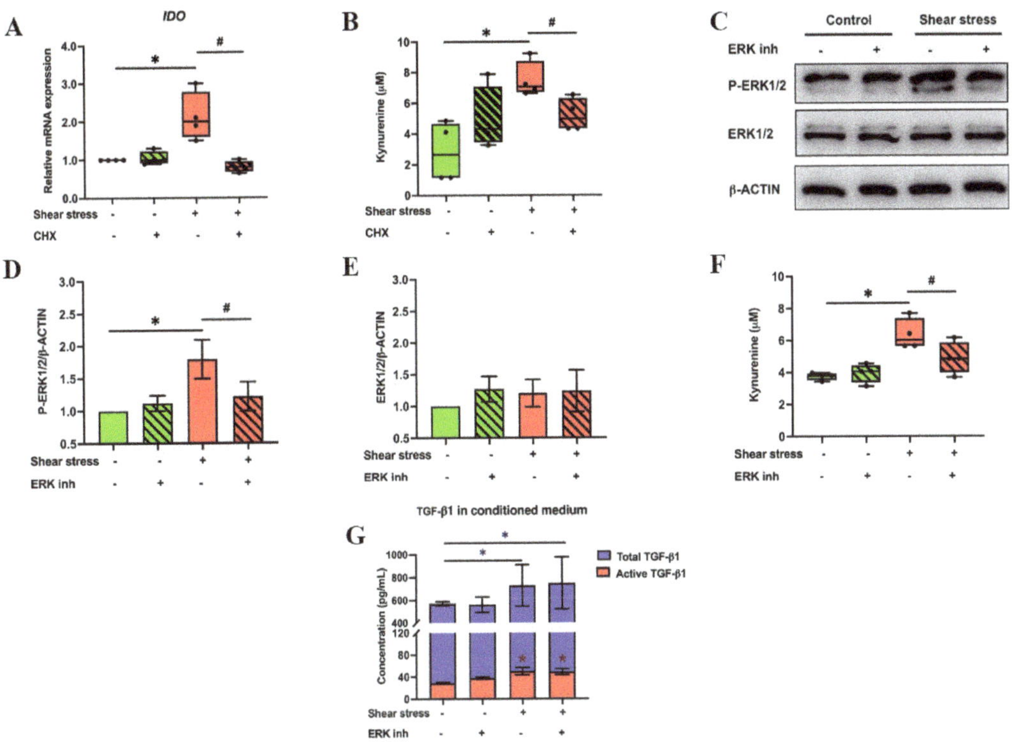

Figure 3. Shear stress enhanced IDO expression and IDO-catabolised kynurenine product in hPDLSCs via ERK1/2 signalling pathway. After shear stress stimulation (at 5 dyn/cm^2), hPDLSCs treated with CHX were examined for *IDO* mRNA expression and kynurenine product (**A**,**B**). The activity of ERK1/2 was examined by Western blot analysis (**C**). The quantitative analysis of Western blot band intensity showed that ERK inhibitor attenuated P-ERK1/2 (**D**), but not ERK1/2 expression in shear stress-induced hPDLSCs (**E**). The addition of an ERK inhibitor inhibits the effect of shear stress-induced kynurenine secretion (**F**). However, shear stress did not affect TGF-β1 secretion (**G**). * $p < 0.05$ vs. non-shear stress without ERK inhibitor group. # $p < 0.05$ vs. control.

3.4. Shear Stress-Derived Conditioned Medium (SS-CM) Induced Regulatory T Cell Differentiation

To further investigate whether SS-CM induces the development of regulatory T (Treg) cells, activated T cells were cultured with SS-CM for 5 days. The Treg cells were characterised by Treg-specific gene markers, *FOXP3* and *IL-10*. The results showed that the mRNA expression of *FOXP3* significantly increased in SS-CM compared to activated T cells and CTL-CM (Figure 4B). The expression of *IL-10* in CTL-CM and SS-CM was upregulated compared to activated T cells. However, the mRNA expression of *IL-10* was not different between CTL-CM and SS-CM (Figure 4C). A previous study indicated that CD4$^+$CD25^{hi}CD127$^{lo/-}$ are potent purity markers of the functional Treg population [53,56,57]. Our study showed that SS-CM increased the percentage of CD4$^+$CD25^{hi}CD127$^{lo/-}$ Treg cells compared to activated T cells and CTL-CM (Figure 4D,E). Therefore, these outcomes suggested that a conditioned medium derived from shear stress-induced hPDLSCs enhances mRNA expression of Treg-specific markers (*FOXP3* and *IL-10*) and increases the Treg cell population.

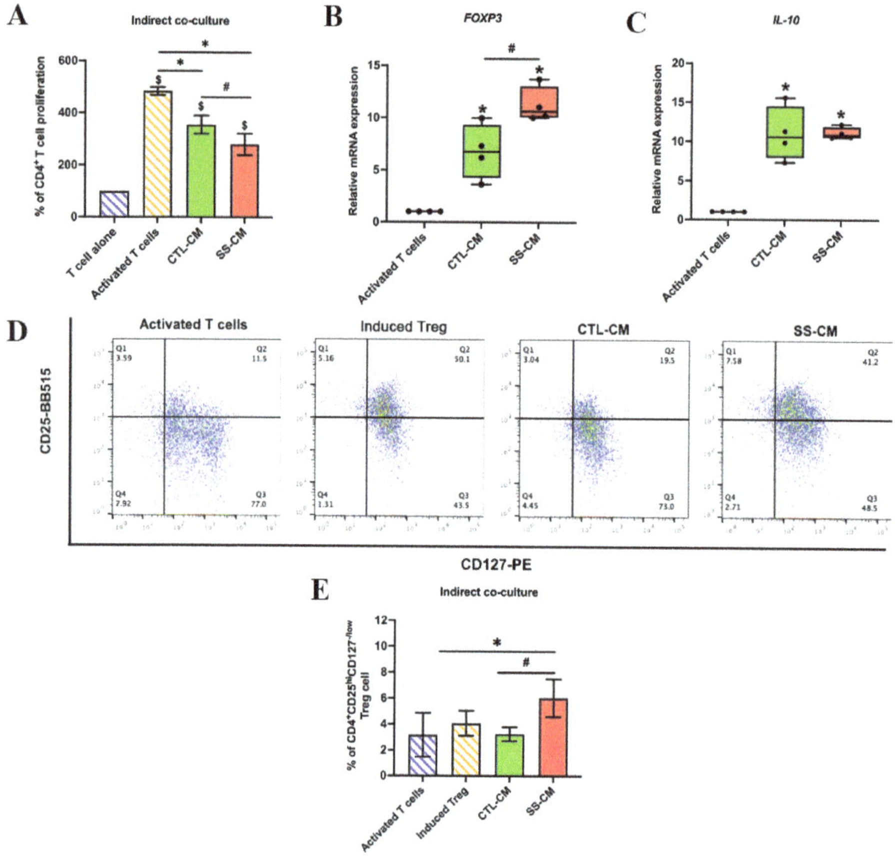

Figure 4. Shear stress attenuated CD4$^+$ T cell proliferation and promoted Treg differentiation. All hPDLSC-derived conditioned media were indirectly co-cultured with CD4$^+$ T cells. The proliferation of CD4$^+$ T cells was assessed using a resazurin assay. SS-CM decreased the percentage of T cell proliferation (**A**). The mRNA expressions of *FOXP3* (**B**) and *IL-10* (**C**) were upregulated in SS-CM-treated CD4$^+$ T cells. The flow cytometry analysis showed an increase in the percentage of CD4$^+$CD25^{hi}CD127$^{lo/-}$ Treg population in SS-CM compared to CTL-CM and activated T cells (**D**,**E**). \$ $p < 0.05$ vs. T cell alone group, * $p < 0.05$ vs. activated T cells group, # $p < 0.05$ vs. CTL-CM group.

4. Discussion

Here, our study showed that shear stress enhanced the secretion of TGF-β1 and IDO-catabolised kynurenine in a conditioned medium derived from shear stress-induced hPDLSCs. These conditioned media potentially inhibited CD4$^+$ T cell proliferation while promoting CD4$^+$CD25^{hi}CD127$^{lo/-}$ Treg differentiation.

TGF-β1 is an immunosuppressive molecule that induces Treg differentiation and inhibits T cell proliferation [17,20–22]. Our data showed that the shear stress at 5 dyn/cm^2 enhanced the secretion of TGF-β1 and the active form of TGF-β1. These findings confirmed that shear stress might be one factor in regulating TGF-β1 activation, similar to platelets [58]. Kynurenine is a product from IDO-metabolised tryptophan. IDO and kynurenine are important factors regulating the T cell function [25]. The decreasing of tryptophan by IDO activity inhibited T cell growth [24]. Moreover, increased kynurenine induced immune cell apoptosis and Treg differentiation [26,27,59]. Interestingly, our study also showed that shear stress activated the expression and activity of the IDO enzyme. Although the amount of kynurenine detected in SS-CM was much lower than that of the exogenous kynurenine experiment (Figure S4B), SS-CM markedly suppressed T cell immune activity. A previous study showed that the aryl hydrocarbon receptor (AHR) on T cells is needed for kynurenine-induced Treg cell differentiation. Moreover, the AHR can be increased when treated with TGF-β1, promoting Treg cell differentiation [60]. Our data suggest that an increased amount of TGF-β1 in SS-CM may amplify the effect of kynurenine on immune suppression. However, further studies on how shear stress-activated IDO and TGF-β1 regulate immunosuppressive properties of hPDLSCs may unveil other mechanisms.

Additionally, shear stress-induced IDO mRNA expression and kynurenine product were inhibited by CHX, the protein synthesis inhibitor, suggesting the involvement of intermediate activators during shear stress stimulation. Studies indicate that IFN-γ and PGE2 are potent inducers to activate IDO expression [31,32,61]. We speculated that shear stress might regulate immunosuppressive properties of hPDLSCs via IFN-γ or PGE2-independent IDO activation. This speculation is based on our previous study showing that TLR3 enhanced the immunosuppressive capacity of hPDLSCs by suppressing proliferation of T cells and promoting expression of *FOXP3* mRNA via IFN-γ-independent IDO activation [7]. Moreover, TGF-β1 had been shown to induce IDO expression in dendritic cells via the P(I)3K signalling pathway, resulting in a positive feedback loop for generating TGF-β1 and IDO [34,62]. We speculate that shear stress may also regulate IDO expression of hPDLSCs via the TGF-β1 signalling pathway.

We further investigated the mechanotransduction-mediated signalling pathway of hPDLSCs after shear stress stimulation. ERK1/2, as a subset of the MAPK signalling pathway, has been reported to be involved in shear stress-mediated mechanotransduction, leading to regulation of the expression of target genes [52,63]. Our study found that shear stress activated the ERK1/2 signalling pathway by increasing the phosphorylation of ERK1/2. The activity of ERK1/2 usually results in increased immunosuppressive effects of dendritic cells via increased TGF-β1 secretion [56]. Our study showed no difference in TGF-β1 secretion after adding ERK inhibitor, indicating that shear stress-induced TGF-β1 secretion in hPDLSCs was not regulated by ERK1/2 activity. Additionally, P38, as a subset of MAPK kinase, is upstream of IDO expression when activated by viral stimulation [64]. Herein, our study showed that kynurenine product in SS-CM was attenuated in the presence of an ERK inhibitor. These data suggested that shear stress enhances kynurenine production in hPDLSCs via the ERK1/2 signalling pathway.

Immunosuppressive properties of PDL cells can be activated by pro-inflammatory cytokines. Conditioned medium from PDL cells exposed to IFN-γ and TLR3 could inhibit T cell proliferation and induce *FOXP3* mRNA expression [1,7]. Here, our study demonstrated that shear stress potentially enhanced the immunosuppressive properties of hPDLSCs by suppressing T cell proliferation and stimulating Treg cell differentiation. Treg cell features were investigated by gene expression of specific markers (*FOXP3* and *IL10*) and protein expression of specific cell surface markers ($CD4^{+}CD25^{hi}CD127^{lo/-}$) [65–67]. Moreover, flow cytometry of viable $CD4^{+}CD25^{hi}CD127^{lo/-}$ Treg cells and sublocalisation of FOXP3 protein via western blot can be further analysed to investigate Treg cell activity [68,69].

In conclusion, our findings demonstrated that shear stress enhanced kynurenine in hPDLSCs via the ERK1/2 signalling pathway. In response to shear stress, hPDLSCs secreted active TGF-β1 and kynurenine, thereby suppressing T cell proliferation and promoting Treg differentiation (Figure 5). Our findings contribute to a better understanding of the immunosuppressive properties of hPDLSCs in response to mechanical stimuli such as the ones generated during tooth movement. We believe that the paracrine-mediated immunoregulatory function of hPDLSCs may be a promising cell-free approach for clinical applications, especially for the case of allogeneic cell therapy.

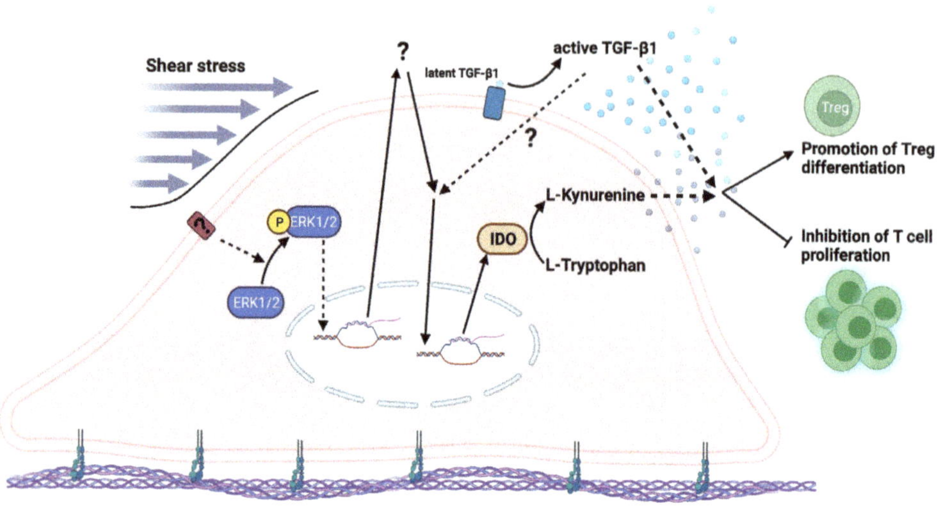

Figure 5. Schematic diagram of shear stress activates an immunosuppressive ability of hPDLSCs via ERK-induced IDO and total and active TGF-β1. Shear stress enhances IDO-dependent kynurenine in hPDLSCs via ERK1/2 activation and increases total and active TGF-β1 in the extracellular matrix or conditioned medium. Increased kynurenine and TGF-β1 secretion inhibit the CD4$^+$ T cell proliferation and promote Treg cell differentiation. Created with Biorender.com.

Supplementary Materials: The following supporting information can be downloaded at: https://www.mdpi.com/article/10.3390/ijms23137119/s1.

Author Contributions: Conceptualisation, R.S., N.K. and P.P.; methodology, R.S., U.N. and P.P.; software, J.N.F., P.P. and T.O.; validation, R.S. and U.N.; formal analysis, R.S.; investigation, R.S., N.K. and P.P.; resources, P.P. and T.O.; data curation, P.P. and T.O.; writing—original draft preparation, R.S. and N.K.; writing—review and editing, N.K., W.N., J.N.F. and T.O.; visualisation, R.S.; supervision, N.K., P.P. and T.O.; project administration, R.S. and U.N.; funding acquisition R.S., N.K. and T.O. All authors have read and agreed to the published version of the manuscript.

Funding: This study was supported by Chulalongkorn University and Thailand Science Research and Innovation Fund, Chulalongkorn University (CU_FRB65_hea(10)_016_32_11) (to N.K.) and Faculty Research Grant (DRF60019), Faculty of Dentistry, Chulalongkorn University (to N.K.) and the NSRF via the Program Management Unit for Human Resources & Institutional Development, Research, and Innovation (B16F640118) (to T.O.). R.S. was supported by the Graduate School, Chulalongkorn University to commemorate the 72nd Anniversary of His Majesty the King Bhumibol Aduladej.

Institutional Review Board Statement: The human PDL cell isolation protocol was approved by the Ethics Committee of the Faculty of Dentistry, Chulalongkorn University (HRBC-DCU 2022-010).

Informed Consent Statement: Informed consent was obtained from all healthy donors involved in the study.

Data Availability Statement: The data presented in this study are available on request from the corresponding author.

Acknowledgments: The Scholarship from the Graduate School, Chulalongkorn University to commemorate the 72nd anniversary of His Majesty the King Bhumibol Aduladej is gratefully acknowledged. The authors gratefully acknowledge Sirikool Thamnium and Worachot Saengha for their assistance with the laboratory work, Daneeya Chaikeawkaew for helping with apotome microscopy and Noppadol Sa-Ard-lam and the Immunology Research Center, Faculty of Dentistry, Chulalongkorn University for handling the flow cytometry. The authors thank the Faculty of Dentistry, Chulalongkorn University for proofreading of the manuscript and the staff of the Center of Excellence for Regenerative Dentistry for their collaboration.

Conflicts of Interest: The authors declare no conflict of interest.

Abbreviations

AHR	Aryl hydrocarbon receptor
BCA	Bicinchoninic acid
CHX	Cycloheximide
COX2	Cyclooxygenase 2
CTL-CM	Non-shear stress-induced hPDLSC-derived conditioned medium
DI	Deionised water
ELISA	Enzyme-linked immunosorbent assay
ERK1/2	Extracellular signalling-regulated kinase1/2
FOXP3	Forkhead box P3
GAPDH	Glyceraldehyde-3-phosphate dehydrogenase
hPDLSCs	Human periodontal ligament stem cells
IDO	Indoleamine 2,3dioxygenase
IFN-γ	Interferon gamma
IL-1β	Interleukin-1 beta
IL-2	Interleukin-2
IL-12	Interleukin-12
IL-10	Interleukin-10
Kyn	Kynurenine
MSCs	Mesenchymal stem cells
PBMCs	Peripheral blood mononucleated cells
P-ERK1/2	Phosphorylated extracellular signalling-regulated kinase ½
PGE2	Prostaglandin E2
PI	Protease inhibitor
P(I)3K	Phosphoinositide 3-kinase
RIPA	Radioimmunoprecipitation assay
RT	Room temperature
SS-CM	Shear stress-induced hPDLSC-derived conditioned medium
TCA	Trichloroacetic acid
TLRs	Toll-like receptors
Treg	Regulatory T cell
TGF-β1	Transforming growth factor beta 1

References

1. Wada, N.; Menicanin, D.; Shi, S.; Bartold, P.M.; Gronthos, S. Immunomodulatory properties of human periodontal ligament stem cells. *J. Cell. Physiol.* **2009**, *219*, 667–676. [CrossRef]
2. Huang, C.-Y.C.; Pelaez, D.; Bendala, J.D.; Garcia-Godoy, F.; Cheung, H.S. Plasticity of stem cells derived from adult periodontal ligament. *Regen. Med.* **2009**, *4*, 809–821. [CrossRef]
3. Ding, G.; Liu, Y.; Wang, W.; Wei, F.; Liu, D.; Fan, Z.; An, Y.; Zhang, C.; Wang, S. Allogeneic Periodontal Ligament Stem Cell Therapy for Periodontitis in Swine. *Stem Cells* **2010**, *28*, 1829–1838. [CrossRef]
4. Liu, Y.; Zheng, Y.; Ding, G.; Fang, D.; Zhang, C.; Bartold, P.M.; Gronthos, S.; Shi, S.; Wang, S. Periodontal Ligament Stem Cell-Mediated Treatment for Periodontitis in Miniature Swine. *Stem Cells* **2008**, *26*, 1065–1073. [CrossRef]
5. Mrozik, K.M.; Wada, N.; Marino, V.; Richter, W.; Shi, S.; Wheeler, D.L.; Gronthos, S.; Bartold, P.M. Regeneration of periodontal tissues using allogeneic periodontal ligament stem cells in an ovine model. *Regen. Med.* **2013**, *8*, 711–723. [CrossRef]

6. Isaka, J.; Ohazama, A.; Kobayashi, M.; Nagashima, C.; Takiguchi, T.; Kawasaki, H.; Tachikawa, T.; Hasegawa, K. Participation of periodontal ligament cells with regeneration of alveolar bone. *J. Periodontol.* **2001**, *72*, 314–323. [CrossRef]
7. Chaikeawkaew, D.; Everts, V.; Pavasant, P. TLR3 activation modulates immunomodulatory properties of human periodontal ligament cells. *J. Periodontol.* **2020**, *91*, 1225–1236. [CrossRef]
8. Liu, O.; Xu, J.; Ding, G.; Liu, D.; Fan, Z.; Zhang, C.; Chen, W.; Ding, Y.; Tang, Z.; Wang, S. Periodontal ligament stem cells regulate B lymphocyte function via programmed cell death protein 1. *Stem Cells* **2013**, *31*, 1371–1382. [CrossRef]
9. Konermann, A.; Stabenow, D.; Knolle, P.A.; Held, S.A.E.; Deschner, J.; Jäger, A. Regulatory role of periodontal ligament fibroblasts for innate immune cell function and differentiation. *Innate Immun.* **2012**, *18*, 745–752. [CrossRef]
10. Konermann, A.; Beyer, M.; Deschner, J.; Allam, J.P.; Novak, N.; Winter, J.; Jepsen, S.; Jäger, A. Human periodontal ligament cells facilitate leukocyte recruitment and are influenced in their immunomodulatory function by Th17 cytokine release. *Cell. Immunol.* **2012**, *272*, 137–143. [CrossRef]
11. Shin, C.; Kim, M.; Han, J.A.; Choi, B.; Hwang, D.; Do, Y.; Yun, J.H. Human periodontal ligament stem cells suppress T-cell proliferation via down-regulation of non-classical major histocompatibility complex-like glycoprotein CD1b on dendritic cells. *J. Periodontal Res.* **2017**, *52*, 135–146. [CrossRef]
12. Liu, J.; Chen, B.; Bao, J.; Zhang, Y.; Lei, L.; Yan, F. Macrophage polarization in periodontal ligament stem cells enhanced periodontal regeneration. *Stem Cell Res. Ther.* **2019**, *10*, 320. [CrossRef]
13. Wada, N.; Tomokiyo, A.; Gronthos, S.; Bartold, P.M. Immunomodulatory Properties of PDLSC and Relevance to Periodontal Regeneration. *Curr. Oral Health Rep.* **2015**, *2*, 245–251. [CrossRef]
14. Li, H.; Shen, S.; Fu, H.; Wang, Z.; Li, X.; Sui, X.; Yuan, M.; Liu, S.; Wang, G.; Guo, Q. Immunomodulatory Functions of Mesenchymal Stem Cells in Tissue Engineering. *Stem Cells Int.* **2019**, *2019*, 967–1206. [CrossRef]
15. Kode, J.A.; Mukherjee, S.; Joglekar, M.V.; Hardikar, A.A. Mesenchymal stem cells: Immunobiology and role in immunomodulation and tissue regeneration. *Cytotherapy* **2009**, *11*, 377–391. [CrossRef]
16. Harrell, C.R.; Djonov, V.; Volarevic, V. The Cross-Talk between Mesenchymal Stem Cells and Immune Cells in Tissue Repair and Regeneration. *Int. J. Mol. Sci.* **2021**, *22*, 2472. [CrossRef]
17. English, K.; Barry, F.P.; Field-Corbett, C.P.; Mahon, B.P. IFN-γ and TNF-α differentially regulate immunomodulation by murine mesenchymal stem cells. *Immunol. Lett.* **2007**, *110*, 91–100. [CrossRef]
18. Clark, D.A.; Coker, R. Molecules in focus Transforming growth factor-beta (TGF-β). *Int. J. Biochem. Cell Biol.* **1998**, *30*, 293–298. [CrossRef]
19. de Araújo Farias, V.; Carrillo-Gálvez, A.B.; Martín, F.; Anderson, P. TGF-β and mesenchymal stromal cells in regenerative medicine, autoimmunity and cancer. *Cytokine Growth Factor Rev.* **2018**, *43*, 25–37. [CrossRef]
20. Groh, M.E.; Maitra, B.; Szekely, E.; Koç, O.N. Human mesenchymal stem cells require monocyte-mediated activation to suppress alloreactive T cells. *Exp. Hematol.* **2005**, *33*, 928–934. [CrossRef]
21. Kehrl, J.H.; Wakefield, L.M.; Roberts, A.B.; Jakowlew, S.; Alvarez-Mon, M.; Derynck, R.; Sporn, M.B.; Fauci, A.S. Production of transforming growth factor beta by human T lymphocytes and its potential role in the regulation of T cell growth. *J. Exp. Med.* **1986**, *163*, 1037–1050. [CrossRef] [PubMed]
22. Chen, W.; Jin, W.; Hardegen, N.; Lei, K.-J.; Li, L.; Marinos, N.; McGrady, G.; Wahl, S.M. Conversion of peripheral $CD4^+$ $CD25^-$ naive T cells to $CD4^+$ $CD25^+$ regulatory T cells by TGF-β induction of transcription factor Foxp3. *J. Exp. Med.* **2003**, *198*, 1875–1886. [CrossRef]
23. Konkel, J.E.; Zhang, D.; Zanvit, P.; Chia, C.; Zangarle-Murray, T.; Jin, W.; Wang, S.; Chen, W. Transforming Growth Factor-β Signaling in Regulatory T Cells Controls T Helper-17 Cells and Tissue-Specific Immune Responses. *Immunity* **2017**, *46*, 660–674. [CrossRef]
24. Mellor, A.L.; Munn, D.; Chandler, P.; Keskin, D.; Johnson, T.; Marshall, B.; Jhaver, K.; Baban, B. Tryptophan catabolism and T cell responses. *Adv. Exp. Med. Biol.* **2003**, *527*, 27–35. [CrossRef] [PubMed]
25. Mellor, A.L.; Keskin, D.B.; Johnson, T.; Chandler, P.; Munn, D.H. Cells Expressing Indoleamine 2,3-Dioxygenase Inhibit T Cell Responses. *J. Immunol.* **2002**, *168*, 3771. [CrossRef]
26. Fallarino, F.; Grohmann, U.; Vacca, C.; Bianchi, R.; Orabona, C.; Spreca, A.; Fioretti, M.C.; Puccetti, P. T cell apoptosis by tryptophan catabolism. *Cell Death Differ.* **2002**, *9*, 1069–1077. [CrossRef]
27. Fallarino, F.; Grohmann, U.; You, S.; McGrath, B.C.; Cavener, D.R.; Vacca, C.; Orabona, C.; Bianchi, R.; Belladonna, M.L.; Volpi, C.; et al. The Combined Effects of Tryptophan Starvation and Tryptophan Catabolites Down-Regulate T Cell Receptor ζ-Chain and Induce a Regulatory Phenotype in Naive T Cells. *J. Immunol.* **2006**, *176*, 6752. [CrossRef] [PubMed]
28. Zunszain, P.A.; Anacker, C.; Cattaneo, A.; Choudhury, S.; Musaelyan, K.; Myint, A.M.; Thuret, S.; Price, J.; Pariante, C.M. Interleukin-1β: A new regulator of the kynurenine pathway affecting human hippocampal neurogenesis. *Neuropsychopharmacology* **2012**, *37*, 939–949. [CrossRef]
29. Issaranggun Na Ayuthaya, B.; Satravaha, P.; Pavasant, P. Interleukin-12 modulates the immunomodulatory properties of human periodontal ligament cells. *J. Periodontal Res.* **2017**, *52*, 546–555. [CrossRef]
30. Andrukhov, O.; Hong, J.S.-A.; Andrukhova, O.; Blufstein, A.; Moritz, A.; Rausch-Fan, X. Response of human periodontal ligament stem cells to IFN-γ and TLR-agonists. *Sci. Rep.* **2017**, *7*, 12856. [CrossRef]

31. Jung, I.D.; Jeong, Y.-I.; Lee, C.-M.; Noh, K.T.; Jeong, S.K.; Chun, S.H.; Choi, O.H.; Park, W.S.; Han, J.; Shin, Y.K.; et al. COX-2 and PGE2 signaling is essential for the regulation of IDO expression by curcumin in murine bone marrow-derived dendritic cells. *Int. Immunopharmacol.* **2010**, *10*, 760–768. [CrossRef]
32. Braun, D.; Longman, R.S.; Albert, M.L. A two-step induction of indoleamine 2,3 dioxygenase (IDO) activity during dendritic-cell maturation. *Blood* **2005**, *106*, 2375–2381. [CrossRef]
33. Carlin, J.M.; Borden, E.C.; Sondel, P.M.; Byrne, G.I. Biologic-response-modifier-induced indoleamine 2,3-dioxygenase activity in human peripheral blood mononuclear cell cultures. *J. Immunol.* **1987**, *139*, 2414–2418.
34. Chen, W. IDO: More than an enzyme. *Nat. Immunol.* **2011**, *12*, 809–811. [CrossRef]
35. Jiang, N.; He, D.; Ma, Y.; Su, J.; Wu, X.; Cui, S.; Li, Z.; Zhou, Y.; Yu, H.; Liu, Y. Force-Induced Autophagy in Periodontal Ligament Stem Cells Modulates M1 Macrophage Polarization via AKT Signaling. *Front. Cell Dev. Biol.* **2021**, *9*. [CrossRef]
36. Wang, Z.; Maruyama, K.; Sakisaka, Y.; Suzuki, S.; Tada, H.; Suto, M.; Saito, M.; Yamada, S.; Nemoto, E. Cyclic Stretch Force Induces Periodontal Ligament Cells to Secrete Exosomes That Suppress IL-1β Production Through the Inhibition of the NF-κB Signaling Pathway in Macrophages. *Front. Immunol.* **2019**, *10*, 1310. [CrossRef]
37. Diaz, M.F.; Vaidya, A.B.; Evans, S.M.; Lee, H.J.; Aertker, B.M.; Alexander, A.J.; Price, K.M.; Ozuna, J.A.; Liao, G.P.; Aroom, K.R.; et al. Biomechanical Forces Promote Immune Regulatory Function of Bone Marrow Mesenchymal Stromal Cells. *Stem Cells* **2017**, *35*, 1259–1272. [CrossRef] [PubMed]
38. Chukkapalli, S.S.; Lele, T.P. Periodontal cell mechanotransduction. *Open Biol.* **2018**, *8*, 180053. [CrossRef]
39. Aveic, S.; Craveiro, R.B.; Wolf, M.; Fischer, H. Current Trends in vitro Modeling to Mimic Cellular Crosstalk in Periodontal Tissue. *Adv. Healthc. Mater.* **2020**, *10*, 2001269. [CrossRef]
40. McCulloch, C.A.; Lekic, P.; McKee, M.D. Role of physical forces in regulating the form and function of the periodontal ligament. *Periodontology* **2000**, *24*, 56–72. [CrossRef]
41. Afanador, E.; Yokozeki, M.; Oba, Y.; Kitase, Y.; Takahashi, T.; Kudo, A.; Moriyama, K. Messenger RNA expression of periostin and Twist transiently decrease by occlusal hypofunction in mouse periodontal ligament. *Arch. Oral Biol.* **2005**, *50*, 1023–1031. [CrossRef] [PubMed]
42. Manokawinchoke, J.; Limjeerajarus, N.; Limjeerajarus, C.; Sastravaha, P.; Everts, V.; Pavasant, P. Mechanical Force–induced TGF-*beta*1 Increases Expression of SOST/POSTN by hPDL Cells. *J. Dent. Res.* **2015**, *94*, 983–989. [CrossRef] [PubMed]
43. Kim, S.G.; Kim, S.G.; Viechnicki, B.; Kim, S.; Nah, H.D. Engineering of a periodontal ligament construct: Cell and fibre alignment induced by shear stress. *J. Clin. Periodontol.* **2011**, *38*, 1130–1136. [CrossRef] [PubMed]
44. Ferrier, J.M.; Dillon, E.M. The water binding capacity of the periodontal ligament and its role in mechanical function. *J. Periodontal Res.* **1983**, *18*, 469–473. [CrossRef] [PubMed]
45. Nath, S.C.; Day, B.; Harper, L.; Yee, J.; Hsu, C.Y.; Larijani, L.; Rohani, L.; Duan, N.; Kallos, M.S.; Rancourt, D.E. Fluid shear stress promotes embryonic stem cell pluripotency via interplay between β-catenin and vinculin in bioreactor culture. *Stem Cells* **2021**, *39*, 1166–1177. [CrossRef] [PubMed]
46. Wittkowske, C.; Reilly, G.C.; Lacroix, D.; Perrault, C.M. In Vitro Bone Cell Models: Impact of Fluid Shear Stress on Bone Formation. *Front. Bioeng. Biotechnol.* **2016**, *4*, 87–87. [CrossRef]
47. Russo, T.A.; Banuth, A.M.M.; Nader, H.B.; Dreyfuss, J.L. Altered shear stress on endothelial cells leads to remodeling of extracellular matrix and induction of angiogenesis. *PLoS ONE* **2020**, *15*, e0241040. [CrossRef]
48. Arora, S.; Srinivasan, A.; Leung, A.; Toh, Y.-C. Bio-Mimicking Shear Stress Environments for Enhancing Mesenchymal Stem Cell Differentiation. *Curr. Stem Cell Res. Ther.* **2020**, *15*, 414–427. [CrossRef]
49. Qi, L.; Zhang, Y. The microRNA 132 Regulates Fluid Shear Stress-Induced Differentiation in Periodontal Ligament Cells through mTOR Signaling Pathway. *Cell. Physiol. Biochem.* **2014**, *33*, 433–445. [CrossRef]
50. Zheng, L.; Chen, L.; Chen, Y.; Gui, J.; Li, Q.; Huang, Y.; Liu, M.; Jia, X.; Song, W.; Ji, J.; et al. The effects of fluid shear stress on proliferation and osteogenesis of human periodontal ligament cells. *J. Biomech.* **2016**, *49*, 572–579. [CrossRef]
51. Tang, M.; Peng, Z.; Mai, Z.; Chen, L.; Mao, Q.; Chen, Z.; Chen, Q.; Liu, L.; Wang, Y.; Ai, H. Fluid shear stress stimulates osteogenic differentiation of human periodontal ligament cells via the extracellular signal-regulated kinase 1/2 and p38 mitogen-activated protein kinase signaling pathways. *J. Periodontol.* **2014**, *85*, 1806–1813. [CrossRef]
52. Zheng, L.; Huang, Y.; Song, W.; Gong, X.; Liu, M.; Jia, X.; Zhou, G.; Chen, L.; Li, A.; Fan, Y. Fluid shear stress regulates metalloproteinase-1 and 2 in human periodontal ligament cells: Involvement of extracellular signal-regulated kinase (ERK) and P38 signaling pathways. *J. Biomech.* **2012**, *45*, 2368–2375. [CrossRef]
53. Van Der Pauw, M.T.M.; Klein-Nulend, J.; Van Den Bos, T.; Burger, E.H.; Everts, V.; Beertsen, W. Response of periodontal ligament fibroblasts and gingival fibroblasts to pulsating fluid flow: Nitric oxide and prostaglandin E2 release and expression of tissue non-specific alkaline phosphatase activity. *J. Periodontal Res.* **2000**, *35*, 335–343. [CrossRef]
54. Limjeerajarus, N.; Keawprachum, B.; Pliankum, M.; Pavasant, P.; Limjeerajarus, C.N. Numerical data on the shear stress distribution generated by a rotating rod within a stationary ring over a 35-mm cell culture dish. *Data Brief* **2018**, *21*, 2253–2258. [CrossRef]
55. Livak, K.J.; Schmittgen, T.D.J.M. Analysis of relative gene expression data using real-time quantitative PCR and the $2^{-\Delta\Delta CT}$. *Method* **2001**, *25*, 402–408. [CrossRef]
56. Arce, F.; Breckpot, K.; Stephenson, H.; Karwacz, K.; Ehrenstein, M.; Collins, M.; Escors, D. Selective ERK Activation Differentiates Mouse and Human Tolerogenic Dendritic Cells, Expands Antigen-Specific Regulatory T Cells, and Suppresses Experimental Inflammatory Arthritis. *Arthritis Rheum.* **2011**, *63*, 84–95. [CrossRef]

57. Martkamchan, S.; Onlamoon, N.; Wang, S.; Pattanapanyasat, K.; Ammaranond, P. The Effects of Anti-CD3/CD28 Coated Beads and IL-2 on Expanded T Cell for Immunotherapy. *Adv. Clin. Exp. Med.* **2016**, *25*, 821–828. [CrossRef]
58. Kouzbari, K.; Hossan, M.R.; Arrizabalaga, J.H.; Varshney, R.; Simmons, A.D.; Gostynska, S.; Nollert, M.U.; Ahamed, J. Oscillatory shear potentiates latent TGF-β1 activation more than steady shear as demonstrated by a novel force generator. *Sci. Rep.* **2019**, *9*, 6065. [CrossRef]
59. Forteza, M.J.; Polyzos, K.A.; Baumgartner, R.; Suur, B.E.; Mussbacher, M.; Johansson, D.K.; Hermansson, A.; Hansson, G.K.; Ketelhuth, D.F.J. Activation of the Regulatory T-Cell/Indoleamine 2,3-Dioxygenase Axis Reduces Vascular Inflammation and Atherosclerosis in Hyperlipidemic Mice. *Front. Immunol.* **2018**, *9*, 950. [CrossRef]
60. Mezrich, J.D.; Fechner, J.H.; Zhang, X.; Johnson, B.P.; Burlingham, W.J.; Bradfield, C.A. An interaction between kynurenine and the aryl hydrocarbon receptor can generate regulatory T cells. *J. Immunol.* **2010**, *185*, 3190–3198. [CrossRef]
61. Taylor, M.W.; Feng, G. Relationship between interferon-γ, indoleamine 2, 3-dioxygenase, and tryptophan catabolism. *FASEB J.* **1991**, *5*, 2516–2522. [CrossRef]
62. Pallotta, M.T.; Orabona, C.; Volpi, C.; Vacca, C.; Belladonna, M.L.; Bianchi, R.; Servillo, G.; Brunacci, C.; Calvitti, M.; Bicciato, S. Indoleamine 2, 3-dioxygenase is a signaling protein in long-term tolerance by dendritic cells. *Nat. Immunol.* **2011**, *12*, 870–878. [CrossRef]
63. Jo, H.; Sipos, K.; Go, Y.-M.; Law, R.; Rong, J.; McDonald, J. Differential Effect of Shear Stress on Extracellular Signal-regulated Kinase and N-terminal Jun Kinase in Endothelial Cells Gi2- and Gβ/*gamma*-Dependent Signaling Pathways. *J. Biol. Chem.* **1997**, *272*, 1395–1401. [CrossRef]
64. Liu, W.L.; Lin, Y.H.; Xiao, H.; Xing, S.; Chen, H.; Chi, P.D.; Zhang, G. Epstein-Barr virus infection induces indoleamine 2,3-dioxygenase expression in human monocyte-derived macrophages through p38/mitogen-activated protein kinase and NF-κB pathways: Impairment in T cell functions. *J. Virol.* **2014**, *88*, 6660–6671. [CrossRef]
65. Liu, W.; Putnam, A.L.; Xu-Yu, Z.; Szot, G.L.; Lee, M.R.; Zhu, S.; Gottlieb, P.A.; Kapranov, P.; Gingeras, T.R.; Fazekas de St Groth, B.; et al. CD127 expression inversely correlates with FoxP3 and suppressive function of human CD4(+) T reg cells. *J. Exp. Med.* **2006**, *203*, 1701–1711. [CrossRef]
66. Yu, N.; Li, X.; Song, W.; Li, D.; Yu, D.; Zeng, X.; Li, M.; Leng, X.; Li, X. CD4(+)CD25 (+)CD127 (low/-) T cells: A more specific Treg population in human peripheral blood. *Inflammation* **2012**, *35*, 1773–1780. [CrossRef]
67. Seddiki, N.; Santner-Nanan, B.; Martinson, J.; Zaunders, J.; Sasson, S.; Landay, A.; Solomon, M.; Selby, W.; Alexander, S.I.; Nanan, R.; et al. Expression of interleukin (IL)-2 and IL-7 receptors discriminates between human regulatory and activated T cells. *J. Exp. Med.* **2006**, *203*, 1693–1700. [CrossRef]
68. de Zoeten, E.F.; Lee, I.; Wang, L.; Chen, C.; Ge, G.; Wells, A.D.; Hancock, W.W.; Ozkaynak, E. Foxp3 processing by proprotein convertases and control of regulatory T cell function. *J. Biol. Chem.* **2009**, *284*, 5709–5716. [CrossRef]
69. Kondělková, K.; Vokurková, D.; Krejsek, J.; Borská, L.; Fiala, Z.; Ctirad, A. Regulatory T cells (TREG) and their roles in immune system with respect to immunopathological disorders. *Acta Med.* **2010**, *53*, 73–77. [CrossRef]

MDPI
St. Alban-Anlage 66
4052 Basel
Switzerland
www.mdpi.com

International Journal of Molecular Sciences Editorial Office
E-mail: ijms@mdpi.com
www.mdpi.com/journal/ijms

www.ingramcontent.com/pod-product-compliance
Lightning Source LLC
Chambersburg PA
CBHW062044220926
51850CB00010B/2653
* 9 7 8 3 7 2 5 8 1 0 6 6 6 *